Vaccinology and Methods in Vaccine Research

Developments in Immunology Series

Vaccinology and Methods in Vaccine Research

Edited by

Rebecca Ashfield
Jenner Institute and Department of Biochemistry, University of Oxford, Oxford, United Kingdom; Ducentis Biotherapeutics Ltd, Oxford, United Kingdom

Angus Nnamdi Oli
Department of Pharmaceutical Microbiology and Biotechnology, Faculty of Pharmaceutical Sciences, Nnamdi Azikiwe University, Awka, Nigeria

Charles Esimone
Department of Pharmaceutical Microbiology and Biotechnology, Faculty of Pharmaceutical Sciences, Nnamdi Azikiwe University, Awka, Nigeria

Linda Anagu
Department of Pharmaceutical Microbiology and Biotechnology, Faculty of Pharmaceutical Sciences, Nnamdi Azikiwe University, Awka, Nigeria

ELSEVIER

ACADEMIC PRESS
An imprint of Elsevier

Academic Press is an imprint of Elsevier
125 London Wall, London EC2Y 5AS, United Kingdom
525 B Street, Suite 1650, San Diego, CA 92101, United States
50 Hampshire Street, 5th Floor, Cambridge, MA 02139, United States
The Boulevard, Langford Lane, Kidlington, Oxford OX5 1GB, United Kingdom

ISBN: 978-0-323-91146-7

For Information on all Academic Press publications
visit our website at https://www.elsevier.com/books-and-journals

Publisher: Stacy Masucci
Acquisitions Editor: Linda Versteeg-Buschman
Editorial Project Manager: Sara Pianavilla
Production Project Manager: Omer Mukthar
Cover Designer: Mark Rogers

Typeset by MPS Limited, Chennai, India

Contents

List of contributors

Linda Onyeka Anagu Department of Pharmaceutical Microbiology and Biotechnology, Nnamdi Azikiwe University, Awka, Nigeria

Nana Efua Andoh Department of Parasitology, Noguchi Memorial Institute for Medical Research, University of Ghana, Accra, Ghana

Rebecca Ashfield Jenner Institute and Department of Biochemistry, University of Oxford, Oxford, United Kingdom; Ducentis Biotherapeutics Ltd, Oxford, United Kingdom

Camila R.R. Barbosa Centre for Inflammation Biology and Cancer Immunology (CIBCI), Peter Gorer Department of Immunobiology, King's College London, London, United Kingdom; Francis Crick Institute, London, United Kingdom

Paulo J.G. Bettencourt Faculty of Medicine, Catholic University of Portugal, Lisbon, Portugal; Center for Interdisciplinary Research in Health, Catholic University of Portugal, Lisbon, Portugal

Sebnem Ercelen Ceylan Genetic Engineering and Biotechnology Institute, Kocaeli, Turkey

Dennis Christensen Statens Serum Institut, Copenhagen, Denmark

Rebecca Chinyelu Chukwuanukwu Department of Medical Laboratory Sciences, Nnamdi Azikiwe University, Awka, Nigeria

George Corby University of Oxford, Oxford, United Kingdom

Alfred Friday Ehiaghe Department of Medical Laboratory Sciences, Nnamdi Azikiwe University, Awka, Nigeria

Katie Ewer Jenner Institute, University of Oxford, Oxford, United Kingdom

Joannah Fergusson Jenner Institute, University of Oxford, Oxford, United Kingdom

Sveinbjorn Gizurarson Faculty of Pharmaceutical Sciences, University of Iceland, Reykjavik, Iceland; Pharmacy Department, Kamuzu University of Health Sciences, Chichiri, Malawi

Sabina Górska Hirszfeld Institute of Immunology and Experimental Therapy Polish Academy of Sciences, Wroclaw, Poland

Adrian V.S. Hill Jenner Institute, University of Oxford, Oxford, United Kingdom

Kefei Hu R&D, Inari Consult ApS, Copenhagen, Denmark; R&D Unit, Croda Denmark, Frederikssund, Denmark

Frances Iseghohi Department of Medical Biotechnology, National Biotechnology Development Agency, Abuja, Nigeria

Erik B. Lindblad R&D, Inari Consult ApS, Copenhagen, Denmark

Stephanie Longet Wellcome Centre for Human Genetics, University of Oxford, Oxford, United Kingdom

Marina Luchner Technical University of Munich, Munich, Germany

Julia L. Marshall The University of Oxford, Oxford, United Kingdom

Anita Milicic Jenner Institute, University of Oxford, Oxford, United Kingdom

Bror Morein DVM, Savacc AB, Uppsala, Sweden

Slavcho Mrenoshki Ss. Cyril and Methodius University, Faculty of Veterinary Medicine, Skopje, North Macedonia

Ekta Mukhopadhyay Jenner Institute, University of Oxford, Oxford, United Kingdom

Andrés Noé Jenner Institute, University of Oxford, Oxford, United Kingdom; Murdoch Children's Research Institute, Parkville, VIC, Australia

Ezinne Janefrances Nwonu Medical Biotechnology Department, National Biotechnology Development Agency, Abuja, Nigeria

Angus Nnamdi Oli Department of Pharmaceutical Microbiology and Biotechnology, Faculty of Pharmaceutical Sciences, Nnamdi Azikiwe University, Awka, Nigeria

Agnieszka Razim Hirszfeld Institute of Immunology and Experimental Therapy Polish Academy of Sciences, Wroclaw, Poland

Sören Reinke Jenner Institute, University of Oxford, Oxford, United Kingdom

Adekunle Babajide Rowaiye Department of Medical Biotechnology, National Biotechnology Development Agency, Abuja, Nigeria

Ahmed M. Salman Jenner Institute, University of Oxford, Oxford, United Kingdom

Aneesh Thakur University of Copenhagen, Copenhagen, Denmark

Malachy Chigozie Ugwu Department of Pharmaceutical Microbiology and Biotechnology, Faculty of Pharmaceutical Sciences, Nnamdi Azikiwe University, Awka, Nigeria

Aondona Priscilla Yahemba Department of Medical Biotechnology, National Biotechnology Development Agency, Abuja, Nigeria

Preface

This book was written during the Covid-19 pandemic. Globally, as of 24th January 2022, the WHO estimates over five and a half million deaths from the SARS-CoV-2 virus have been reported to the WHO. This worldwide health crisis has highlighted the need for the rapid development of effective, affordable vaccines that can be quickly deployed to all countries, especially low- and middle-income countries (LMICs).

Healthcare companies and academic institutions across the world answered this need with remarkable speed and ingenuity, with a plethora of vaccine candidates (over 100) in clinical development during 2021, and a handful already approved for mass immunization programs. Regulatory authorities worked closely with those developing new vaccines to facilitate approval processes, and activities that would normally be carried out sequentially, for example, scaling up good manufacturing practice (GMP) manufacture and different phases of clinical testing, were parallel tracked tracked to reduce overall timelines.

The majority of Covid vaccines target the SARS-CoV-2 surface Spike (S) protein, which is an important target antigen for neutralizing antibodies. A variety of approaches for subunit vaccines have been applied, including RNA, DNA, viral vectors, protein in adjuvant, and virus-like particles (VLP). The more traditional "whole cell" approach has also been used, with both inactivated and live-attenuated viral vaccines.

In the 21st century, outbreaks of several viruses including Ebola, MERS (Middle East respiratory syndrome), SARS (severe acute respiratory syndrome), influenza, Zika, and Chikungunya have affected numerous countries worldwide. Many researchers are aiming for a platform technology approach, with a single vaccine design (e.g., viral vector or RNA formulated in liposomes), which can be rapidly adapted to create vaccines against emerging pathogens or new pathogen strains.

It is clear that several challenges remain in the rapid international deployment of effective vaccines. These include the required scale-up in GMP manufacture of new vaccines to produce the billions of doses required in a pandemic situation. Manufacturing capacity is rate limiting, especially in LMICs that lack the expensive facilities required for production of vaccines and other biological medicines. It is also challenging to deploy vaccines internationally, especially those that need to be stored at low temperatures, as many regions of the world lack freezer equipment and electricity supply cannot be guaranteed.

This book is a collaboration between scientists based in several different countries including Nigeria and the UK, and is representative of the many ongoing international collaborations in vaccine development, and more generally in the fight against infectious diseases. The editors intend the book to be used by scientists who are developing new vaccines and teaching the concepts of vaccinology to students.

Early chapters review the basic immunology that underpins the concept of immunization and the varied approaches that have been taken to develop vaccines. One chapter focuses on practical methods that can be implemented in evaluating antigens as vaccine candidates, including immunomonitoring in clinical trials. Later chapters explore recent developments in vaccine research, including the use of immunopeptidomics to identify target T-cell epitopes, and approaches to developing cancer vaccines, a rapidly expanding area of research due to the discovery of tumor-specific neoepitopes that can be targeted by an immune response while sparing normal tissue.

We have included a comprehensive review of adjuvants, which can greatly improve the immune response to subunit vaccines, especially soluble proteins and VLPs. Finally, there is a practical guide to the principles of GMP manufacture and to the planning and execution of clinical trials.

As the Covid-19 pandemic has illustrated, vaccines can have a global impact in controlling disease and saving lives. Ground-breaking science and international, multidisciplinary collaboration are key in advancing vaccine development. There remains much to be done, in preparing for future outbreaks, and in developing more effective vaccines against diseases with high mortality rates in certain areas of the world, for example, Ebola hemorrhagic fever, malaria, HIV, and tuberculosis.

Acknowledgements

Dedicated to the memory of Shahid Khan, a highly respected scientist and enthusiastic teacher, who was the inspiration for this book. Shahid's work resulted in the first genetically attenuated malaria vaccine to be clinically tested in humans, an important contribution to the ongoing and challenging search for a vaccine against malaria.

Also dedicated to all those who lost their lives as a result of the deprivation of medical services following the immense pressure due to the Covid-19 pandemic. In loving memory of Princess Ifeoma Emmanuella Onaku and Chief Silas Nwachukwu Anagu.

Chapter 1

Immunology and the concept of vaccination

Rebecca Chinyelu Chukwuanukwu[1], Alfred Friday Ehiaghe[1], Adekunle Babajide Rowaiye[2] and Angus Nnamdi Oli[3]
[1]Department of Medical Laboratory Sciences, Nnamdi Azikiwe University, Awka, Nigeria, [2]Department of Medical Biotechnology, National Biotechnology Development Agency, Abuja, Nigeria, [3]Department of Pharmaceutical Microbiology and Biotechnology, Faculty of Pharmaceutical Sciences, Nnamdi Azikiwe University, Awka, Nigeria

Manipulation of the immune system by vaccination

Daily, millions of people worldwide benefit from modulation of the immune system via immunotherapy. Unfavorable immune responses can be controlled to treat autoimmunity, allergy, and transplant rejection, and conversely responses can be stimulated to treat cancer, or induce protective immune responses against pathogens using vaccines. Immunization (also known as vaccination) is the most widely practiced form of immunotherapy. It is recognized as perhaps the greatest triumph of modern medicine, attributed originally to Edward Jenner who was the first to use a scientific approach to immunization in 1798. He utilized observations made about two decades earlier by a farmer, Benjamin Jesty, that individuals exposed to cowpox were protected against smallpox. The term vaccination originates from the Latin word "vacca" for cow after Jenner successfully demonstrated that inoculation with cowpox could protect against the often-fatal smallpox virus. It is interesting to note that Jenner knew nothing of the infectious agents that cause disease. Neither did he understand the basis for the consequent immunity that resulted following his vaccination. Therefore, despite the huge success of his vaccinations in offering protection against smallpox, findings from his work remained unexploited for almost a century.

This narrative changed when Louis Pasteur, together with Camile Guérin and Albert Calmette, demonstrated that weakening a pathogen could stimulate protective immune responses against that specific pathogen without causing disease. This is termed attenuation and was the basis for production of the Bacille Calmette-Guerin (BCG) tuberculosis vaccine, which was first

Vaccinology and Methods in Vaccine Research. DOI: https://doi.org/10.1016/B978-0-323-91146-7.00005-6

used in 1921, and remains in use today. Pasteur made the germ theory of disease famous and also opined that vaccination could actually prevent infectious diseases. For this reason, Louis Pasteur is considered to be the founder of modern immunology. Robert Koch built on the work of Pasteur and was able to prove that infectious diseases are caused by microorganisms. These discoveries extended Jenner's vaccination strategy to other diseases.

The essential strategy for vaccination is to prepare an innocuous form of a pathogen, or individual pathogen antigens, that are immunogenic and can establish memory cells and protective immunity. This can be achieved using killed or live attenuated organisms, purified microbial components or recombinant antigens. In the case of Edward Jenner's landmark experiment, cowpox was a closely related but much less dangerous pathogen that was able to offer cross-protection against the virulent and dangerous smallpox. Approaches to vaccine design are described in more detail in Chapter 2, Vaccine Types and Reverse Vaccinology.

Overview of immunity and the immune system

The concept of immunity

It is the role of the immune system to provide a defense against the various pathogens that the host encounters on a daily basis. The two cardinal events for immunity are *recognition* and *defense*, which are mediated by the cells of the immune system, including polymorphonuclear and mononuclear leukocytes. To identify a foreign entity or pathogen, the immune cells must recognize foreign configurations or structures, known as antigens. This is possible because higher animals have evolved the ability to distinguish between themselves (self) and other species (nonself). This capacity to discriminate between self and nonself, and to react to nonself, is the basis of immunity.

There exists an enormous variety of infectious pathogens: bacteria, viruses, fungi, protozoa, and parasites, with various methods of transmission and infection. Therefore, no single recognition and defense strategy is effective against all pathogens. Consequently, a wide variety of cellular and secreted components are required for effective immune protection.

Millions of microorganisms are encountered daily, a number of which are capable of causing disease. Yet, most healthy individuals succumb to infection only occasionally, because potential disease-causing organisms (pathogens) are detected and destroyed before an infection can establish, often within minutes of a challenge.

The first line of defense against foreign invasion consists of physical and chemical barriers. The physical protective barrier includes the skin, hairs, cilia, epithelial cells, and the mucous membranes which line and protect the respiratory, digestive, and reproductive tracts. In addition, chemical barriers comprise secreted antimicrobial proteins including enzymes in tears and saliva (lysozymes), and pulmonary surfactants. When these barriers are breached and pathogens gain entry into the system, the next lines of defense come into play: the innate and adaptive immune systems. These work together to provide robust, protective immune responses.

The innate immune response

The innate immune system consists of cellular components and secreted mediators that defend against pathogens. It is termed innate (natural) because it is present at birth, is nonspecific, i.e., does not recognize pathogen-specific antigens, and lacks memory (so is not improved by prior exposure to any particular microorganism). Innate immune responses are induced rapidly following exposure to pathogens, to limit spread and proliferation of the infectious agent within the body.

The innate immune system has evolved over millennia to recognize chemical structures that are generally characteristic of infectious pathogens. These foreign molecular patterns can be recognized and bound by pattern recognition receptors (PRRs), which are encoded in the germline, and include toll-like receptors and mannan-binding lectins. PRRs recognize microbial proteins, lipids, carbohydrates, and nucleic acids collectively known as pathogen-associated molecular patterns (PAMPs). After recognition of a foreign invader, the innate immune system is activated to defend the host via cellular components, including polymorphonuclear and mononuclear phagocytes, and secreted mediators (Table 1.1).

Cells of the innate immune system

Cells of the innate immune system are derived from the *myeloid* lineage, which differentiates from myeloid progenitor cells in the bone marrow,

TABLE 1.1 Cells and secreted mediators of the innate immune system.

Cells	Secreted mediators
Neutrophils	Complement proteins
Eosinophils	Lysozymes
Basophils	Cytotoxins
Mast cells	Cytokines
Monocytes	Eicosanoids
Macrophages	Pentraxins
Dendritic cells	Collectins
Natural killer cells	Defensins

whereas the adaptive immune system comprises *lymphocytes*, which also derive from bone marrow stem cells. There is also a family of innate lymphoid cells, including natural killer (NK) cells, a subset of lymphocytes that do not recognize pathogen-specific antigens and are hence considered to be part of the innate immune system.

Monocytes and macrophages

Monocytes can differentiate into dendritic cells (DCs), which are professional antigen-presenting cells (APCs), or into phagocytic macrophages, which can exhibit either pro- or antiinflammatory effects (Atri et al., 2018).

Macrophages

These are specialized innate immune cells engaged in the detection, ingestion, and destruction of pathogens and other harmful particles. They play an important role in host defense and tissue homeostasis (Hirayama et al., 2017). Tissue macrophages comprise tissue-resident macrophages, and those are derived from circulating monocytes that invade tissues as part of the inflammatory response. Like DCs, they can present antigens to cells of the adaptive immune system, and secrete cytokines. Macrophages can polarize into proinflammatory (M1) or tolerogenic (M2) phenotypes.

Dendritic cells (DCs)

DCs are specialized APCs found in barrier tissues and are essential in induction of pathogen-specific adaptive immune responses. They engulf pathogens and migrate to lymph nodes, where they display pathogen-derived peptides on the cell surface, which are recognized by T lymphocytes. DCs are thus key in linking innate and adaptive immune systems, and are crucial in generating vaccine-specific immune responses. Peptide presentation by MHC (major histocompatibility complex) molecules and T-cell activation is described in greater detail in the section on adaptive immunity, and also in Chapter 5, Immunopeptidomics for the Identification of T-Cell Vaccine Targets. DCs are described in detail in Chapter 3, Dendritic Cells as Vaccine Targets.

Neutrophils

Neutrophils are bone marrow-derived cells of the myeloid lineage, a member of the granulocyte subfamily (with basophils, eosinophils, and mast cells). Neutrophils moderate early immune responses to infection, and are recruited by chemokines and cytokines released by macrophages following pathogen recognition (McDonald et al., 2010). Neutrophils also express PRRs, which upon activation by PAMPs, trigger the release of reactive oxygen species (ROS), and neutrophil extracellular traps, primarily composed of DNA, in order to immobilize pathogens (Delgado-Rizo et al., 2017; Li et al., 2020).

Although not considered to be professional APCs, neutrophils are able to migrate from peripheral sites to lymph nodes (Beauvillain et al., 2011), and can present antigens to T cells (Leliefeld, Koenderman, & Pillay, 2015).

Eosinophils

Eosinophils are prominent in the blood during allergic conditions, parasitic infection, and cancer. Upon activation, eosinophils secrete cytokines, ROS, enzymes, and lipid mediators including eicosanoids. An elevated concentration of eosinophils in plasma is referred to as blood eosinophilia, and tissue eosinophilia at the site of an infection or during inflammation.

Basophils

Basophils are activated during fungal, bacterial, and viral infections. As a granulocyte, they can release enzymes from intracellular granules that combat pathogenic bacteria, for example, lysozyme (attacking the bacterial cell wall) and proteolytic enzymes.

Natural killer cells

Fully developed NK cells have a half-life of 7–10 days, and kill target cells without the requirement for presentation of pathogen-specific peptides by MHC. They are present in peripheral blood, spleen, uterus, and liver, and are activated by exposure to macrophage-derived cytokines. NK cells are triggered by viral infection, with a response that peaks in 3–4 days, before humoral and/or T-cell responses develop. They are considered to be part of the innate immune system, while differentiating from the lymphocyte rather than myeloid cell lineage. NK cells are able to distinguish between healthy and infected or cancerous cells, by assessing the balance between activating and inhibitory receptor signals on the target cell. All healthy human cells express MHC class I proteins, which send inhibitory signals to NK cells, preventing their destruction. This signal transduction pathway is mediated by a family of NK receptors, for example, NKG2A. Downregulation of cell surface MHC proteins, characteristic of infected or cancerous cells, leads to activation of NK cells.

Natural killer T cells

Natural killer T cells are similar to T cells of the adaptive immune system in that they express TCRs (T-cell receptors) on the cell surface. They can identify microbial lipid antigens displayed by CD1d, and can be activated rapidly following infection, producing copious amounts of proinflammatory chemokines and cytokines (Terabe & Berzofsky, 2008).

Secreted proteins

Cytokines

Cytokines, a group of proteins produced by both myeloid cells and lymphocytes, regulate the immune response to pathogens and facilitate communication between cells (Berraondo et al., 2019).

Interferons

Aristizábal and González (2013) and Kotredes and Gamero (2013) showed that the interferon INF-α can activate NK cells and DCs, inhibiting the growth of cancer cells.

Interleukins

Interleukins (ILs) modulate intercellular cross-talk and are crucial in generating an inflammatory response.

Chemokines

Chemokines comprise a family of soluble chemoattractant cytokines mediating cell migration through venules, and between blood and tissues, in response to a chemical gradient.

The complement system

In addition to immune cells, several classes of soluble molecules are present in the blood and extracellular fluids that are capable of killing or rendering pathogens more susceptible to destruction by other immune mechanisms. These include complement proteins, antimicrobial peptides, and enzymes. Complement proteins have the ability to lyse pathogens and also render them more susceptible to phagocytosis by phagocytic cells (Stevens, 2010).

The complement system is one of the most important secreted mediators of innate immunity. It comprises a group of heat labile soluble serum proteins, which are usually inactive unless activated by a trigger such as detection of antigen–antibody complexes or microorganisms. The system produces a rapid and amplified response to one of these triggers, via a chain of events (Rastogi, 2008). Three different pathways (Fig. 1.1) have been described for complement activation, namely, the classical pathway, involving antigen–antibody complex formation; the alternative pathway, which is antibody independent; and the lectin pathway, also antibody independent (Stevens, 2010).

When activated, complement molecules:

1. Act as opsonins: extracellular proteins that bind to substances or cells to aid their clearance by phagocytosis;
2. Amplify T cell metabolism, and immune complex clearance;

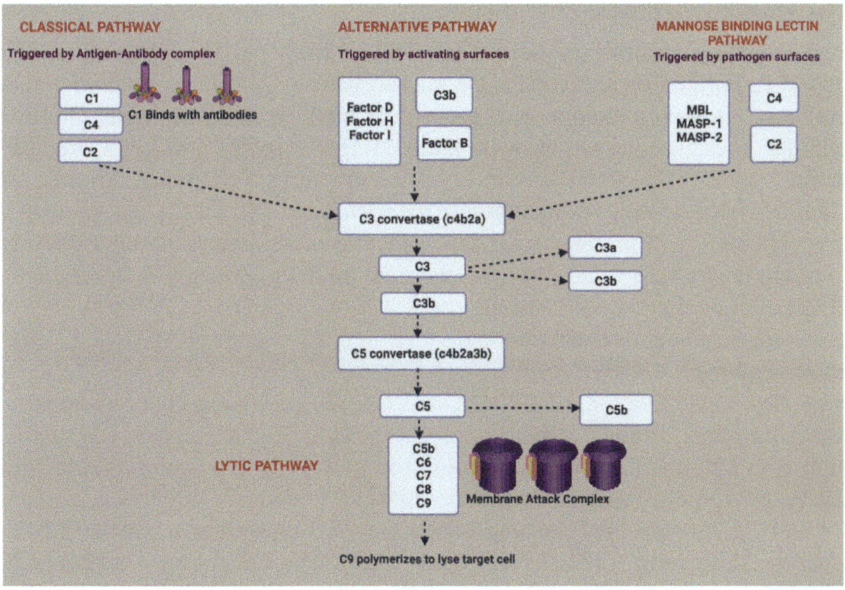

FIGURE 1.1 The complement system. *Source: Illustrations were made using Biorender software.*

3. Facilitate the uptake of immune complexes in the spleen;
4. Enhance vascular permeability;
5. Attract monocytes and neutrophils to the regions with elevated antigen levels;
6. Stimulate the secretion of immunoregulatory molecules that amplify immune responses.

Complement activation promotes phagocytosis and inflammation: The C3b generated by the classical and alternative pathways binds to specific receptor proteins on macrophages and neutrophils. This enhances the ability of these cells to phagocytose microbial cells onto which C3b has attached (Rastogi, 2008).

The activation of the complement system via the classical pathway inhibits the formation of immune complexes that may precipitate in the plasma, while complement activation via the alternative pathway solubilizes immune complexes that have already precipitated. C3-bearing immune complexes are readily removed from tissues and circulation by macrophages and other phagocytes (Stevens, 2010).

Complement is also involved in cytotoxic and cytolytic damage: foreign cells (e.g., bacteria) are lysed by membrane attack complexes formed from complement proteins (Idris, Oyeyinka, Arinola, & Okafor, 2016).

The classical pathway

This pathway is antibody dependent, and is triggered by IgM and/or IgG bound to antigen on the surface of microorganisms. IgM is the most potent antibody class in activation of the classical pathway because it has multiple binding sites; subclasses of IgG, namely, IgG1, IgG2, and IgG3 are also active, with IgG3 being the most efficient (Idris, Oyeyinka, Arinola, & Okafor, 2016). Pathogens including *Escherichia coli*, viruses, a few protozoa, mycoplasmas, and proteins such as C-reactive protein can also bind directly to complement C1q to activate the classical pathway (Massey & Mcpherson, 2007).

Complement activation can be divided into three stages (Fig. 1.1): (1) fixation (binding) of C1 (C1q, C1r, and C1s) to triggers such as IgG, IgM, or microbes; (2) activation, of C4, C2, and C3; (3) formation of the membrane attack complex through activation of C5 through C9, leading to lysis of foreign cells (Stevens, 2010). The recognition unit (C1q, C1r, and C1s) binds to the Fc section of immunoglobulin moieties. C1s becomes stimulated, causing C4 and C2 cleavage and forming C4b2a (i.e., C3 convertase), which, in turn, cleaves C3 to form C4b2a3b (C5 convertase). The C5 convertase splits C5, which then assembles C6, C7, C8, and C9 to form the membrane attack complex. When C9 polymerizes, it causes lysis of the target cell (Stevens, 2010).

The alternative pathway

The alternative complement activation pathway is independent of either IgM or IgG. It involves activation of C3 by a variety of triggers that include bacterial toxins and capsular polysaccharides. Upon activation, C3 is hydrolyzed to produce C3b, which then binds factor B. The complex splits, forming C3Bb—an enzyme that possesses C3 convertase activity. On stabilization by properdin, it causes the cleavage of more C3. Any further molecule of C3 complexed to the C3bBbP enzyme (C3bBb3bP) triggers the cleavage of C5 that further cleaves C5b. C5b draws together C6, C7, C8, and C9 to form the membrane attack complex, as for the classical pathway, resulting in bacterial lysis (Rastogi, 2008; Stevens, 2010).

The lectin pathway

This is another antibody-independent pathway through which complement is activated. It employs nonspecific recognition of the carbohydrate component of microbial cell walls (Beutler, 2004; Eisen & Minchinton, 2003). Mannose-binding or mannan-binding lectin (MBL) binds mannose in a calcium-dependent manner in order to activate complement (Arnold, Dwek, Rudd, & Sim, 2006). Structurally, MBL resembles C1q and is reported to associate with three MBL-serine proteases (MASPs): MASP-1, MASP-2, and MASP-3, which together with MBL bind mannose on bacterial surfaces leading to the formation of an activated C1-like complex. MASP-2 cleaves

C2 and C4, and activation proceeds along the classical pathway (Kerr et al., 2008).

Regulation of complement activation

A system of regulatory proteins modulates the activity of complement. These proteins prevent tissue damage that may otherwise result from spontaneous binding of stimulated complement components to host cells, or inadvertent stimulation of plasma complement components (Arlaud, Barlow, Gaboriaud, Gros, & Narayana, 2007; Idris, Oyeyinka, Arinola, & Okafor, 2016; Schwaeble & Reid, 1999).

Complement activation is regulated efficiently by the following factors:

1. The labile nature of activated complement factors;
2. C8-binding protein: this prevents binding of C9 component to C5b678, so preventing membrane puncture and lysis of target cells;
3. Serum protein Vitronectin (Protein S) that inhibits C5b-7 membrane binding, and prevents C9 polymerization;
4. Presence of C3b inactivator in serum;
5. C1 esterase inhibitor, which inhibits C1 and its subunits;
6. Factor I, acting on free C3b fragment in solution to cleave the molecule;
7. Factor H, which binds to C3b and destroys it.

Adaptive immune response

The innate immune system can respond and clear an infection rapidly. Only after the innate defenses have failed is an adaptive immune response necessary: this is achieved through the interaction of innate and adaptive immune cells (Fig. 1.2), and coordination of the two immune responses, for example, through cytokine secretion. The adaptive immune system can adapt to protect vertebrates from almost any pathogen. Unlike innate immunity, it takes time to be activated: days rather than minutes or hours. However, when activated, it is efficient and specific.

B and T lymphocytes

The adaptive arm of the immune system is composed of B and T lymphocytes, which express B-cell receptors (BCRs) and TCRs, respectively (Fig. 1.3). B cells give rise to humoral immunity via secretion of antibodies (derived from their BCRs) in response to recognition of native, unprocessed antigen. T cells comprise the cellular adaptive immune response: TCRs recognize processed peptide fragments in the context of MHC molecules, leading to stimulation of both cytotoxic and helper T cells. A huge diversity of BCRs and TCRs (millions per individual) are generated through genetic rearrangements of germline gene segments, described in more detail below. Each receptor recognizes, on average, a single target antigen—this ensures that the

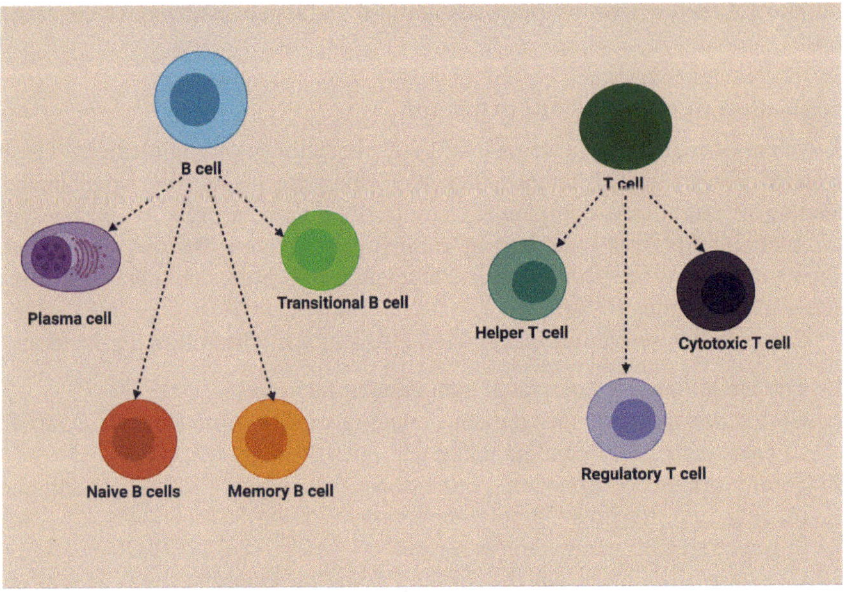

FIGURE 1.2 Cells of the adaptive immune response. *Source: Illustrations were made using Biorender software.*

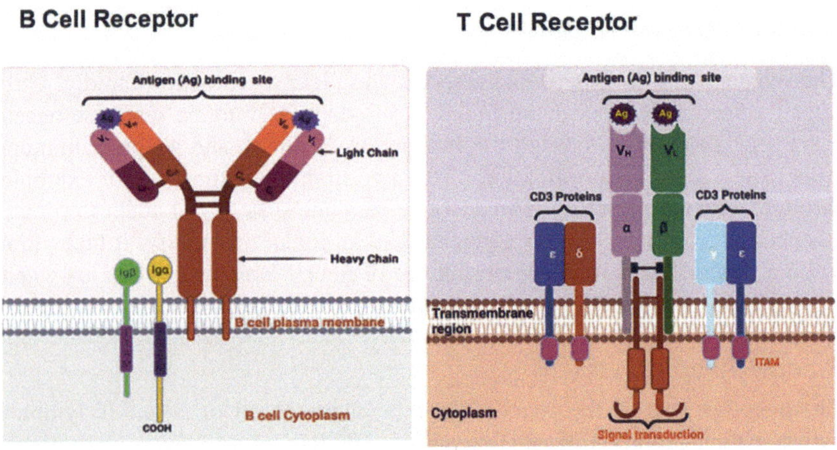

FIGURE 1.3 Structure of the B- and T-cell receptors. *Source: Illustrations were made using Biorender software.*

adaptive immune response is pathogen-specific. Following antigen recognition and binding, the lymphocyte proliferates to produce exact clones of the B or T cells, a process known as clonal expansion.

To generate an effective immune response, a number of different immune cells need to become activated, and interact with each other. These interactions take place in organized tissues collectively known as the lymphoid system, which contain lymphocytes at varying developmental stages, and are categorized into primary and secondary lymphoid organs.

Primary lymphoid organs are the site of lymphoid development (lymphopoiesis). It is here that the progenitor (primitive immature) cells mature. T lymphocytes develop in the thymus hence the nomenclature T lymphocytes, while B lymphocytes develop in the bone marrow (and fetal liver).

Secondary lymphoid organs are the sites of lymphocyte interaction with each other, and with antigens. Following contact with their cognate (specific) antigen, they become effector cells and subsequently disseminate to sites of infection. The major secondary lymphoid organs are the lymph nodes, spleen, and the mucosa-associated lymphoid tissues (e.g., the tonsils). Cell trafficking between different tissues is controlled by a family of soluble chemoattractant proteins (chemokines), and chemokine receptors on the surface of immune cells.

Cell-mediated adaptive immune responses

T lymphocytes

Cellular adaptive immune responses are elicited when antigens reach the secondary lymphoid organs, typically through the action of APCs. APCs, mainly DCs, engulf pathogens and foreign proteins, then process them and present peptide fragments to T cells. Naïve T-cell lymphocytes recognize these antigenic fragments, become activated, proliferate, and differentiate into effector cells that eliminate cells expressing their cognate antigen.

T lymphocytes have both effector and regulatory functions. Effector functions include secretion of proinflammatory cytokines and direct killing of infected cells; effector T cells are also responsible for delayed hypersensitivity responses and rejection of tissue grafts. In humans, the developing thymus becomes infiltrated with prelymphocytes after 3 months of fetal development. The prelymphocytes multiply within the thymus and rearrange their TCR α and β chain gene segments, resulting in expression of a single TCR specificity on the cell surface; this genetic recombination is similar to the process that produces antibodies, described in detail below. The cells then undergo positive selection to remove cells bearing TCRs which fail to recognize the host's MHC alleles.

Survivors of the positive selection process express either CD4 or CD8, along with a functional TCR; they migrate to the thymic medulla. The mean-residence time of single positive T cells in the medulla is about 12 days—a period within which they undergo negative selection, resulting in apoptosis of any cell bearing a TCR that recognizes self-antigen. This process is facilitated by nuclear transcription factor AIRE, which induces expression of many proteins not usually expressed in the thymus. However, negative selection does not remove all self-reactivity, so the induction of regulatory T cells is important to ensure peripheral tolerance, that is, to prevent autoimmunity.

Classes of T cell

CD8 + T cells interact with peptides in complex with MHC class I proteins, while CD4 + T cells recognize antigen presented by MHC class II. The T-cell coreceptor (CD8 or CD4) binds to MHC (class I and II, respectively) to increase the binding avidity of the TCR–MHC interaction. CD4 + cells are typically T helper and T regulatory cells, while CD8 + cells are cytotoxic or "killer" T cells.

Cytotoxic T cells (CD8 T cells) selectively kill target cells expressing a specific antigen using cytotoxic effector proteins. These proteins include perforins that trigger the formation of pores in the target cell membrane, allowing granzymes to move into the cell and cause death by apoptosis (programmed cell death). Cytotoxic T cells also release proinflammatory cytokines. As cytotoxic T cells recognize peptide epitopes processed from antigens within the cell, they are able to kill host cells infected with intracellular pathogens including viruses, parasites (e.g., malaria) and intracellular bacteria, for example, TB.

Helper T cells (CD4 helper T cells, Fig. 1.4) recognize peptides presented by MHC class II on specialized APC surfaces including DCs and macrophages. The peptides are loaded onto MHC following phagocytosis of pathogens, and processing of proteins inside endosomes. Subsequently, the helper

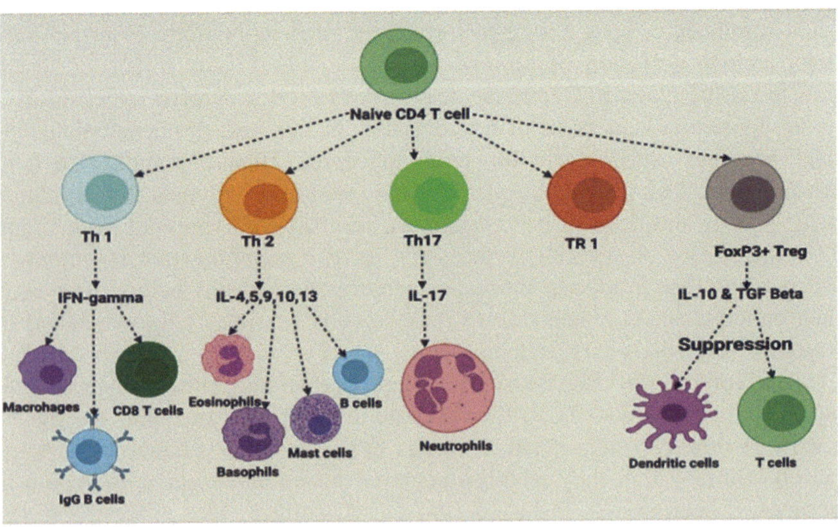

FIGURE 1.4 The clonal expansion of helper T cells. *Source: Illustrations were made using Biorender software.*

T cell is stimulated to secrete cytokines that activate a variety of immune cells including macrophages, cytotoxic T cells, and antibody-secreting B cells. There are several subclasses of helper T cell including Th1, Th2, Th17, and the more recently described (and less well characterized) Th22. Th1 cells produce proinflammatory cytokines including interferon gamma (IFN-γ) and tumor necrosis factor-beta (TNF-β), that target intracellular pathogens and cancerous cells. Th2 cells produce antiinflammatory cytokines including IL-4, IL-5, IL-10, and IL-13. Th2 cells function to help B cells generate immunoglobulins against extracellular pathogens, and in some cases to regulate proinflammatory Th1-mediated immune responses; Th2 responses can become pathogenic, for example, in allergy. Th17 cells are generally proinflammatory, and secrete a variety of cytokines including IL-17, IL-21, and IL-22.

Regulatory T cells (Tregs) are usually CD4 + , although CD8 Treg cells are also known (Yu et al., 2018). The cells suppress immune reactions by a number of different mechanisms, including secretion of tolerogenic cytokines (IL-10 and TGFβ) to block the differentiation of DCs and promote apoptosis of effector T cells (Curotto de Lafaille & Lafaille, 2009; Stadhouders, Lubberts, & Hendriks, 2018). Tregs also make direct cell–cell contacts with APCs and decrease effector T-cell activation. Treg cells express the transcription factor FoxP3 and the cell surface marker CD25 (part of the high affinity IL-2 receptor complex). Tregs can be antigen dependent or independent, and play an important role in controlling inflammation and maintaining peripheral tolerance to self-antigens.

On completion of thymic development and selection, T cells recirculate between the blood and lymphoid organs via the lymphatic system. These recirculating lymphocytes that have not yet encountered their cognate antigen are known as *naïve T cells.* On contact with their cognate antigen on the surface of APCs, they are induced to proliferate and differentiate into *effector T cells* that can respond quickly and efficiently when they encounter cognate antigen on infected cells. Following antigen recognition, *memory T cells* are also produced. These are long-lived cells whose function is to provide protection from the same pathogen on subsequent challenge.

The major histocompatibility complex (MHC) locus

MHC is a wide-ranging DNA locus found in the genome of vertebrates. It comprises a group of closely related polymorphic genes that express multisubunit proteins, many of which display short antigenic peptide fragments (epitopes) on the cell surface for recognition by T cells. The peptides could be derived from self-proteins, grafts, or from pathogens (Janeway, Travers, Walport, & Shlomchik, 2001a, 2001b).

The MHC locus is situated on Chromosome 6 in humans (Fig. 1.5), spanning approximately 30MB (from 6p22.1 to 6p21.3), and containing 224

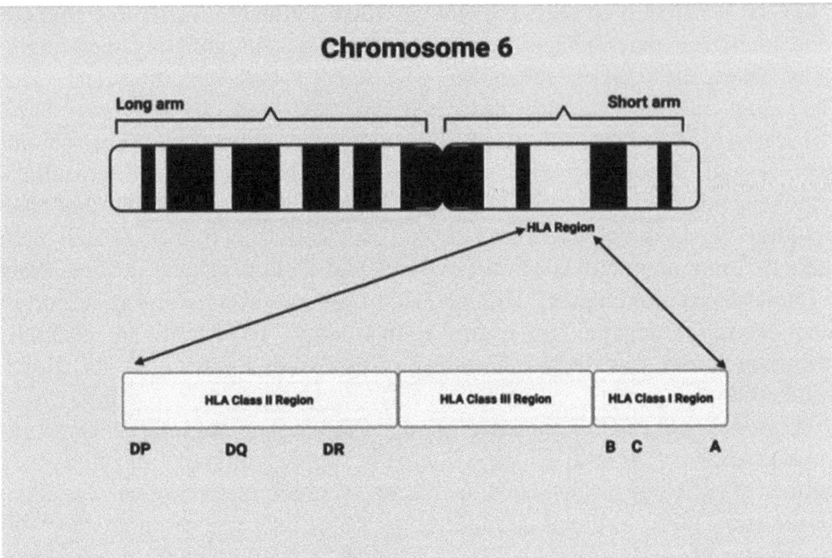

FIGURE 1.5 Human MHC locus located on human chromosome 6. *Source: Illustrations were made using Biorender software.*

genes (Gao et al., 2010). These DNA sequences code for the MHC proteins which are often called human leukocyte antigens (HLA) in humans. MHC alleles have a codominant phenotypic expression. They are highly diverse in the population, with the genes exhibiting the greatest degree of polymorphism in the human genome (Tsuji et al., 2017).

Types of MHC protein and antigen presentation

There are three main MHC subgroups: MHC class I, II, and III (Table 1.2). While MHC class I and II proteins are involved in the display of antigenic fragments on the cell surface, MHC III genes code for other immune proteins such as cytokines, compliment factors, and heat shock proteins (Cruz-Tapias, Castiblanco, & Anaya, 2013).

MHC I

The MHC I complex comprises classical and nonclassical MHC I molecules. Classical MHC class I includes the A, B, and C genes that code for HLA-A, B, and C proteins, respectively. These are highly diverse cell surface proteins that, together with β2 microglobulin, form an antigenic peptide–presenting structure for recognition by CD8 T lymphocytes (Krovi & Gapin, 2016). There are many hundreds of alleles encoding HLA-A, B, and C proteins each with different peptide-binding specificities, ensuring a diverse spectrum of peptide presentation within a population. A given individual, if

TABLE 1.2 Summary of structure, location, and function of MHC proteins.

Properties	MHC I	MHC II	MHC III
Cells expressed	Every cell with a nucleus	Professional APC: B lymphocytes, macrophages, dendritic cells	Hepatocytes, mononuclear phagocytes
Cellular location (classical molecules)	Cell surface	Cell surface	Cytoplasmic
Classical proteins expressed	HLA (A, B, and C)	HLA (DRA, DRB1, DQA1, DQB1, DPA1, and DPB1)	Some complement factors, cytokines, and heat shock proteins
Components of peptide–MHC complex	Alpha chains (comprising α1, α2, and α3 domains) and $beta_2$ microglobulin. Peptide binds to α1 and α2 domains.	Alpha chains (α1 and α2 domains) and beta chains (β1 and β2 domains). Peptide binds to α1 and β1 domains	Do not bind peptides
T lymphocytes that bind MHC protein	CD8 + lymphocytes	CD4 + lymphocytes	None
Source of antigen	Cytoplasmic proteins from self-antigens, tumors or intracellular pathogens	Ingested from extracellular pathogens	Do not process antigen
Protein degrading enzymes	Cytoplasmic proteasomes	Proteases from endosomes and lysosomes	Antigens are not degraded
Loading site of peptide fragments	Endoplasmic reticulum	Specialized vesicular compartment	Peptides are not loaded
Molecules involved in transportation and loading of peptides	TAP, Tapasin, and other chaperones	Invariant chain (Ii), DM	Peptides are not loaded
Antigen fragment size	Mostly 9–10 residues, range of 8–14	13–18 residues	Do not process antigen

heterozygous, will possess two different A, B, and C alleles, that is, a total of six class I HLA molecules.

The nonclassical MHC class I molecules (MHC-Ib) are nonpolymorphic molecules, including CD1, HLA-E, HLA-F, HLA-G, HLA-H, MHC chain-related gene A, and neonatal Fc-receptor for IgG (FcRn) (Halenius, Gerke, & Hengel, 2015). While some MHC-Ib molecules are antigen-presenting structures that are recognized by specific subsets of TCRs, others are ligands for innate immune receptors. MHC-Ib molecules are classified as a component of the innate or nonspecific immune system because they elicit a more rapid response (Krovi & Gapin, 2016).

MHC I proteins are found on platelets and all cells with nuclei; they present peptide epitopes to cytotoxic T lymphocytes (CTL), which express CD8 protein in addition to TCR on their cell surface. MHC I triggers a cell-mediated immune response to kill tumorigenic cells and those infected with viruses and other intracytoplasmic pathogens such as *Mycoplasma* and *Rickettsia* (Gulzar & Copeland, 2004). Apart from binding CD8 receptors of T cells, MHC I also binds to receptors of NK cells (Rosenberg & Huang, 2018).

Structurally, MHC I (Fig. 1.6) has a variable heavy α-subunit and a small invariant β2 microglobulin subunit, which are about 45 and 12 KDa, respectively, in humans. The alpha subunit contains the binding groove for the peptide epitope, while the β2 microglobulin subunit provides structural support

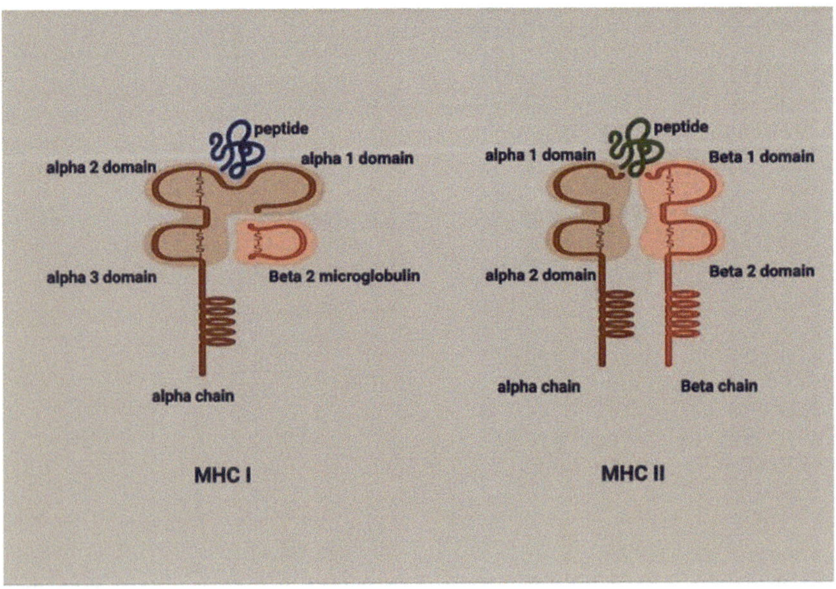

FIGURE 1.6 The class I MHC–peptide and class II MHC–peptide complexes. *Source: Illustrations were made using Biorender software.*

and stability (Khan, 2016). The alpha subunit comprises a small cytoplasmic tail, transmembrane helix, and an N-terminal extracellular region consisting of alpha 1, alpha 2, and alpha 3 domains (Wieczorek et al., 2017). The alpha 1 and alpha 2 domains form the deep epitope-binding pocket that binds a peptide that is 8−14 amino acids in length, more typically 9 or 10 amino acids. The floor of the groove is formed by eight β sheets, situated between two long α-helices. The peptide binds noncovalently to the MHC I−binding groove with two or more "anchor residues" fitting into pockets found in the floor of the groove (Wieczorek et al., 2017). The alpha 3 domain is invariant between alleles, and binds to the CD8 coreceptor (Khan, 2016).

Peptides loaded onto the MHC I complex can either be obtained from endogenous or degraded exogenous proteins, the latter in a process known as cross-presentation (Michalek, Grant, Gramm, Goldberg & Rock, 1993). Endogenously derived self-peptides are obtained from protein degradation and ribosomal artifacts, for example, mis-translated polypeptides. Proteins from tumors, viruses, and intracellular microbes are also degraded by proteasomes into small peptides. The Class I antigen presentation pathway is shown in Fig. 1.8 and described in detail in Chapter 5, Immunopeptidomics for the Identification of T-Cell Vaccine Targets, so will not be repeated here.

CD8

CD8 is a TCR coreceptor, and specifically binds to HLA class I protein (Gao & Jakobsen, 2000). It is found on T cells, DCs, cortical thymocytes, and NK cells (Leong, Cooper, & Leong, 2003). To function effectively, the CD8 coreceptor forms a dimer comprising two chains (alpha and beta). Both chains belong to the immunoglobulin superfamily having intracellular, transmembrane, and immunoglobulin variable (IgV)-like extracellular domains.

The α3 domain of Class I MHC interacts with the extracellular IgV-like domain of CD8-α triggering a signal transduction event in which the cytoplasmic tail of the CD8 coreceptor interacts with lymphocyte-specific protein tyrosine kinase (Lck) (Devine, Sun, Barr, & Kavathas, 1999). Upon binding, Lck phosphorylates the CD3 alpha, beta, and zeta chains of the TCR. The signal transduction process culminates in the stimulation of transcription factors such as NFAT, NF-κB, and AP-1. Overall, the CD8 coreceptor plays a major role in CTL activation by increasing the sensitivity to antigen by up to 1 million-fold (Cole et al., 2012).

MHC class II

Each MHC class II protein is dimeric, with an alpha chain and a beta chain (Fig. 1.6). The classical MHC II locus has three gene clusters: DP, DQ, and DR. These gene clusters code for the alpha and beta chains, namely DP alpha 1 (HLA-DPA1), DP beta 1 (HLA-DPB1), DQ alpha 1 (HLA-DQA1), DQ beta 1 (HLA-DQB1), DR alpha (HLA-DRA), and DR beta 1 (HLA-DRB1).

Genes for two nonclassical MHC class II-like proteins are also encoded in this locus, HLA-DM and HLA-DO (Wosen, Mukhopadhyay, Macaubas, & Mellins, 2018). These molecules are localized in endosomes/lysosomes, and facilitate the loading of peptide fragments onto class II MHC class molecules (Alfonso & Karlsson, 2000). As for class I MHC, there are hundreds of alleles encoding DP, DQ and DR chains, with differing peptide-binding properties. Each individual, if heterozygous, will express two different DP, DQ, and DR proteins (a total of 6).

During inflammation, MHC II molecules can be found on all cell types, but they are predominantly found on professional APCs such as B cells, DCs, and macrophages (Unanue, Turk, & Neefjes, 2016). The professional APCs are specialized in eliminating extracellular microbes, or those that can survive in the phagosomes of macrophages (Gominet, Compain, Beloin, & Lebeaux, 2017). In DCs and macrophages, microbes are ingested by phagocytosis into phagosomes, and in B cells through endocytosis into endosomes. After ingestion, antigens are degraded in the endocytic compartments and peptide epitopes displayed on the surface for recognition by CD4 + T helper lymphocytes (Rock, Reits & Neefjes, 2016).

Class II MHC proteins are heterodimers with polymorphic α and β subunits. Unlike the class I MHC protein, where the antigenic peptide-binding groove (Fig. 1.7) is shaped by the alpha 1 and alpha 2 domains, in the class II MHC molecule it is shaped by the alpha 1 and beta 1 domains. Beta sheets

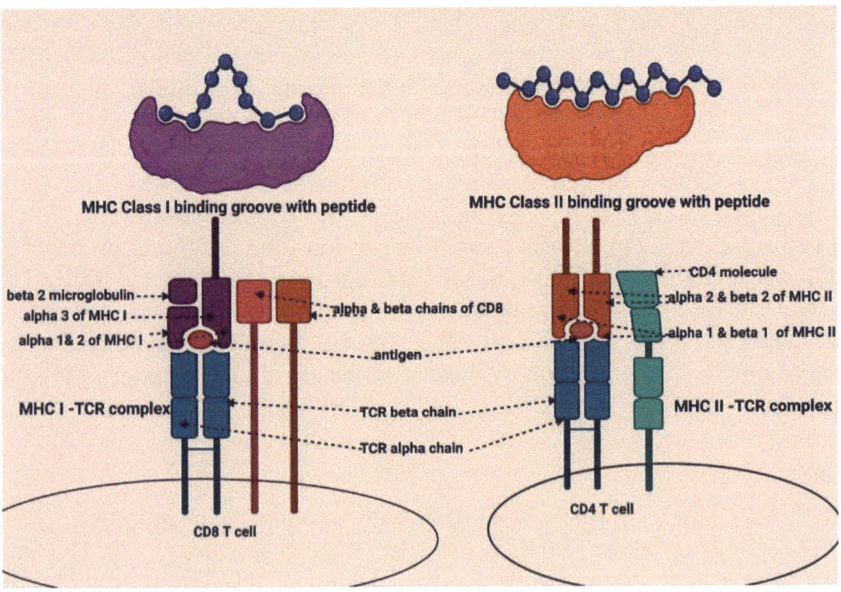

FIGURE 1.7 Binding grooves of MHC I and II and MHC–TCR complexes. *Source: Illustrations were made using Biorender software.*

form the floor of the peptide-bending cleft and the sides are formed by alpha helices, leaving both ends of the binding groove open. With this open conformation, class II MHC molecules present longer peptide fragments (13–18 amino acids) than MHC I (Stern et al., 1994). In addition, the class II immunoglobulin domains alpha 2 and beta 2 bind the CD4 coreceptor, which increases binding avidity of the TCR–MHC complex in a similar fashion to CD8 and class I MHC (Wang et al., 2011).

The antigen processing pathway for class II MHC peptides resembles that for class I MHC peptides (Fig. 1.8). The folding and assembly of class II MHC proteins with the invariant chain (Ii) takes place in the ER (Roche & Cresswell, 1990). Acting as a pseudopeptide, the invariant chain Ii fills the peptide-binding pocket of the MHC II molecule through its cytosolic dileucine motif, and targets it into the endosomal pathway (Neefjes, Jongsma, Paul & Bakke, 2011). In the endosomal location known as MHC class II compartment (MIIC), the MHC II molecules bound to the invariant chain Ii encounter peptide fragments generated from antigens by the resident proteases (Neefjes, 1999). Proteases including cathepsin L and S also degrade the invariant chain (Unanue et al., 2016), leaving a small fragment of the invariant chain known as the class II–associated invariant chain peptide (CLIP), which lies within the peptide-binding groove of the MHC II molecule (Ghosh, Amaya, Mellins, & Wiley, 1995).

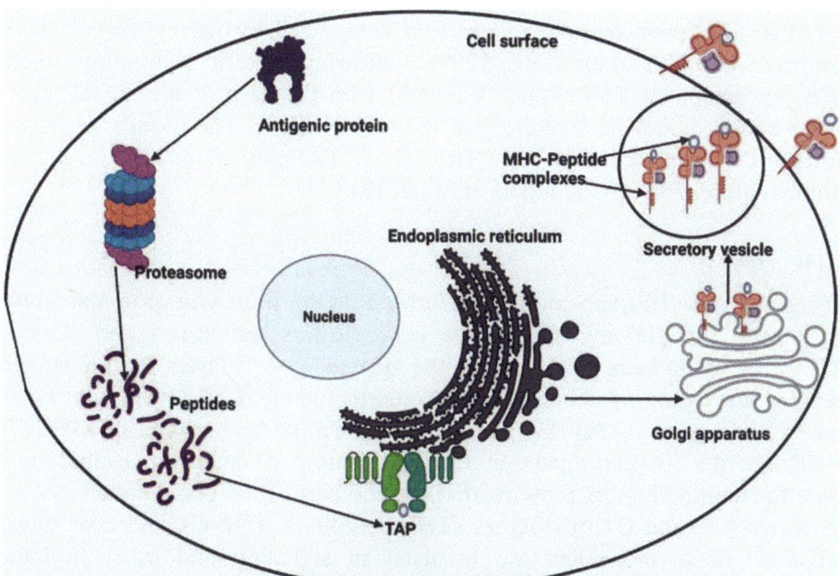

FIGURE 1.8 The class I MHC antigen-processing pathway. *Source: Illustrations were made using Biorender software.*

Immunodominant peptides with higher-binding affinity for MHC II then displace the CLIP fragments, with the aid of the DM chaperone (Denzin & Cresswell, 1995). DM releases the low affinity CLIP peptide and replaces it with higher affinity peptide fragments from degraded antigens (Blum, Wearsch & Cresswell, 2013). From the MIIC, the MHC II-peptide (pMHC-II) complex moves to the plasma membrane through tubules or vesicles (Boes et al., 2002; Kleijmeer et al., 2001).

CD4

The CD4 coreceptor is expressed on DCs, T helper cells, monocytes, and macrophages (Claeys & Vermeire, 2019). Like CD8, CD4 belongs to the immunoglobulin superfamily with four immunoglobulin domains (D1−D4) in the extracellular portion (Barclay, Brady, Davis, & Lange, 1993). The α2 and β2 membrane-proximal (nonvariable) domains of class II MHC interact with the D1 domain of CD4. This causes the Lck kinase attached to the cytoplasmic tail of CD4 to trigger tyrosine phosphorylation of the immunoreceptor tyrosine activation motifs (ITAM) on the cytoplasmic domains of CD3 (Barber et al., 1989; Rudd, Trevillyan, Dasgupta, Wong & Schlossman, 2010). ITAM phosphorylation amplifies the TCR signal, and further activates SH2 domain-containing tyrosine kinases, such as ZAP70. The entire signal transduction process culminates in the stimulation of transcription factors such as NFAT, NF-κB, and AP-1, ultimately resulting in T lymphocyte activation (Owen et al., 2013).

MHC II-peptide complexes are first recognized by naïve (Th0) CD4 T lymphocyte cells (Cunha-Neto, 1999). Following specific recognition, there is an expansion of Th0 clones (Fig. 1.4) with the help of the signal transducer and activator of transcription (STAT) proteins. These cells are functionally differentiated into Th1, Th2, Th17, TR1, and Foxp3 + Treg cells (Cunha-Neto, 1999; Stadhouders et al., 2018).

MHC class III

The MHC class III gene complex is located on the short arm of human chromosome 6 (6p21.3) in between the gene clusters for class I and class II (Gruen & Weissman, 2001). It is the densest gene cluster in the human genome consisting of about 61 genes, approximately 27% of the total number of MHC genes (224) (Hurt et al., 2004). Unlike MHC I and II, MHC III molecules do not participate in the presentation of antigens. Rather they code for immunological proteins that include part of the complement system (B factor, C2, and C4), cytokines (TNF-α, TNF-β, TNF-C), and heat shock proteins. These molecules are involved in signaling and other immune responses (Relle & Schwarting, 2012; Xie et al., 2003).

MHC class III proteins are secreted by hepatocytes and phagocytes such as monocytes, macrophages, and DCs (Colten, 1984).

Humoral (antibody mediated) adaptive immune response

Antibodies, or immunoglobulins, are serum proteins secreted by the terminally differentiated cells of the B cell lineage, plasma cells. They bind specifically to native antigens that are accessible on the surface of pathogens encountered in the blood, peripheral organs, or mucosal surfaces. As with T cells, naïve B cells are also activated on contact with their cognate antigen; they usually require T-cell help for this activation. Once activated, B cells will differentiate to antibody-producing plasma cells and memory B cells.

Antibodies provide protection in three major ways: *opsonization*, which is coating the surface of the pathogen to enhance engulfment by phagocytes; *neutralization*, which involves preventing the pathogen from entering the cell by disrupting binding to specific host cell receptors; and *complement activation*.

Somatic mutation: When a B cell becomes activated, it undergoes several rounds of mutation and selection for high-affinity immunoglobulin secreting plasma cells. This is a process known as somatic hypermutation, and occurs in the germinal centers of lymph nodes with the help of Tfh (T follicular helper) cells. The selected high-affinity B cells continue to proliferate and mature with T-cell help, while cells with low-affinity receptors die by apoptosis.

B lymphocytes: origin and differentiation

B cells are derived from a lymphoid-myeloid precursor that can differentiate into a common myeloid progenitor or an early lymphocyte progenitor (ELP). Depending on the type of cytokine present, ELP can form B cells, T cells, NK cells, or other innate lymphocyte subsets. B-cell precursors remain in the microenvironment of bone marrow stromal cells. Several transcription and growth factors such as early B-cell factor, interleukin-7, and paired box protein 5 are involved in the differentiation of common lymphoid progenitors into pre-B cells. During this maturation process, the gene segments coding the heavy and light chain of the antibody molecule are rearranged. Immature B cells are identifiable with distinctive markers, namely, surface antigens CD19, CD45R, CD43, CD24, terminal deoxyribonucleotide transferase, recombination-activating genes (RAG-1 and RAG-2), and c-Kit. During maturation in the bone marrow, both positive and negative selection occurs, as for T cells in the thymus.

In the spleen, immature B cells further develop into either follicular or marginal zone B cells. All matured naïve (unstimulated) B cells express IgM on their surface. Naive B cells have a life span of only a few days; however, if they encounter antigen in the secondary lymphoid organs, they are activated and differentiate to form memory cells and antibody-secreting plasma cells. Activated B cells expresses markers such CD25 (high-affinity interleukin-2 receptor subunit).

Structure of immunoglobulins

Antibody molecules have a common basic structure comprising two identical light (L) chains and two identical heavy (H) chains. Both the heavy and light chains have variable (V) and constant (C) regions, with the tip of the V regions containing the hyper-variable CDRs (complementarity determining regions) that bind antigens: three CDRs are contributed by the H chain and three by the L chain (Fig. 1.9). L and H chains are held together by inter-chain disulfide bonds formed between cysteine residues, forming immunoglobulin fold domains. The Fc antibody region is formed by the interaction of the two H chains, again by disulfide bonding to form immunoglobulin domains, with further stabilization provided by the presence of N-linked glycosylation. Fc domains differ according to antibody class, leading to the triggering of different effector functions following antigen binding.

Antibody diversity: The germline DNA in a human encodes 30–65 V (variable) gene segments for each L/H chain, as well as 27 D (diversity) segments for the H chain, and 4–6 J (joining) segments per chain. Genetic recombination events in the bone marrow combine random V, D, and J segments together to create a huge diversity of antigen-binding domains: recombination can create over 10^{10} different BCRs per individual, although in practice this number will be limited by the total number of B cells in circulation. There are two sets of L chain gene segments, kappa and lambda, only one of which will be rearranged per B cell to ensure clonality of the BCR.

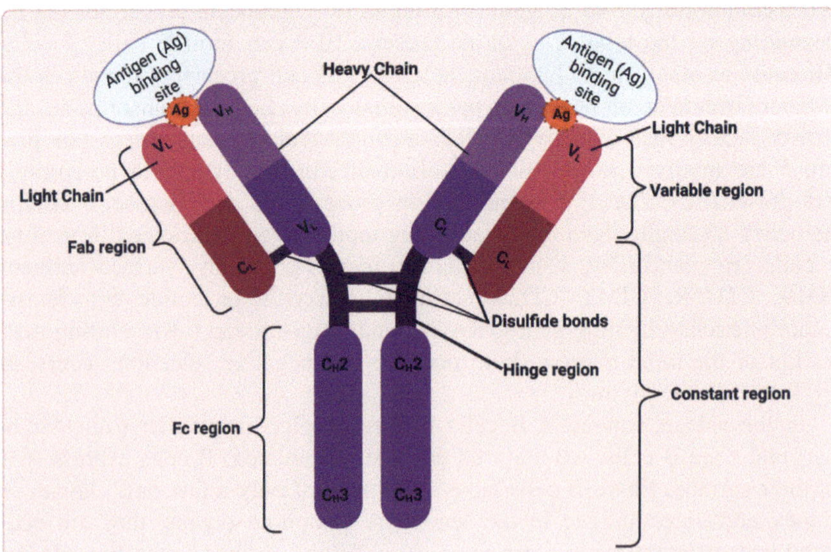

FIGURE 1.9 Structure of an immunoglobulin (antibody). *Source: Illustrations were made using Biorender software.*

Further diversity is created by somatic hypermutation (which follows antigen recognition and B-cell activation) in germinal centers of lymph nodes, which creates point mutations in CDRs leading to an overall increase in antigen affinity.

Immunoglobulin classes

Humans have five isotypes of immunoglobulin, namely, immunoglobulin G (IgG), immunoglobulin A (IgA), immunoglobulin M (IgM), immunoglobulin D (IgD), and immunoglobulin E (IgE). These differ in structure, function, and heavy-chain constant regions while retaining the V domains that determine antigen specificity. IgM is expressed as a surface-bound receptor (BCR) on the surface of naïve B cells, and following antigen recognition and B-cell activation, a phenomenon known as class-switching occurs. Class switching involves genetic recombination events that sequentially removes exons encoding H chain Fc domains and replaces them with an exon encoding another domain. The order of exons encoding isotype-specific gene sequences is μ (IgM), δ (IgD), $\gamma 3$ (IgG3), $\gamma 1$ (IgG1), $\alpha 1$ (IgA1), $\gamma 2$ (IgG2) $\gamma 4$ (IgG4), ε (IgE), and $\alpha 2$ (IgA2).

IgM antibodies dominate the primary immune response. A nonpolymerized (singlet) form of IgM with a membrane-bound sequence is the antibody receptor used by B lymphocytes to identify antigen. Postantigen recognition, an alternative splicing event removes the transmembrane exon leading to secretion of soluble, pentameric IgM, with 10 antigen-binding sites. Due to its large size, it is restricted to the intravascular blood stream and to some extent the lymph, but not peripheral tissues. It is very efficient in activating complement.

IgG has a single Fc domain and is divalent in terms of antigen binding. It constitutes 75% of total serum immunoglobulin and is also found in nonmucosal tissues. It directly inactivates pathogens by interacting with effector moieties, that is, complement and Fc receptors found on the surface of phagocytic cells. It is the major immunoglobulin of the secondary immune response and is important for vaccine-induced immunity. IgG has four subclasses: IgG1, IgG2, IgG3, and IgG4. These subclasses have functional differences, for example, IgG3 has the highest capacity to activate complement, while IgG1 has the highest affinity for Fc gamma-receptors and is therefore the best mediator of opsonization. IgG is the only immunoglobulin class that can transport across the placenta.

IgA is found as IgA1 and IgA2 in humans. IgA2 differs from IgA1 in that it possesses a shorter hinge and resists bacterial proteases better. IgA molecules are mostly dimers of two immunoglobulin molecules linked by an extra J chain polypeptide, and so are tetravalent. Secretory IgA found in mucosa is made of dimeric (tetravalent) IgA plus an extra protein known as the secretory component. It is involved in shielding the body's mucosal surfaces from infectious agents. Dimeric IgA is common in seromucous secretions across epithelia: tears, saliva, breast milk, respiratory, gastrointestinal, and genitourinary secretions. Monomeric and small amounts of dimeric IgA are found in the serum where they can link pathogens to effector cells via Fc receptors specific for IgA.

IgD accounts for less than 1% of total circulating immunoglobulin. It is bound to the surface of B cells and has a very short half-life. It is not known to possess any of the functional activity of other immunoglobulin classes, but may have a role as an antigen receptor on the surface of B cells. It is monomeric and has a long hinge region.

IgE is a dimeric monomer, like IgG, and is present in very low concentrations in blood and extracellular fluid. It binds to receptors on eosinophils and is important in immunity to parasitic worms. When it binds to high-affinity IgE receptors on mast cells and basophils, it can trigger inflammatory mediators such as histamine, provoking effects such as sneezing, coughing, and vomiting that can expel infectious agents. In atopic individuals, allergic reactions are triggered when allergens bind to IgE on the surface of mast cells.

Human monoclonal antibodies as therapeutic agents

Mouse monoclonal antibodies administered to human subjects are significantly immunogenic; the human antimouse antibodies (HAMA) that form can accelerate clearance of a monoclonal antibody from the blood so reducing its efficacy, and in addition can potentially cause hypersensitivity reactions.

More recently, several approaches have been taken to produce fully human antibodies. First, production of human antibody-secreting hybridomas, by fusing human B cells with an immortal B cell line. Immortalized Epstein-Barr virus (EBV)–transformed B cell lines have also been used as a source of human monoclonal antibodies. EBV-transformed lines tend to lose their antibody secretory potential upon prolonged culturing. A better approach is to isolate the genes encoding a particular antibody, and to stably transfect these into an immortalized cell line (e.g., HEK293 or CHO cells) to generate cells that secrete large amounts of the recombinant antibody. Production of monoclonal antibodies using this technology can be easily scaled up for manufacture of clinical-grade material, and antibodies targeting Ebola, HIV, and SARS-CoV-2 have been produced in this way.

Prophylactic and therapeutic vaccines

Vaccines can be either prophylactic or therapeutic. A prophylactic (preventive) vaccine acts by inducing immunity to pathogen-specific antigens prior to natural exposure to that pathogen. Most vaccines for infectious agents are prophylactic. Therapeutic vaccines boost an immune response post-disease induction, and have been used for some pathogens (Rabies, for example), cancer, and in an experimental setting for chronic diseases including dementia.

Immunologic memory and vaccination

Vaccines activate lymphocytes, which are the effector cells of the adaptive immune system and mediate immunological memory. These memory cells (B and T cells) are able to respond quickly when challenged with the "real" infectious agent, to prevent an infection becoming established. This recall response is much more rapid than the activation of new B and T cells during a natural infection in unvaccinated individuals, which takes weeks to develop. These unvaccinated, infected individuals will also generate a memory B and T cell response, and therefore can become protected against future infections, providing that the infection does not prove fatal in the meantime.

There are numerous issues that have prevented the production of effective vaccines against some infectious agents, for example, high pathogen mutation rates giving rise to multiple strains, some of which are able to escape vaccine-induced immunity. There is also a large variation in the level of serum antibodies that is required for neutralization of different pathogen species. In order to improve vaccine efficacy, it is possible to combine vaccine antigens with chemical agents known as adjuvants, which can increase the vaccine response by stimulating various cells of the immune system (known as immunopotentiation), or, in the case of alum, act as an antigen depot. More information on specific adjuvants is given in Chapter 9, Adjuvants, Immunomodulators, and Adaptogens.

References

Alfonso, C., & Karlsson, L. (2000). Nonclassical MHC class II molecules. *Annual Review of Immunology, 18*, 113–142. Available from https://doi.org/10.1146/annurev.immunol.18.1.113.

Aristizábal, B., & González, Á. (2013). Innate immune system. In J. M. Anaya, Y. Shoenfeld, A. Rojas-Villarraga, et al. (Eds.), *Autoimmunity: From bench to bedside [Internet]*. Bogota: El Rosario University Press. Chapter 2https://www.ncbi.nlm.nih.gov/books/NBK459455/.

Arlaud, G. J., Barlow, P. N., Gaboriaud, C., Gros, P., & Narayana, S. V. (2007). Deciphering complement mechanisms: The contributions of structural biology. *Molecular Immunology, 44*(16), 3809–3822. Available from https://doi.org/10.1016/j.molimm.2007.06.147.

Arnold, J. N., Dwek, R. A., Rudd, P. M., & Sim, R. B. (2006). Mannan binding lectin and its interaction with immunoglobulins in health and in disease. *Immunology Letters, 106*(2), 103–110. Available from https://doi.org/10.1016/j.imlet.2006.05.007.

Atri, C., Guerfali, F. Z., & Laouini, D. (2018). Role of human macrophage polarization in inflammation during infectious diseases. *International Journal of Molecular Sciences, 19*(6), 1801. Available from https://doi.org/10.3390/ijms19061801. Available from 29921749, PMCID: PMC6032107.

Barber, E. K., Dasgupta, J. D., Schlossman, S. F., Trevillyan, J. M., & Rudd, C. E. (1989). The CD4 and CD8 antigens are coupled to a protein-tyrosine kinase (p56lck) that phosphorylates the CD3 complex. *Proceedings of the National Academy of Sciences of the United States of America*, *86*(9), 3277–3281. Available from https://doi.org/10.1073/pnas.86.9.3277.

Barclay, A. N., Brady, R. L., Davis, S. J., & Lange, G. (1993). CD4 and the immunoglobulin superfamily. *Philosophical Transactions of the Royal Society of London. Series B, Biological Sciences*, *342*(1299), 7–12. Available from https://doi.org/10.1098/rstb.1993.0129.

Beauvillain, C., Cunin, P., Doni, A., Scotet, M., Jaillon, S., Loiry, M. L., Magistrelli, G., Masternak, K., Chevailler, A., Delneste, Y., & Jeannin, P. (2011). CCR7 is involved in the migration of neutrophils to lymph nodes. *Blood*, *117*(4), 1196–1204. Available from https://doi.org/10.1182/blood-2009-11-254490.

Berraondo, P., Sanmamed, M. F., Ochoa, M. C., Etxeberria, I., Aznar, M. A., Pérez-Gracia, J. L., Rodríguez-Ruiz, M. E., Ponz-Sarvise, M., Castañón, E., & Melero, I. (2019). Cytokines in clinical cancer immunotherapy. *British Journal of Cancer*, *120*(1), 6–15. Available from https://doi.org/10.1038/s41416-018-0328-y, Epub 2018 Nov 9.

Beutler, B. (2004). Innate immunity: An overview. *Molecular Immunology*, *40*(12), 845–859. Available from https://doi.org/10.1016/j.molimm.2003.10.005.

Blum, J. S., Wearsch, P. A., & Cresswell, P. (2013). Pathways of antigen processing. *Annual Review of Immunology*. *31*, 443–473. Available from: https://doi.org/10.1146/annurev-immunol-032712-095910. Epub 2013 Jan 3. PMID: 23298205; PMCID: PMC4026165.

Boes, M., Cerny, J., Massol, R., Op den Brouw, M., Kirchhausen, T., Chen, J., & Ploegh, H. L. (2002). T-cell engagement of dendritic cells rapidly rearranges MHC class II transport. *Nature*, *418*(6901), 983–988. Available from https://doi.org/10.1038/nature01004.

Claeys, E., & Vermeire, K. (2019). The CD4 receptor: An indispensable protein in T cell activation and a promising target for immunosuppression. *Archives of Microbiology & Immunology*, *3*(3), 133–150.

Cole, D. K., Laugel, B., Clement, M., Price, D. A., Wooldridge, L., & Sewell, A. K. (2012). The molecular determinants of CD8 co-receptor function. *Immunology*, *137*(2), 139–148. Available from https://doi.org/10.1111/j.1365-2567.2012.03625.x.

Colten, H. R. (1984). Expression of the MHC class III genes. *Philosophical Transactions of the Royal Society of London. Series B, Biological Sciences*, *306*(1129), 355–366. Available from https://doi.org/10.1098/rstb.1984.0096.

Cruz-Tapias, P., Castiblanco, J., & Anaya, J. M. (2013). Major histocompatibility complex: Antigen processing and presentation. In: Anaya JM, Shoenfeld Y, Rojas-Villarraga A, et al., editors. Autoimmunity: From Bench to Bedside [Internet]. Bogota (Colombia): El Rosario University Press; Chapter 10. Available from: https://www.ncbi.nlm.nih.gov/books/NBK459467/

Cunha-Neto, E. (1999). MHC-restricted antigen presentation and recognition: constraints on gene, recombinant and peptide vaccines in humans. *Brazilian Journal of Medical and Biological Research = Revista Brasileira de Pesquisas Medicas e Biologicas / Sociedade Brasileira de Biofisica*, *32*(2), 199–205. Available from https://doi.org/10.1590/s0100-879x1999000200008.

Curotto de Lafaille, M. A., & Lafaille, J. J. (2009). Natural and adaptive foxp3 + regulatory T cells: More of the same or a division of labor? *Immunity*, *30*(5), 626–635. Available from https://doi.org/10.1016/j.immuni.2009.05.002.

Delgado-Rizo, V., Martínez-Guzmán, M. A., Iñiguez-Gutierrez, L., García-Orozco, A., Alvarado-Navarro, A., & Fafutis-Morris, M. (2017). Neutrophil extracellular traps and its implications in inflammation: An overview. *Frontiers in Immunology*, *8*, 81. Available from https://doi.org/10.3389/fimmu.2017.00081.

Denzin, L. K., & Cresswell, P. (1995). HLA-DM induces CLIP dissociation from MHC class II alpha beta dimers and facilitates peptide loading. *Cell*, *82*(1), 155–165. Available from https://doi.org/10.1016/0092-8674(95)90061-6.

Devine, L., Sun, J., Barr, M. R., & Kavathas, P. B. (1999). Orientation of the Ig domains of CD8 alpha beta relative to MHC class I. *Journal of Immunology*, *162*(2), 846–851.

Eisen, D. P., & Minchinton, R. M. (2003). Impact of mannose-binding lectin on susceptibility to infectious diseases. *Clinical Infectious Diseases: An Official Publication of the Infectious Diseases Society of America*, *37*(11), 1496–1505. Available from https://doi.org/10.1086/379324.

Gao, G. F., & Jakobsen, B. K. (2000). Molecular interactions of coreceptor CD8 and MHC class I: The molecular basis for functional coordination with the T-cell receptor. *Immunology Today*, *21*(12), 630–636. Available from https://doi.org/10.1016/s0167-5699(00)01750-3.

Gao, J., Liu, K., Liu, H., Blair, H. T., Li, G., Chen, C., Tan, P., & Ma, R. Z. (2010). A complete DNA sequence map of the ovine major histocompatibility complex. *BMC Genomics*, *11*, 466. Available from https://doi.org/10.1186/1471-2164-11-466.

Ghosh, P., Amaya, M., Mellins, E., & Wiley, D. C. (1995). The structure of an intermediate in class II MHC maturation: CLIP bound to HLA-DR3. *Nature*, *378*(6556), 457–462. Available from https://doi.org/10.1038/378457a0.

Gominet, M., Compain, F., Beloin, C., & Lebeaux, D. (2017). Central venous catheters and biofilms: Where do we stand in 2017? *Acta Pathologica, Microbiologica, et Immunologica Scandinavica*, *125*(4), 365–375. Available from https://doi.org/10.1111/apm.12665.

Gruen, J. R., & Weissman, S. M. (2001). Human MHC class III and IV genes and disease associations. *Frontiers in Bioscience: A Journal and Virtual Library*, *6*, D960–D972. Available from https://doi.org/10.2741/gruen.

Gulzar, N., & Copeland, K. F. (2004). CD8 + T-cells: Function and response to HIV infection. *Current HIV Research*, *2*(1), 23–37. Available from https://doi.org/10.2174/1570162043485077.

Halenius, A., Gerke, C., & Hengel, H. (2015). Classical and non-classical MHC I molecule manipulation by human cytomegalovirus: so many targets—but how many arrows in the quiver? *Cellular & Molecular Immunology*, *12*(2), 139–153. Available from https://doi.org/10.1038/cmi.2014.105.

Hirayama, D., Iida, T., & Nakase, H. (2017). The phagocytic function of macrophage-enforcing innate immunity and tissue homeostasis. *International Journal of Molecular Sciences*, *19*(1), 92. Available from https://doi.org/10.3390/ijms19010092. Available from 29286292, PMCID: PMC5796042.

Hurt, P., Walter, L., Sudbrak, R., Klages, S., Müller, I., Shiina, T., Inoko, H., Lehrach, H., Günther, E., Reinhardt, R., & Himmelbauer, H. (2004). The genomic sequence and comparative analysis of the rat major histocompatibility complex. *Genome Research*, *14*(4), 631–639. Available from https://doi.org/10.1101/gr.1987704.

Idris, M., Oyeyinka, G. O., Arinola, O. G., & Okafor, G. O. (2016). In L. S. Salimonu (Ed.), *Basic immunology* (3rd ed., pp. 89–97). College Press Ibadan, ISBN: 978-978-54640-3-0.

Janeway, C. A., Jr, Travers, P., Walport, M., & Shlomchik, M. J. (2001a). *Immunobiology: The immune system in health and disease. The major histocompatibility complex and its functions* (5th ed.). New York: Garland Science. Available from https://www.ncbi.nlm.nih.gov/books/NBK27156/.

Janeway, C. A., Jr, Travers, P., Walport, M., & Shlomchik, M. J. (2001b). *Immunobiology: The immune system in health and disease. The components of the immune system* (5th ed.). New York: Garland Science. Available from https://www.ncbi.nlm.nih.gov/books/NBK27092/.

Kerr, F. K., Thomas, A. R., Wijeyewickrema, L. C., Whisstock, J. C., Boyd, S. E., Kaiserman, D., Matthews, A. Y., Bird, P. I., Thielens, N. M., Rossi, V., & Pike, R. N. (2008). Elucidation of the substrate specificity of the MASP-2 protease of the lectin complement pathway and identification of the enzyme as a major physiological target of the serpin, C1-inhibitor. *Molecular Immunology, 45*(3), 670–677. Available from https://doi.org/10.1016/j.molimm.2007.07.008.

Khan, M. A. (2016). T regulatory cell-mediated immunotherapy for solid organ transplantation: A clinical perspective. *Molecular Medicine (Cambridge, Mass.), 22*, 892–904. Available from https://doi.org/10.2119/molmed.2016.00050.

Kleijmeer, M., Ramm, G., Schuurhuis, D., Griffith, J., Rescigno, M., Ricciardi-Castagnoli, P., Rudensky, A. Y., Ossendorp, F., Melief, C. J., Stoorvogel, W., & Geuze, H. J. (2001). Reorganization of multivesicular bodies regulates MHC class II antigen presentation by dendritic cells. *The Journal of Cell Biology, 155*(1), 53–63. Available from https://doi.org/10.1083/jcb.200103071.

Kotredes, K. P., & Gamero, A. M. (2013). Interferons as inducers of apoptosis in malignant cells. *Journal of Interferon & Cytokine Research: The Official Journal of the International Society for Interferon and Cytokine Research, 33*(4), 162–170. Available from https://doi.org/10.1089/jir.2012.0110.

Krovi, S. H., & Gapin, L. (2016). Structure and function of the non-classical major histocompatibility complex molecule MR1. *Immunogenetics, 68*(8), 549–559. Available from https://doi.org/10.1007/s00251-016-0939-5.

Leliefeld, P. H., Koenderman, L., & Pillay, J. (2015). How neutrophils shape adaptive immune responses. *Frontiers in Immunology, 6*, 471. Available from https://doi.org/10.3389/fimmu.2015.00471.

Leong, A. S., Cooper, K., & Leong, F. J. W. (2003). *Manual of diagnostic antibodies for immunohistology*. Cambridge University Press.

Li, T., Zhang, Z., Li, X., Dong, G., Zhang, M., Xu, Z., & Yang, J. (2020). Neutrophil extracellular traps: Signaling properties and disease relevance. *Mediators of Inflammation, 2020*, 9254087. Available from https://doi.org/10.1155/2020/9254087.

Massey, H. D., & Mcpherson, R. A. (2007). Mediators of inflammation: Complement, cytokines, and adhension molecules. In P. A. McPherson, & M. R. Pincus (Eds.), *Henry's clinical diagnosis and management by laboratory methods* (21st ed., pp. 850–875). Philadelphia, PA: Saunders Elsevier.

McDonald, B., Pittman, K., Menezes, G. B., Hirota, S. A., Slaba, I., Waterhouse, C. C., Beck, P. L., Muruve, D. A., & Kubes, P. (2010). Intravascular danger signals guide neutrophils to sites of sterile inflammation. *Science (New York, N.Y.), 330*(6002), 362–366. Available from https://doi.org/10.1126/science.1195491, Erratum in: Science. 2011 Mar 25;331 (6024):1517.

Michalek, M. T., Grant, E. P., Gramm, C., Goldberg, A. L., & Rock, K. L. (1993). A role for the ubiquitin-dependent proteolytic pathway in MHC class I-restricted antigen presentation. *Nature, 363*(6429), 552–554. Available from https://doi.org/10.1038/363552a0.

Neefjes, J. (1999). CIIV, MIIC and other compartments for MHC class II loading. *European Journal of Immunology, 29*(5), 1421–1425. Available from https://doi.org/10.1002/(SICI)1521-4141(199905)29:051421:AID-IMMU14213.0.CO;2-C.

Neefjes, J., Jongsma, M. L., Paul, P., & Bakke, O. (2011). Towards a systems understanding of MHC class I and MHC class II antigen presentation. *Nature Reviews Immunology, 11*(12), 823–836. Available from https://doi.org/10.1038/nri3084.

Owen, C., Lees, E. K., Grant, L., Zimmer, D. J., Mody, N., Bence, K. K., & Delibegović, M. (2013). Inducible liver-specific knockdown of protein tyrosine phosphatase 1B improves glucose and lipid homeostasis in adult mice. *Diabetologia*, *56*(10), 2286–2296. Available from https://doi.org/10.1007/s00125-013-2992-z.

Rastogi, S. C. (2008). *Element of immunology* (5th ed., pp. 97–107). New Dehhi: Satish Kumar Jain for CBS Publishers.

Relle, M., & Schwarting, A. (2012). Role of MHC-linked susceptibility genes in the pathogenesis of human and murine lupus. *Clinical & Developmental Immunology*, *2012*, 584374. Available from https://doi.org/10.1155/2012/584374.

Roche, P. A., & Cresswell, P. (1990). Invariant chain association with HLA-DR molecules inhibits immunogenic peptide binding. *Nature*, *345*(6276), 615–618. Available from https://doi.org/10.1038/345615a0.

Rock, K. L., Reits, E., & Neefjes, J. (2016). Present yourself! By MHC class I and MHC class II molecules. *Trends in Immunology*, *37*(11), 724–737. Available from https://doi.org/10.1016/j.it.2016.08.010.

Rosenberg, J., & Huang, J. (2018). CD8 + T cells and NK cells: Parallel and complementary soldiers of immunotherapy. *Current Opinion in Chemical Engineering*, *19*, 9–20. Available from https://doi.org/10.1016/j.coche.2017.11.006.

Rudd, C. E., Trevillyan, J. M., Dasgupta, J. D., Wong, L. L., & Schlossman, S. F. (2010). Pillars article: The CD4 receptor is complexed in detergent lysates to a protein-tyrosine kinase (pp58) from human T lymphocytes. 1988. *Journal of Immunology*, *185*(5), 2645–2649.

Schwaeble, W. J., & Reid, K. B. (1999). Does properdin crosslink the cellular and the humoral immune response? *Immunology Today*, *20*(1), 17–21. Available from https://doi.org/10.1016/s0167-5699(98)01376-0.

Stadhouders, R., Lubberts, E., & Hendriks, R. W. (2018). A cellular and molecular view of T helper 17 cell plasticity in autoimmunity. *Journal of Autoimmunity*, *87*, 1–15. Available from https://doi.org/10.1016/j.jaut.2017.12.007.

Stern, L. J., Brown, J. H., Jardetzky, T. S., Gorga, J. C., Urban, R. G., Strominger, J. L., & Wiley, D. C. (1994). Crystal structure of the human class II MHC protein HLA-DR1 complexed with an influenza virus peptide. *Nature*, *368*(6468), 215–221. Available from https://doi.org/10.1038/368215a0.

Stevens, C. D. (2010). *Clinical immunology and serology: A laboratory perspective*. (3rd ed.). Philadelphia, PA: F. A. Davis Company.

Terabe, M., & Berzofsky, J. A. (2008). The role of NKT cells in tumour immunity. *Advances in Cancer Research*, *101*, 277–348.

Tsuji, H., Taniguchi, Y., Ishizuka, S., Matsuda, H., Yamada, T., Naito, K., & Iwaisaki, H. (2017). Structure and polymorphisms of the major histocompatibility complex in the Oriental stork, Ciconia boyciana. *Scientific Reports*, *7*, 42864. Available from https://doi.org/10.1038/srep42864.

Unanue, E. R., Turk, V., & Neefjes, J. (2016). Variations in MHC class II antigen processing and presentation in health and disease. *Annual Review of Immunology*, *34*, 265–297. Available from https://doi.org/10.1146/annurev-immunol-041015-055420.

Wang, X. X., Li, Y., Yin, Y., Mo, M., Wang, Q., Gao, W., Wang, L., & Mariuzza, R. A. (2011). Affinity maturation of human CD4 by yeast surface display and crystal structure of a CD4-HLA-DR1 complex. *Proceedings of the National Academy of Sciences of the United States of America*, *108*(38), 15960–15965. Available from https://doi.org/10.1073/pnas.1109438108.

Wieczorek, M., Abualrous, E. T., Sticht, J., Álvaro-Benito, M., Stolzenberg, S., Noé, F., & Freund, C. (2017). Major histocompatibility complex (MHC) class I and MHC class II proteins: Conformational plasticity in antigen presentation. *Frontiers in Immunology*, *8*, 292. Available from https://doi.org/10.3389/fimmu.2017.00292.

Wosen, J. E., Mukhopadhyay, D., Macaubas, C., & Mellins E. D. (2018). Epithelial MHC class II expression and its role in antigen presentation in the gastrointestinal and respiratory tracts. *Frontiers in Immunology*, *9*, 2144. Available from: https://doi.org/10.3389/fimmu.2018.02144. PMID: 30319613; PMCID: PMC6167424.

Xie, T., Rowen, L., Aguado, B., Ahearn, M. E., Madan, A., Qin, S., Campbell, R. D., & Hood, L. (2003). Analysis of the gene-dense major histocompatibility complex class III region and its comparison to mouse. *Genome Research*, *13*(12), 2621–2636. Available from https://doi.org/10.1101/gr.1736803.

Yu, Y., Ma, X., Gong, R., Zhu, J., Wei, L., & Yao, J. (2018). Recent advances in CD8 + regulatory T cell research. *Oncology Letters*, *15*(6), 8187–8194. Available from https://doi.org/10.3892/ol.2018.8378.

Chapter 2

Vaccine types and reverse vaccinology

Angus Nnamdi Oli[1] and Adekunle Babajide Rowaiye[2]
[1]Department of Pharmaceutical Microbiology and Biotechnology, Faculty of Pharmaceutical Sciences, Nnamdi Azikiwe University, Awka, Nigeria, [2]Department of Medical Biotechnology, National Biotechnology Development Agency, Abuja, Nigeria

Introduction

Vaccinology entails scientific inquiry into vaccine development and utilization. It cuts across several fields of medicine including immunology and infectious diseases, epidemiology, public health, and pediatrics. The journey of a new vaccine (Fig. 2.1) usually begins with defining the pathogen that causes a particular disease, and then identifying target antigens within that pathogen. The selected antigen, or in some cases an entire pathogen, is formulated into a final vaccine and passes through preclinical and clinical studies to evaluate its safety and efficacy. Vaccine development is usually a long and complex process with a combination of public and private involvement. However, the timeline for production of novel COVID-19 vaccines was truncated as several activities such as engaging modern platform technologies, GMP manufacture, regulatory approvals, concurrent Phase I–II trials, and efficacy data acquisition from Phase IIb/III trials were fast-tracked, and

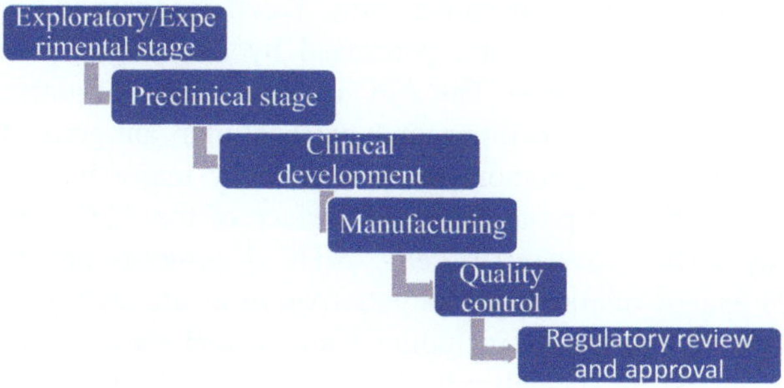

FIGURE 2.1 Cycle of vaccine development.

Vaccinology and Methods in Vaccine Research. DOI: https://doi.org/10.1016/B978-0-323-91146-7.00013-5

performed in parallel rather than sequentially (Khuroo et al., 2020). Vaccine safety was not compromised during this process, as volunteer numbers in Phase II/III trials exceeded those in a typical vaccine development program.

Clinical Vaccinology deals with scientific inquiry into vaccine utilization. It provides a wide range of information about vaccines, specifically the part they play in reducing the tide of vaccine preventable infectious diseases and cancers, and vaccine safety. Second to clean water, vaccines remain the most cost-effective and successful instrument for promoting and maintaining public health and preventing infectious diseases (Andre et al., 2008; Oli et al., 2017).

Fundamentals of vaccine immunology

This section provides a short summary of how the immune system responds to pathogens, and to vaccines. More details are provided in Chapter 1, Immunology and the concept of vaccination.

The immune system has two major subcomponents—the innate and the adaptive immune systems, with functional cross-talk between them. While the innate immune system offers a rapid, nonspecific, broad spectrum first line of defense, the adaptive system is slower and specific for antigenic proteins found on or inside pathogens (Rowaiye et al., 2020). Generally, the innate system recognizes conserved molecular patterns associated with microbes, which differ from those of host cells. As it lacks memory, the reaction of the innate immune system does not improve with repeated infections (Clem, 2011). The innate system comprises anatomic barriers (skin and mucus membranes); physiological barriers (fever, gastric acidity, lysozyme in tears, interferons, and collectins); the complement pathway (classical, alternative or properdin, and lectin pathways); inflammatory responses mediated by myeloid cells (granulocytes, monocytes, and natural killer cells), and pattern recognition receptors which include C-type lectin receptors and Toll-like receptors (Clem, 2011).

In the adaptive immune system, T cells originate from the Thymus, and the B (antibody producing) cells are from the bone marrow. While B cells can recognize antigens in their native form, T cell recognition and activation require that the antigens are first processed by an antigen processing cell (APC) (Margolick et al., 2006). The APC uses two types of antigen presentation mechanisms, for intracellular and extracellular antigens. Intracellular antigens from viruses and tumors are presented by major histocompatibility complex class I (MHC I) proteins on the surface of the APC, where they are recognized by CD8 cytotoxic T cells. MHC I proteins are found on all somatic cells except mature erythrocytes (Andre et al., 2008). On the other hand, extracellular pathogens including bacteria and parasites are phagocytosed and processed by specialized APC in endosomal compartments, where pathogenic peptide fragments combine with MHC II proteins. Once on the cell surface, MHC II-peptide complexes bind to CD4 helper T cells (Alberts et al., 2002; Oli et al., 2020). CD4 cells comprise several subsets including Th1 cells that stimulate cell-mediated immunity, and Th2 cells that can stimulate antibody-mediated immunity directed against extracellular pathogens (Kaiko et al., 2008). MHC proteins are alternatively known as human leukocyte antigens (HLAs) in humans.

Following T cell activation, there is a clonal expansion to produce firstly effector T cells specific for the present infection, and secondly memory T cells customized for responding to future infections (Hirahara & Nakayama, 2016). B cell activation is considered more important than T cell induction for the majority of vaccines, due to the fact that antibodies have the ability to neutralize pathogens early in infection and prevent disease progression (Forthal, 2014). Different diseases can be prevented by distinct levels of serum antibodies known as "correlates of protection," which vaccinations must be able to exceed. For example, for Diphtheria this value is 0.1 IU/L, and for Rabies virus $>$ 0.5 UI/mL (Heilmann et al., 2010; Zhang et al., 2019).

B cells are activated by recognition of a specific epitope on the surface of an antigen by the Fab region of the B cell receptor (BCR; i.e., membrane bound antibody), and by stimulation from cytokines produced by Th2 cells (Goldman & Prabhakar, 1996.). Somatic hypermutation of antibody complementary determining regions, located at the tip of the Fab arms, allows the B cells to develop greater binding affinity for the antigen. BCR recognition stimulates a B cell to transform into plasma cells and culminates in the secretion of neutralizing antibodies specific for the target antigen (Tonomura et al., 2003). Different classes of antibody vary in their Fc region, which mediates antibody effector functions, for example by binding to complement or to Fc-receptors leading to cell-mediated cytotoxicity (lysis) or phagocytosis.

The first product of the clonal expansion of B cells is IgM (Blum et al., 2013), followed by IgG, which is a better neutralizing antibody because it facilitates higher affinity Fc-receptor binding and opsonization (Goldsby et al., 2003). Other antibodies produced by plasma cells include IgA which is found in saliva, breast milk, tears, and mucus; IgE, produced in response to parasitic infections and allergic reactions; and IgD, found on mature B cells. IgG is recognized as the most important antibody class in vaccinology (Blum et al., 2013).

Immunization is the process of stimulating the immune system with an immunogen, for the purpose of developing immunological memory against future infections. It can either be natural (passive) or artificial (active). Passive immunization is transferred to unimmunized individuals via the administration of antibodies (immunoglobulins); however, the approach only confers temporary immunity. Antibodies can be derived from natural sources such as milk, colostrum, and placenta, or isolated from infected individuals, or manufactured as recombinant proteins. Artificial immunization is achieved through medical intervention such as vaccination, or natural exposure to infectious disease (Slifka & Amanna, 2018).

Classes of vaccines

Vaccines can be divided into two broad types: whole-pathogen vaccines (or whole cell vaccines) and subunit vaccines. The latter contain one or more target antigens, usually selected on the basis that they will induce a neutralizing antibody response—typically these antigens will be expressed on the pathogen surface. Whole-pathogen vaccines comprise live-attenuated vaccines and inactivated vaccines, while subunit vaccines include soluble proteins, recombinant DNA or RNA, polysaccharide conjugate vaccines, virus-like particles, viral vectors, and Toxoid vaccines (Table 2.1).

Whole-pathogen vaccines

Live-attenuated vaccines

In 1937, Max Theiler, a Professor of Epidemiology and Public Health, developed the first live-attenuated vaccine (LAV). He passed the virulent Yellow Fever virus many times through laboratory mice until it no longer caused disease in humans but could induce immunity (Frierson, 2010). LAVs have weakened (mutated) copies of the original pathogen, which means that they simulate an attenuated version of the natural infection. As LAVs are able to replicate in humans, they stimulate all arms of the immune system, and thus induce strong, long lasting antibody, and cell-mediated responses (Clem, 2011). Live viral vectors that are not able to replicate in humans are also used as vaccine delivery systems. As few as one or two doses of LAV can confer a lifetime protection against the disease-causing pathogen. Due to smaller genome sizes, it is easier to attenuate viruses than bacteria (Clem, 2011). Other examples of LAV include the Measles, Mumps, and Rubella (MMR) combined vaccine, Chickenpox, Rotavirus, and Smallpox vaccines.

The LAV approach remains an important means of combating many viral infections. In the 20th century, large-scale vaccination programs with LAV recorded a remarkable improvement in human and animal health (Pandey et al., 2016). The successful eradication of smallpox, polio, and measles in many parts of the world is largely due to the success of LAV. While the approach has worked well against acute viral infections, it has improved less effective with chronic infections; an example of a LAV for chronic infections is the BCG vaccine for TB, which is partially effective (Minor, 2015). A major limitation of LAVs is the fact that refrigeration is required to maintain vaccine potency. This storage issue constitutes a formidable logistic challenge in mass vaccination campaigns in rural African communities that lack electricity (Dairo & Osizimete, 2016). In addition, LAVs contain live,

TABLE 2.1 Types of vaccines.

PROPERTIES	VACCINE TYPES					
	Live attenuated	Inactivated	Sub-unit	Toxoid	Conjugate	Nucleic acids/ Viral Vectors
First vaccine produced	Yellow Fever, 1937	Polio, 1955	Hepatitis B Virus-Like Particles (VLPs), 1981 (USA)	Tetanus, 1924	Hib, 1987	West Nile Virus, 2005 (Horses); COVID-19 human vaccines 2020
Method of preparation	Attenuation	Inactivated of pathogens	A portion of the pathogen is used (in the case of Hepatitis and Human Papillomavirus Vaccine (HPV) these self-assemble into VLPs)	Toxins of pathogens used	Polysaccharides conjugated to proteins	Genetic material that codes for antigen placed in a vector
Mechanism of action	Humoral and Cell mediated immunity	Mainly humoral immunity	Epitopes bind to antibodies or T-cells: mainly humoral immunity	T cell response	B and T cell activation	*In situ* antigen production
Immunogenicity	Strong	Weaker than LAV	Poorer immunogenic response than LAV	Weaker than LAV	Enhanced immunogenicity compared to polysaccharide alone	May have lower immunogenicity than LAV (especially DNA vaccines)

(*Continued*)

TABLE 2.1 (Continued)

Immunopotentiator	Adjuvant not required	Adjuvants/ additional booster doses	Booster doses, adjuvants, carriers	Adjuvants, booster doses	Adjuvants	Adjuvants, carriers, etc.
Commercially approved	Chickenpox, Rotavirus, MMR, Smallpox	Hepatitis A, Influenza, Polio, Rabies, Typhoid, Cholera, Plague, Pertussis	Hepatitis B, HPV, Shingles, Pneumococcal, Meningococcal	Diphtheria Tetanus	*Haemophilus influenzae* (Hib), *Neisseria meningitidis* (Nm) and *Streptococcus pneumoniae* (Sp)	COVID-19 Vaccines
Limitations	Can trigger the actual disease e.g. Polio; Require refrigeration; Less efficacious for chronic infections; Easier approach for viruses than bacteria	Weaker immune response than LAV; Requires manufacture of intact pathogen	Weaker immune response than LAV; Requires identification of specific, immunogenic antigens;Potential denaturation of antigens	Weaker immune response than LAV; Applies to Bacteria only	Must be coupled with proteins	Integration to aberrant genomic sites for DNA vaccine; Reactogenicity (RNA vaccines in particular)
Advantages	Strong long lasting immunity	Safe compared to LAV; Less stringent storage requirements than LAV as not live	Safer than LAV; Specific	Safer than LAV	Safer than LAV; Vaccines against multiple strains can be combined in the same injection	Easy to produce, longer shelf life (DNA Vaccines only)

replication competent pathogens and can therefore not be administered to individuals with compromised immune systems. Such individuals include patients with an organ transplant, chronic illnesses, and immunodeficiencies.

LAVs have the potential of reverting back to their original virulent form and can trigger the actual disease, for example, some polio cases are caused by the attenuated vaccine (albeit at a low frequency, in fewer than one in a million vaccinated children) (Arvas, 2014). Though there are safety concerns, LAVs are more effective than inactivated vaccines. For example, in young chickens, a single dose of live-attenuated influenza vaccine provided stronger innate and mucosal IgA protection than a single dose of influenza inactivated vaccine (Jang et al., 2018). LAVs elicit strong humoral and cell-mediated immunity; hence, adjuvants are not required. Because the whole pathogen is used, LAVs contain naturally occurring immunopotentiating molecules (Karch & Burkhard, 2016).

The objective of attenuation is to inhibit pathogen replication in the host cells to some degree, so that they evoke a strong immune response but do not cause disease. This can be achieved either by classical or targeted attenuation. The classical attenuation of live pathogens (mostly viruses) is achieved by passing them multiple times through cultured cells or live animals which are not the target species for the vaccine. Targeted attenuation involves the use of genetic engineering approaches such as deletion or mutation of genes that are responsible for replication and assembly, altered replication fidelity, codon deoptimization, RNA interference, or zinc finger nuclease control (Lauring et al., 2010). However, as with the classical approach to attenuation, the major challenge with targeted attenuation is that live-attenuated viruses can revert back to virulence through mutation and recombination (Hanley, 2011).

Inactivated vaccines

The first inactivated vaccine was against Polio virus, made by Jonas Salk, an American virologist. Since its introduction in 1955, the approach has continued to gain ground (Juskewitch et al., 2010). Inactivated vaccines contain pathogens previously destroyed using heat, chemicals, or radiation treatment (Rappuoli & Aderem, 2011; Seo, 2015). The mode of inactivation may affect the immunogenicity produced by the vaccine: for example, irradiated *Listeria monocytogenes* bacteria induced a greater protective T cell immune response in experimental mice than heat killed bacteria (Datta et al., 2006). With inactivated microbes, the vaccine is more stable and hence does not necessarily require freeze drying or refrigeration for storage and transportation.

Inactivated vaccines have been used for over a century and are safe compared to LAV (Sanders et al., 2014). However, they produce weaker immune responses and hence there is a need to coadminister adjuvants and/or additional booster doses to achieve the desired protective effect (Petrovsky & Aguilar, 2004).

Compared to LAV, they elicit mainly humoral immunity, with little to no cell-mediated immunity (Lee & Nguyen, 2015). Inactivated vaccines are noninfectious, and licensed examples include viral and licensed examples target viral and bacterial diseases such as Hepatitis A, influenza, Polio, Rabies, tick-borne encephalitis, typhoid, cholera, plague, and pertussis. Many other experimental vaccines are at different stages of clinical development (Stauffer et al., 2006).

During the inactivation process of the pathogen, care is taken to preserve important epitopes (e.g., targets for neutralizing antibodies) from denaturation (Sanders et al., 2014). To produce high quality antigen, the production process of the vaccine must overcome challenges such as aggregate formation, protein cross linking, denaturation, and degradation (Delrue et al., 2012). BBIBP-CorV is an inactivated vaccine produced by Sinopharm against severe acute respiratory syndrome coronavirus 2 (SARS-CoV-2). Empirical data reveal that this vaccine produces strong neutralizing antibodies against SARS-CoV-2 in experimental animal models (Wang et al., 2020). The same report showed that immunized *Rhesus macaques* dosed intratracheally with BBIBP-CorV were protected against SARS-CoV-2 infection, providing rationale for evaluation of the vaccine in human clinical trials (Wang et al., 2020).

Subunit vaccines

Subunit vaccines contain part of a pathogen which can be targeted by a neutralizing immune response. The targeted antigens can be formulated as highly purified proteins, polysaccharides, glycoproteins, or synthetic peptides; nucleic acids or vectors expressing target antigens; virus like particles (VLPs) displaying antigen on their surface, or viral vectors expressing antigens from their genomes (Foged, 2011). The strength of this approach is that the immune response is targeted to specific antigens, which can then induce neutralization of the pathogen. It is sometimes challenging to identify those antigens that would evoke a neutralizing response—for example, a particular protein may be highly expressed on the surface of a pathogen but if it is highly variable, it may evade vaccine-induced antibodies.

Generally, subunit vaccines have a good safety profile as they contain isolated proteins, protein subfragments, or glycans (Vartak & Sucheck, 2016). Although subunit vaccines are safer and sometimes easier to manufacture than whole pathogens, nonviral vector-based vaccines usually require formulation with adjuvants to elicit strong immunogenicity (Foged, 2011; Moyle & Toth, 2013). The first recombinant protein vaccine for hepatitis B virus, is a VLP derived from hepatitis B virus surface antigen. The genetic code for the viral protein was inserted into the genome of yeast cells, and as the cells multiply the viral protein is expressed and forms VLPs when the yeast is lysed (Ho et al., 2020). Similar recombinant vaccines have been produced against human papillomavirus (two VLP vaccines), shingles (herpes zoster;

recombinant glycoprotein in adjuvant), meningococcus, and pneumococcus (both bacterial glycans conjugated to protein) (James et al., 2018; McCarthy et al., 2018; Reglinski et al., 2018; Sharma & Sharma, 2007).

Although subunit vaccines are safer than LAV, they are less effective. They lack pathogen-associated molecular patterns, which are recognized by the host immune system resulting in a strong innate response (Demento et al., 2011). Immunopotentiation of a subunit vaccine can be achieved by the administration of booster doses or adjuvants, or by using a carrier system, such as a viral vector (Lee & Nguyen, 2015; Stauffer et al., 2006).

A vaccine carrier offers a delivery system, to enhance vaccine bioavailability and potency. Recent studies demonstrated that modulation of the physiochemical properties of these carriers can achieve optimal antigen presentation, endosomal escape, particle biodistribution, and cellular trafficking (Foged, 2011; Vartak & Sucheck, 2016). They can also be modified with adjuvants, to augment humoral and cellular immunological responses against the antigen. Nanoparticle vaccine carriers including liposomes, polymeric nanoparticles, and inorganic nanoparticles are able to transport antigens to immune cells by targeting specific receptors, and are being explored for use in immunotherapy against cancer and infectious diseases (Moyle & Toth, 2013). Carrier systems are described in greater detail in Chapter 9, Adjuvants, immunomodulators, and adaptogens.

Subunit vaccines have shown promise in the prevention of diseases such as human cytomegalovirus, tetanus, malaria, *Salmonella enterica* serovar enteritis infections, and SARS-CoV-2 (Anderholm et al., 2016; Greenwood, 2014; Saylor et al., 2020; Wang et al., 2020).

Toxoid vaccines

Toxoid vaccines are a type of subunit vaccine, first exemplified by inactive tetanus toxoid (TT) vaccine introduced in 1924. It resulted in a 95% decrease in the incidence of Tetanus (Tetanus Vaccines, 2017). Toxoid vaccines are produced from inactivated bacterial toxins, which trigger an immune response, in particular T-helper cell responses (van der Heiden et al., 2017). Toxoid vaccines do not elicit immune responses as strongly as LAV and therefore require the use of multiple doses or adjuvants such as aluminum or calcium salts (Baxter, 2007; Christensen, 2016). Toxoid vaccines do not revert back to virulence and therefore they are considered safe. Formalin has been used for the chemical inactivation of toxins; however, inactivation by alkylation with iodoacetamide under reversibly denaturing conditions retains more of the epitope structure of the native toxin (Jones et al., 2008). Empirical data suggest that genetically inactivated mutant pertussis toxoid is safer and provokes a stronger immune response compared to chemically inactivated toxin (Robbins et al., 2009). Another advantage of toxoid vaccines is their being less susceptible to changes in light, humidity, and temperature than whole pathogens (Baxter, 2007). In the development of toxoid vaccines, the use of nanotoxoids as a particulate delivery system has shown good potential to enhanceimmunogenicity; these comprise either toxoid aggregates combined with polymeric micelles or cell-membrane-coated nanoparticles. The latter approach allows for the quick loading of several kinds of toxins, to confer immunity against several bacterial agents (Angsantikul et al., 2018).

Conjugate vaccines

In 1987, the first glycoconjugate vaccine to prevent *Haemophilus influenzae* type b (Hib), was licensed for human use (Goldblatt, 2000). Antigens or toxoids are linked to polysaccharides from the target pathogen to induce a helper (CD4) T cell response against the protein component, which supports a strong antibody response to the polysaccharide. These vaccines have continued to be very useful in the prevention of bacterial infections especially in children, including *Neisseria meningitidis*, *Streptococcus pneumoniae*, and Hib infections that are the major causes of childhood meningitis (Findlow & Borrow, 2016). These bacteria have a polysaccharide coating which act as virulence determinants. In the absence of the conjugated protein to provide T cell help, the capsular polysaccharides alone activate B cells in a T cell-independent manner by producing mostly IgM and few memory B cells, which are insufficient to protect high-risk individuals (Goldblatt, 2000; Nishat & Andreana, 2016; Pobre et al., 2014).

For vaccines against Hib, *N. meningitidis*, and *S. pneumoniae*, the carrier proteins used are the nontoxic mutant of diphtheria toxin (CRM197), diphtheria toxoid, and TT (Findlow & Borrow, 2016). Advances in conjugate vaccine development have led to more immunogenic T cell-dependent antigenic vaccines as compared with the T cell-independent polysaccharide vaccines of the past (Pichichero, 2013). Another reason for the success of these vaccines is that they help reduce asymptomatic carriage and hence reduce the spread of infection (Mäkelä & Käyhty, 2002).

Nucleic acid vaccines and other experimental approaches

Nucleic acid vaccines supply the host cells with genetic sequences, either RNA or DNA, that code for the target antigen. They usually employ plasmids, nanoparticles, or viruses as their delivery systems. Viral vector-based vaccines use modified viruses to deliver the genetic material into the host cell via infection (Porter & Raviprakash, 2017). Stripped of any disease-causing genes, the viruses can either be replication-competent or replication-defective, where genes crucial for replication have been mutated or deleted.

DNA vaccines

In the 1990s, the in vivo transfection of APCs in mice to induce primary adaptive immune responses initiated the development of DNA vaccines (Hobernik & Bros, 2018). Similarly, DNA administered directly to the skin and muscles of mice generated immune responses to the encoded antigen (Coban et al., 2013; Weiner, 2008). This led to the development and approval of the first veterinary DNA vaccine against Equine West Nile virus (WNV) in 2005. DNA vaccines enter cell nuclei where they are transcribed into RNA, which is then translated into the corresponding antigenic protein(s) in the cytoplasm (Lee & Nguyen, 2015). The antigens can be secreted to generate an antibody response and are also processed by the proteasome, leading to peptide presentation on cell surface MHC molecules and recognition by T cell receptors on naïve CD4 or CD8 T cells. Presentation can either be directly on the surface of transfected cells (somatic cells or professional APCs, e.g., dendritic cells) or cross-presentation following phagocytosis of transfected somatic cells by APCs (Ghaffarifar, 2018; Khan, 2013; Sun et al., 2021).

DNA vaccines also trigger innate immune responses: an unidentified cytosolic DNA sensor binds to transfected plasmids and triggers the TBK1-STING pathway. In this process, type I interferons are produced, which act as an adjuvant (Sun et al., 2021). DNA vaccines are easy and relatively inexpensive to produce, and have a longer shelf life (Hobernik & Bros, 2018). To date, only a few veterinary DNA vaccines have been approved and are listed in Table 2.2. They include vaccines against Equine WNV (De Filette et al., 2012); infectious hematopoietic necrosis (IHNV) in salmon (Corbeil et al., 2000); canine melanoma (Liao et al., 2006); and the CLYNAVTM vaccine for Atlantic salmon (European Food Safety Authority et al., 2017). A conditional approval by the United States Department of Agriculture (USDA) has been given for the first commercial DNA vaccine against the highly pathogenic H5N1 influenza in chickens (Redding & Weiner, 2009).

Beyond prophylaxis, DNA can also be employed in therapeutic cancer vaccines. In this regard, personalized antitumor DNA vaccines can be made directly from biopsies, or synthesized following sequencing of tumor DNA and identification of potential neoepitopes (Lee et al., 2018). This topic is explored in more detail in the chapter on approaches to cancer vaccination. Other applications for this approach include the treatment of autoimmune and allergic diseases (Gutowska-Owsiak & Ogg, 2017; Wraith et al., 2003).

Historically, DNA vaccines have evoked low levels of immunogenicity, illustrated by the first clinical trials of vaccines for Influenza, HIV, malaria, and cancer (Weiner, 2008). Furthermore, immune responses generated in primate and large animal models are lower in comparison to those generated in small animals (Weiner, 2008). Various approaches have been taken to boost host cellular and humoral response to DNA vaccines (Coban et al., 2013; Weiner, 2008).

The efficiency of DNA vaccines can be improved by vector optimization, the introduction of nanocarriers and the attachment of virus-derived nuclear localization sequences. Vector optimization strategies include the use of APC-specific promoters, terminator sequences, inclusion of multiple antigen sequences, and the use of molecular adjuvants (Ghaffarifar, 2018; Lee et al., 2018). Nanocarriers can prevent extracellular DNA degradation, enhance endo/lysosomal escape of DNA, and enable APC targeting (Ghaffarifar, 2018). The use of specific localization sequences in the vaccine facilitates the entry of DNA into the nucleus for integration into the genome (Ghaffarifar, 2018; Karunamoorthi, 2014).

A major concern surrounding DNA vaccines is the possibility that foreign DNA might integrate to aberrant sites in the human genome leading to potential pathologies, including transformation to a cancerous phenotype.

In the first two decades of the 21st century, DNA vaccines against HCV, malaria, Influenza, HIV-1, and cytomegalovirus have undergone various stages of human clinical testing (Weiner, 2008).

Viral vector vaccines

Viral vector vaccines use modified versions of a specific virus (the vector) to deliver antigens to APCs. Vaccines that have been shown to give promising results include the recombinant vesicular stomatitis virus-Zaire Ebola virus (rVSV-ZEBOV) (Choi et al., 2021), the Oxford-AstraZeneca COVID-19 vaccine (AZD1222) (Chung et al., 2020), and the Russian Sputnik V vaccine (Jones et al., 2008). Merck developed the rVSV-ZEBOV vaccine using the replication-competent, live attenuated recombinant vesicular stomatitis virus. This viral vector-based vaccine carries a gene that codes for the EBOV surface glycoprotein (GP1,2) (Marzi et al., 2011). The safety and immunogenicity of this vaccine was evaluated in three clinical studies in Canada, Liberia, Sierra Leone, Spain, and the United States. Though the rVSV-ZEBOV vaccine provides protection against the most lethal Zaire Ebola virus species, it does not confer immunity against other species of Ebola virus (Damon & Frey, 2021). Viral vector vaccines have also been extensively employed against SARS-CoV-2, the virus that causes COVID-19; these are covered in a separate chapter which addresses vaccines targeting SARS-CoV-2.

RNA vaccines

Unlike DNA vaccines, which deliver genetic material into the nucleus, mRNA is delivered into the cytoplasm (Pardi et al., 2018). As for DNA vaccines, ribosomes translate the RNA to generate viral antigens, proteolysis takes place in the proteasome, and epitopes loaded onto Class I MHC proteins are displayed on the cell surface for recognition by CD8 cells. Antibody responses are generated against intact antigen that is secreted from the cell, or expressed on the cell surface if a transmembrane domain is

TABLE 2.2 Approved veterinary DNA vaccines.

Vaccines	Production technology	Indication	Remarks
Equine West Nile virus vaccine (WNV)	Uses a recombinant plasmid coding for the prM and E proteins of WNV	To immunize horses from WNV and from several mosquito-borne diseases, like Eastern equine encephalitis (EEE) and Western equine encephalitis (WEE)	
Infectious hematopoietic necrosis virus vaccine (IHNV)	Uses a recombinant plasmid coding for the glycoprotein of the IHNV	IHNV in salmon and trout fishes	
Infectious pancreatic necrosis virus vaccine (IPNV)	Uses a recombinant plasmid coding for the glycoprotein of the IPNV	IPNV in salmon and trout	
Canine melanoma vaccine	Uses a recombinant plasmid coding for the human melanocyte protein tyrosinase	Tyrosinase positive melanoma, in dogs	
CLYNAVTM vaccine	Uses a recombinant plasmid coding for salmonid alphavirus subtype 3 (SAV3) protein	For Atlantic salmon fish	
Avian influenza virus vaccine	Recombinant DNA vaccine	H5N1 influenza in chickens	Conditional approval by the United States Department of Agriculture (USDA)

included. Secreted protein can be taken up by APCs, degraded, and presented by MHC II to CD4 cells (Pardi et al., 2018). In 1993 it was established that liposome-encapsulated RNA induced in vivo T cell stimulation (Kallen & Theß, 2014; Verbeke et al., 2019).

Extensive research efforts have seen the development of many RNA vaccines that use cationic polymers, nanoparticles, and viral vectors as delivery systems, and are in different phases of clinical development. These include vaccines against Influenza virus (Petsch et al., 2012), rabies virus (Schnee et al., 2016), HIV-1 (Pollard et al., 2013), RSV (Geall et al., 2012), Zika virus (Richner et al., 2017), melanoma (Weide et al., 2009), nonsmall-cell lung cancer (Sebastian et al., 2014), and prostate cancer (Kübler et al., 2015). At the time of writing, two human mRNA vaccines have been approved for use, the Moderna and Pfizer/BioNTech COVID-19 vaccines: more information is provided in chapter 8.

RNA vaccines share the advantages of DNA vaccines over conventional vaccines in that they are noninfectious, and manufacture follows the same pathway for vaccines against multiple indications. However, adverse reactions to RNA vaccines might be seen in individuals susceptible to autoimmunity (Pardi et al., 2018). Reactogenicity has also been reported in clinical trials of COVID-19 RNA vaccines (Wadman, 2020). Manufacture of RNA vaccines is described in the chapter addressing approaches to cancer vaccination.

Immunoinformatics

With the advent of sequencing the entire genomes of humans and animals, there is a huge accumulation of immunologically relevant data. In addition, epidemiological and clinical immunological data can be accessed in publicly available databases (Tomar & De, 2010). To interpret and apply this goldmine of data effectively, researchers need to utilize informatics in a time and cost-effective manner. This has led to the rapid development of immunoinformatics, which stands as a new field interfacing between genomics, computational methods, and experimental immunology (Alberts et al., 2002; Oli et al., 2020). Immunoinformatics makes use of computer-based methods to solve immunological problems. One application is the prediction of B- and T-cell epitopes, thus opening new windows for potential translational research (Fig. 2.2).

One of the principal objectives of immunoinformatics is to predict immunogenicity and antigenicity (Flower, 2007). It involves the use of software, computational tools, databases for T cell and B cell epitope prediction, vaccine design, protein structure comparisons, MHC characterization and gene analysis, immune system modeling, and network analysis of immune system data (Evans, 2008; Tomar & De, 2010, 2014). NERVE, Vaxign, and RANKPEP are software programs widely used for reverse vaccinology (RV) (Moxon et al., 2019; Reche et al., 2002).

Reverse vaccinology (RV) and other applications

Following the publishing of the *H. influenzae* genome in 1995, researchers commenced using genomic information to computationally enhance vaccine

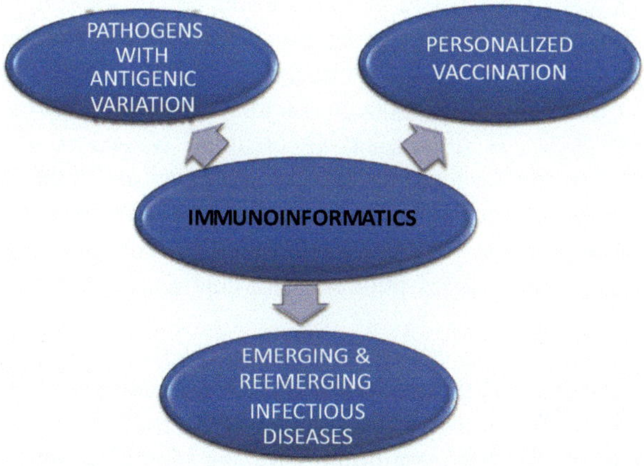

FIGURE 2.2 Applications of immunoinformatics in clinical and genomic medicine.

discovery. In 2000 Rino Rappuoli and his team produced a vaccine against Serogroup B meningococcus, thus championing the birth of RV (Hietalahti & Meri, 2015; Rappuoli, 2001). Subsequently, the technology was used to design vaccines against Group A *Streptococcus*, Group B *Streptococcus*, *Staphylococcus aureus*, and *S. pneumoniae* (Delany et al., 2013; Seib et al., 2012). RV is an important aspect of immunoinformatics and in contrast to classical vaccinology it provides speed, accuracy, and efficiency.

RV provides a more comprehensive approach to vaccine development because it reveals all possible epitopes (B, CD4 +, and CD8 + T cell) in a pathogen, increasing by orders of magnitude potential vaccine targets (Delany et al., 2013; Seib et al., 2012). An initial limitation of RV was that antigens could not be assessed based on their ability to generate antibodies, although advances in structural biology and B cell technologies show some hope in surmounting this challenge (Liljeroos et al., 2015; Van Regenmortel, 2016). One disadvantage of RV is that currently, it can only identify protein antigens but not polysaccharide antigens (Rappuoli & Aderem, 2011).

Following identification of potential candidate antigens by RV, these can be ranked as vaccine candidates using a variety of approaches:

1. Assess antigen variability between pathogen strains, and select the least variable antigens/epitopes (Bidmos et al., 2018).
2. Assess immunogenicity by expressing antigens as soluble proteins using, for example, DNA encoded in viral vectors and injecting into mice, rats, or rabbits (Zheng et al., 2012).
3. Assess the ability of antibodies from immunized animals to protect against infection in in vitro infectivity assays (Straub et al., 2007).
4. Assess the ability of candidate antigens to protect against disease in animal challenge models (Muñoz-Fontela et al., 2020).

With the advent of RV, there is a greater research focus on pathogen antigens which could be developed as subunit vaccines. RV has led to the discovery of pili in Gram-positive bacteria such as pneumococcus and streptococcus, and also meningococcal factor G binding protein which binds to complement factor H in humans (Moballegh Naseri et al., 2020; Seib et al., 2012). Optimal target antigens for vaccines are both necessary for microbial pathogenesis, and highly immunogenic. RV was helpful in the speedy design of various SARS-CoV-2 vaccines, and this approach holds great promise for future global public health challenges (Moxon et al., 2019; Reche et al., 2002).

Sequencing peptide epitopes presented by MHC molecules using mass spectrometry has enabled the design of T-cell inducing vaccines (Backert & Kohlbacher, 2015; Pimenova et al., 2008). This topic is covered in more detail in Chapter 5, Immunopeptidomics for the identification of T cell vaccine targets.

Today, the development of novel vaccines has been made possible through the molecular and mechanistic understanding of the immune system. Data provided by the sequencing of the B cell Receptor repertoire, high-throughput discovery of protective human antibodies, and structural description of the distinctive nature or features of protective antigens and epitopes have contributed to rational vaccine design, for example by identifying epitopes that are recognized by strongly neutralizing antibodies capable of conferring protection against specific pathogens (Rappuoli et al., 2016).

Immunoinformatic data obtained from sequencing a genome, the protein-coding regions of genes in the genome (whole-exome sequencing) and RNA expressed from the genome have made it possible to understand a patient's set of human leukocyte antigen (HLA) alleles. Combined with data from accurate sequencing of peptide epitopes by mass spectrometry, such information enables the design of personalized prophylactic and therapeutic vaccines, and cancer immunotherapies (Backert & Kohlbacher, 2015)(Pimenova et al., 2008). With the advent of next-generation sequencing, it is possible to determine with high accuracy the presence and quantity of mRNA transcripts in an individual's cells, including splice variants and other modifications, and to follow changes in the RNA profile over time.

Personalized cancer vaccines, or neoantigen cancer vaccines (NCV) have been developed for use in for cancer immunotherapy. Neoantigens are important immunogens and are not found in normal cells, as they are created by tumor-specific mutations and are thus potentially seen as foreign by the immune system. They can activate CD4 + and CTLs or CD8 + to induce immunological responses. Immunoinformatic tools can identify these neoantigens as new tumor immunotherapy targets using a combination of different algorithms that can identify and predict the affinity of neoantigen peptides to MHCs. Examples of NCV "include nucleic acid, dendritic cell (DC)-based, tumor cell, and synthetic long peptide (SLP) vaccines" (Peng et al., 2019). Cancer neoantigen vaccines are described in more detail in Chapter 7.

Vaccination has been serving as an indispensable tool in the fight against infectious diseases and cancer. Microbial proteins vital to pathogenesis have been mined through bioinformatics tools. Most important to the immunoinformatician are those conserved proteins that are both necessary for microbial pathogenesis, and that are highly immunogenic. The CadF protein of *C. jejuni* is a good example of a protein identified through a data mining approach: Moballegh Naseri et al. (2020) were able to prove that CadF is conserved, and thus could serve as a good vaccine target against *C. jejuni*.

References

Alberts, B., Johnson, A., Lewis, J., et al. (2002). *T cells and MHC proteins. Molecular biology of the cell* (4th ed). Garland Science Available from. Available from https://www.ncbi.nlm.nih.gov/books/NBK26926/ Accessed 24.12.20.

Anderholm, K. M., Bierle, C. J., & Schleiss, M. R. (2016). Cytomegalovirus vaccines: Current status and future prospects. *Drugs*, *76*(17), 1625–1645. Available from https://doi.org/10.1007/s40265-016-0653-5. Available from 27882457, PMCID: PMC5628252.

Andre, F. E., Booy, R., Bock, H. L., Clemens, J., Datta, S. K., John, T. J., Lee, B. W., Lolekha, S., Peltola, H., Ruff, T. A., Santosham, M., & Schmitt, H. J. (2008). Vaccination greatly reduces disease, disability, death and inequity worldwide. *Bulletin of the World Health Organization*, *86*(2), 140–146. Available from https://doi.org/10.2471/blt.07.040089. Available from 18297169, PMCID: PMC2647387.

Angsantikul, P., Fang, R. H., & Zhang, L. (2018). Toxoid vaccination against bacterial infection using cell membrane-coated nanoparticles. *Bioconjugate Chemistry*, *29*(3), 604–612. Available from https://doi.org/10.1021/acs.bioconjchem.7b00692, Epub 2017 Dec 27. Available from 29241006.

Arvas, A. (2014). Vaccination in patients with immunosuppression. *Turkish Archives of Pediatrics.*, *49*(3), 181–185. Available from https://doi.org/10.5152/tpa.2014.2206. Available from 26078660, PMCID: PMC4462293.

Backert, L., & Kohlbacher, O. (2015). Immunoinformatics and epitope prediction in the age of genomic medicine. *Genome Medicine.*, *7*, 119. Available from https://doi.org/10.1186/s13073-015-0245-0. Available from 26589500, PMCID: PMC4654883.

Baxter, D. (2007). Active and passive immunity, vaccine types, excipients and licensing. *Occupational Medicine (London).*, *57*(8), 552–556. Available from https://doi.org/10.1093/occmed/kqm110. Available from 18045976.

Bidmos, F. A., Siris, S., Gladstone, C. A., & Langford, P. R. (2018). Bacterial vaccine antigen discovery in the reverse vaccinology 2.0 era: Progress and challenges. *Frontiers in Immunology*, *9*, 2315. Available from https://doi.org/10.3389/fimmu.2018.02315. Available from 30349542, PMCID: PMC6187972.

Blum, J. S., Wearsch, P. A., & Cresswell, P. (2013). Pathways of antigen processing. *Annual Review of Immunology*, *31*, 443–473. Available from https://doi.org/10.1146/annurev-immunol-032712-095910, Epub 2013 Jan 3. Available from 23298205. Available from PMC4026165.

Christensen, D. (2016). Vaccine adjuvants: Why and how. *Human Vaccines & Immunotherapeutics.*, *12*(10), 2709–2711. Available from https://doi.org/10.1080/21645515.2016.1219003. Available from 27551808, PMCID: PMC5084984.

Chung, Y. H., Beiss, V., Fiering, S. N., & Steinmetz, N. F. (2020). COVID-19 vaccine frontrunners and their nanotechnology design. *ACS Nano.*, *14*(10), 12522–12537. Available from https://doi.org/10.1021/acsnano.0c07197, Epub 2020 Oct 9. Available from 33034449, PMCID: PMC7553041.

Clem, A. S. (2011). Fundamentals of vaccine immunology. *Journal of Global Infectious Diseases.*, *3*(1), 73–78. Available from https://doi.org/10.4103/0974-777X.77299. Available from 21572612, PMCID: PMC3068582.

Coban, C., Kobiyama, K., Jounai, N., Tozuka, M., & Ishii, K. J. (2013). DNA vaccines: A simple DNA sensing matter? *Human Vaccines & Immunotherapeutics.*, *9*(10), 2216–2221. Available from https://doi.org/10.4161/hv.25893, Epub 2013 Aug 2. Available from 23912600.

Corbeil, S., Kurath, G., & LaPatra, S. E. (2000). Fish DNA vaccine against infectious hematopoietic necrosis virus: Efficacy of various routes of immunisation. *Fish & Shellfish Immunology*, *10*(8), 711–723. Available from https://doi.org/10.1006/fsim.2000.0286. Available from 11185755.

Dairo, D. M., & Osizimete, O. E. (2016). Factors affecting vaccine handling and storage practices among immunization service providers in Ibadan, Oyo State, Nigeria. *African Health Sciences*, *16*(2), 576–583. Available from https://doi.org/10.4314/ahs.v16i2.27. Available from 27605974, PMCID: PMC4994548.

Damon, I. K., & Frey, S. E. (2021). Use of Ebola vaccine: Recommendations of the advisory committee on immunization practices, United States, 2020. *MMWR. Recommendations and Reports: Morbidity and Mortality Weekly Report. Recommendations and Reports*, *70*(1), 1–12. Available from https://doi.org/10.15585/mmwr.rr7001a1. Available from 33417593, PMCID: PMC7802368.

Datta, S. K., Okamoto, S., Hayashi, T., Shin, S. S., Mihajlov, I., Fermin, A., Guiney, D. G., Fierer, J., & Raz, E. (2006). Vaccination with irradiated Listeria induces protective T cell immunity. *Immunity*, *25*(1), 143–152. Available from https://doi.org/10.1016/j.immuni.2006.05.013. Available from 16860763.

De Filette, M., Ulbert, S., Diamond, M., & Sanders, N. N. (2012). Recent progress in West Nile virus diagnosis and vaccination. *Veterinary Research*, *43*(1), 16. Available from https://doi.org/10.1186/1297-9716-43-16. Available from 22380523, PMCID: PMC3311072.

Delany, I., Rappuoli, R., & Seib, K. L. (2013). Vaccines, reverse vaccinology, and bacterial pathogenesis. *Cold Spring Harbor Perspectives in Medicine.*, *3*(5), a012476. Available from https://doi.org/10.1101/cshperspect.a012476. Available from 23637311, PMCID: PMC3633180.

Delrue, I., Verzele, D., Madder, A., & Nauwynck, H. J. (2012). Inactivated virus vaccines from chemistry to prophylaxis: Merits, risks and challenges. *Expert Review of Vaccines*, *11*(6), 695–719. Available from https://doi.org/10.1586/erv.12.38. Available from 22873127.

Demento, S. L., Siefert, A. L., Bandyopadhyay, A., Sharp, F. A., & Fahmy, T. M. (2011). Pathogen-associated molecular patterns on biomaterials: A paradigm for engineering new vaccines. *Trends in Biotechnology*, *29*(6), 294–306. Available from https://doi.org/10.1016/j.tibtech.2011.02.004, Epub 2011 Apr 2. Available from 21459467.

European Food Safety AuthorityHouston, R., Moxon, S., Nogué, F., Papadopoulou, N., Ramon, M., & Waigmann, E. (2017). Assessment of the potential integration of the DNA plasmid vaccine CLYNAV into the salmon genome. *EFSA Journal.*, *15*(1), e04689. Available from https://doi.org/10.2903/j.efsa.2017.4689. Available from 32625277, PMCID: PMC7009875.

Evans, M. C. (2008). Recent advances in immunoinformatics: Application of in silico tools to drug development. *Current Opinion in Drug Discovery & Development*, *11*(2), 233–241.

Findlow, H., & Borrow, R. (2016). Interactions of conjugate vaccines and co-administered vaccines. *Human Vaccines & Immunotherapeutics.*, *12*(1), 226–230. Available from https://doi.org/10.1080/21645515.2015.1091908. Available from 26619353, PMCID: PMC4962715.

Flower, D. R. (2007). Immunoinformatics and the in silico prediction of immunogenicity. An introduction. *Methods in Molecular Biology.*, *409*, 1–15. Available from https://doi.org/10.1007/978-1-60327-118-9_1. Available from 18449989.

Foged, C. (2011). Subunit vaccines of the future: The need for safe, customized and optimized particulate delivery systems. *Therapeutic Delivery.*, *2*(8), 1057–1077. Available from https://doi.org/10.4155/tde.11.68. Available from 22826868.

Forthal, D. N. (2014). Functions of antibodies. *Microbiology Spectrum.*, *2*(4), 1–17. Available from 25215264, PMCID: PMC4159104.

Frierson, J. G. (2010). The yellow fever vaccine: A history. *The Yale Journal of Biology and Medicine*, *83*(2), 77–85. Available from 20589188, PMCID: PMC2892770.

Geall, A. J., Verma, A., Otten, G. R., Shaw, C. A., Hekele, A., Banerjee, K., Cu, Y., Beard, C. W., Brito, L. A., Krucker, T., O'Hagan, D. T., Singh, M., Mason, P. W., Valiante, N. M., Dormitzer, P. R., Barnett, S. W., Rappuoli, R., Ulmer, J. B., & Mandl, C. W. (2012). Nonviral delivery of self-amplifying RNA vaccines. *Proceedings of the National Academy of Sciences of the United States of America*, *109*(36), 14604–14609. Available from https://doi.org/10.1073/pnas.1209367109, Epub 2012 Aug 20. Available from 22908294.

Ghaffarifar, F. (2018). Plasmid DNA vaccines: Where are we now? *Drugs of Today (Barcelona)*, *54*(5), 315–333. Available from https://doi.org/10.1358/dot.2018.54.5.2807864. Available from 29911696.

Goldblatt, D. (2000). Conjugate vaccines. *Clinical and Experimental Immunology*, *119*(1), 1–3. Available from https://doi.org/10.1046/j.1365-2249.2000.01109.x. Available from 10671089, PMCID: PMC1905528.

Goldman, A. S., & Prabhakar, B. S. (1996). Immunology overview. In S. Baron (Ed.), *Medical microbiology* (4th ed). University of Texas Medical Branch at Galveston, Chapter 1. Available from. Available from https://www.ncbi.nlm.nih.gov/books/NBK7795/, Accessed 24.12.20.

Goldsby, R., Kindt, T. J., Osborne, B. A., & Kuby, J. (2003). *Cells and organs of the immune system. Immunology* (5th ed., pp. 24–56). New York: W. H. Freeman and Company.

Greenwood, B. (2014). The contribution of vaccination to global health: Past, present and future. *Philosophical Transactions of the Royal Society of London. Series B, Biological Sciences*, *369*(1645), 20130433. Available from https://doi.org/10.1098/rstb.2013.0433. Available from 24821919, PMCID: PMC4024226.

Gutowska-Owsiak, D., & Ogg, G. S. (2017). Therapeutic vaccines for allergic disease. *NPJ Vaccines.*, *2*, 12. Available from https://doi.org/10.1038/s41541-017-0014-8. Available from 29263869, PMCID: PMC5604746.

Hanley, K. A. (2011). The double-edged sword: How evolution can make or break a live-attenuated virus vaccine. *Evolution (N Y).*, *4*(4), 635–643. Available from https://doi.org/10.1007/s12052-011-0365-y. Available from 22468165, PMCID: PMC3314307.

Heilmann, C., Budtz-Jørgensen, E., Nielsen, F., Heinzow, B., Weihe, P., & Grandjean, P. (2010). Serum concentrations of antibodies against vaccine toxoids in children exposed perinatally to immunotoxicants. *Environmental Health Perspectives*, *118*(10), 1434–1438. Available from https://doi.org/10.1289/ehp.1001975, Epub 2010 Jun 1. Available from 20562056.

Hietalahti, J., & Meri, S. (2015). Uudet rokotteet B-ryhmän meningokokkia vastaan [New vaccines against group B meningococcal diseases]. *Duodecim; Laaketieteellinen Aikakauskirja*, *131*(6), 525–532, Finnish. Available from 26237895.

Hirahara, K., & Nakayama, T. (2016). CD4 + T-cell subsets in inflammatory diseases: Beyond the Th1/Th2 paradigm. *International Immunology*, *28*(4), 163–171. Available from https://doi.org/10.1093/intimm/dxw006, Epub 2016 Feb 12. Available from 26874355.

Ho, J. K., Jeevan-Raj, B., Netter, H. J., & Hepatitis, B. (2020). Virus (HBV) subviral particles as protective vaccines and vaccine platforms. *Viruses.*, *12*(2), 126. Available from https://doi.org/10.3390/v12020126. Available from 31973017, PMCID: PMC7077199.

Hobernik, D., & Bros, M. (2018). DNA vaccines-how far from clinical use? *International Journal of Molecular Sciences.*, *19*(11), 3605. Available from https://doi.org/10.3390/ijms19113605. Available from 30445702, PMCID: PMC6274812.

James, S. F., Chahine, E. B., Sucher, A. J., & Hanna, C. (2018). Shingrix: The new adjuvanted recombinant herpes zoster vaccine. *The Annals of Pharmacotherapy*, *52*(7), 673–680. Available from https://doi.org/10.1177/1060028018758431, Epub 2018 Feb 18. Available from 29457489.

Jang, H., Elaish, M., Kc, M., Abundo, M. C., Ghorbani, A., Ngunjiri, J. M., & Lee, C. W. (2018). Efficacy and synergy of live-attenuated and inactivated influenza vaccines in young chickens. *PLoS One*, *13*(4), e0195285. Available from https://doi.org/10.1371/journal.pone.0195285. Available from 29624615, PMCID: PMC5889186.

Jones, R. G., Liu, Y., Rigsby, P., & Sesardic, D. (2008). An improved method for development of toxoid vaccines and antitoxins. *Journal of Immunological Methods*, *337*(1), 42–48. Available from https://doi.org/10.1016/j.jim.2008.05.009, Epub 2008 Jun 9. Available from 18571196.

Juskewitch, J. E., Tapia, C. J., & Windebank, A. J. (2010). Lessons from the Salk polio vaccine: Methods for and risks of rapid translation. *Clinical and Translational Science.*, *3*(4), 182–185. Available from https://doi.org/10.1111/j.1752-8062.2010.00205.x. Available from 20718820, PMCID: PMC2928990.

Kaiko, G. E., Horvat, J. C., Beagley, K. W., & Hansbro, P. M. (2008). Immunological decision-making: How does the immune system decide to mount a helper T-cell response? *Immunology*, *123*(3), 326–338. Available from https://doi.org/10.1111/j.1365-2567.2007.02719.x, Epub 2007 Nov 5. Available from 17983439.

Kallen, K. J., & Theß, A. (2014). A development that may evolve into a revolution in medicine: mRNA as the basis for novel, nucleotide-based vaccines and drugs. *Therapeutic Advances in Vaccines.*, *2*(1), 10–31. Available from https://doi.org/10.1177/2051013613508729. Available from 24757523, PMCID: PMC3991152.

Karch, C. P., & Burkhard, P. (2016). Vaccine technologies: From whole organisms to rationally designed protein assemblies. *Biochemical Pharmacology*, *120*, 1–14. Available from https://doi.org/10.1016/j.bcp.2016.05.001, Epub 2016 May 6. Available from 27157411.

Karunamoorthi, K. (2014). Malaria vaccine: A future hope to curtail the global malaria burden. *International Journal of Preventive Medicine.*, *5*(5), 529–538. Available from 24932383, PMCID: PMC4050672.

Khan, K. H. (2013). DNA vaccines: Roles against diseases. *Germs.*, *3*(1), 26–35. Available from https://doi.org/10.11599/germs.2013.1034. Available from 24432284, PMCID: PMC3882840.

Khuroo, M. S., Khuroo, M., Khuroo, M. S., Sofi, A. A., & Khuroo, N. S. (2020). COVID-19 vaccines: A race against time in the middle of death and devastation!. *Journal of Clinical and Experimental Hepatology.*, *10*(6), 610–621. Available from https://doi.org/10.1016/j.jceh.2020.06.003, Epub 2020 Jun 10. Available from 32837093.

Kübler, H., Scheel, B., Gnad-Vogt, U., Miller, K., Schultze-Seemann, W., Vom Dorp, F., Parmiani, G., Hampel, C., Wedel, S., Trojan, L., Jocham, D., Maurer, T., Rippin, G., Fotin-Mleczek, M., von der Mülbe, F., Probst, J., Hoerr, I., Kallen, K. J., Lander, T., &

Stenzl, A. (2015). Self-adjuvanted mRNA vaccination in advanced prostate cancer patients: A first-in-man phase I/IIa study. *Journal for Immunotherapy of Cancer.*, *3*, 26. Available from https://doi.org/10.1186/s40425-015-0068-y. Available from 26082837, PMCID: PMC4468959.

Lauring, A. S., Jones, J. O., & Andino, R. (2010). Rationalizing the development of live attenuated virus vaccines. *Nature Biotechnology*, *28*(6), 573–579. Available from https://doi.org/10.1038/nbt.1635, Epub 2010 Jun 7. Available from 20531338.

Lee, J., Arun Kumar, S., Jhan, Y. Y., & Bishop, C. J. (2018). Engineering DNA vaccines against infectious diseases. *Acta Biomaterialia*, *80*, 31–47. Available from https://doi.org/10.1016/j.actbio.2018.08.033, Epub 2018 Aug 31. Available from 30172933.

Lee, S., & Nguyen, M. T. (2015). Recent advances of vaccine adjuvants for infectious diseases. *Immune Network.*, *15*(2), 51–57. Available from https://doi.org/10.4110/in.2015.15.2.51, Epub 2015 Apr 23. Available from 25922593.

Liao, J. C., Gregor, P., Wolchok, J. D., Orlandi, F., Craft, D., Leung, C., Houghton, A. N., & Bergman, P. J. (2006). Vaccination with human tyrosinase DNA induces antibody responses in dogs with advanced melanoma. *Cancer Immunity: A Journal of the Academy of Cancer Immunology*, *6*, 8. Available from 16626110, PMCID: PMC1976276.

Liljeroos, L., Malito, E., Ferlenghi, I., & Bottomley, M. J. (2015). Structural and computational biology in the design of immunogenic vaccine antigens. *Journal of Immunology Research.*, *2015*, 156241. Available from https://doi.org/10.1155/2015/156241, Epub 2015 Oct 7. Available from 26526043.

Mäkelä, P. H., & Käyhty, H. (2002). Evolution of conjugate vaccines. *Expert Review of Vaccines*, *1*(3), 399–410. Available from https://doi.org/10.1586/14760584.1.3.399. Available from 12901578.

Margolick, J. B., Markham, R. B., & Scott, A. L. (2006). Infectious disease epidemiology: Theory and practice. Chapter 10. In K. E. Nelson, & C. F. Masters (Eds.), *The immune system and host defense against infections* (pp. 317–343). Boston: Jones and Bartlett.

Marzi, A., Ebihara, H., Callison, J., Groseth, A., Williams, K. J., Geisbert, T. W., & Feldmann, H. (2011). Vesicular stomatitis virus-based Ebola vaccines with improved cross-protective efficacy. *The Journal of Infectious Diseases*, *204*(Suppl. 3), S1066–S1074. Available from https://doi.org/10.1093/infdis/jir348. Available from 21987743, PMCID: PMC3203393.

McCarthy, P. C., Sharyan, A., & Sheikhi Moghaddam, L. (2018). Meningococcal vaccines: Current status and emerging strategies. *Vaccines (Basel).*, *6*(1), 12. Available from https://doi.org/10.3390/vaccines6010012. Available from 29495347, PMCID: PMC5874653.

Minor, P. D. (2015). Live attenuated vaccines: Historical successes and current challenges. *Virology*, *479-480*, 379–392. Available from https://doi.org/10.1016/j.virol.2015.03.032, Epub 2015 Apr 8. Available from 25864107.

Moballegh Naseri, M., Shams, S., Moballegh Naseri, M., & Bakhshi, B. (2020). In silico analysis of epitope-based CadF vaccine design against *Campylobacter jejuni*. *BMC Research Notes.*, *13*(1), 518. Available from https://doi.org/10.1186/s13104-020-05364-z. Available from 33168057, PMCID: PMC7652678.

Moxon, R., Reche, P. A., & Rappuoli, R. (2019). Editorial: Reverse vaccinology. *Frontiers in Immunology*, *10*, 2776. Available from https://doi.org/10.3389/fimmu.2019.02776. Available from 31849959, PMCID: PMC6901788.

Moyle, P. M., & Toth, I. (2013). Modern subunit vaccines: Development, components, and research opportunities. *ChemMedChem.*, *8*(3), 360–376. Available from https://doi.org/10.1002/cmdc.201200487, Epub 2013 Jan 11. Available from 23316023.

Muñoz-Fontela, C., Dowling, W. E., Funnell, S. G. P., Gsell, P. S., Riveros-Balta, A. X., Albrecht, R. A., Andersen, H., Baric, R. S., Carroll, M. W., Cavaleri, M., Qin, C., Crozier, I., Dallmeier, K., de Waal, L., de Wit, E., Delang, L., Dohm, E., Duprex, W. P., Falzarano, D., … Barouch, D. H. (2020). Animal models for COVID-19. *Nature*, *586*(7830), 509–515. Available from https://doi.org/10.1038/s41586-020-2787-6, Epub 2020 Sep 23. Available from 32967005.

Nishat, S., & Andreana, P. R. (2016). Entirely carbohydrate-based vaccines: An emerging field for specific and selective immune responses. *Vaccines (Basel)*, *4*(2), 19. Available from https://doi.org/10.3390/vaccines4020019. Available from 27213458, PMCID: PMC4931636.

Oli, A. N., Agu, R. U., Ihekwereme, C. P., & Esimone, C. O. (2017). An evaluation of the cold chain technology in South-East, Nigeria using Immunogenicity study on the measles vaccines. *Pan African Medical Journal.*, *27*(Suppl 3), 28. Available from https://doi.org/10.11604/pamj.supp.2017.27.3.11491, eCollection 2017. PubMed. Available from 29296163.

Oli, A. N., Obialor, W. O., Ifeanyichukwu, M. O., Odimegwu, D. C., Okoyeh, J. N., Emechebe, G. O., Adejumo, S. A., & Ibeanu, G. C. (2020). Immunoinformatics and vaccine development: An overview. *Immunotargets and Therapy.*, *9*, 13–30. Available from https://doi.org/10.2147/ITT.S241064, eCollection 2020. Review. PubMed. Available from 32161726.

Pandey, A., Cabello, A., Akoolo, L., Rice-Ficht, A., Arenas-Gamboa, A., McMurray, D., Ficht, T. A., & de Figueiredo, P. (2016). The case for live attenuated vaccines against the neglected zoonotic diseases brucellosis and bovine tuberculosis. *PLoS Neglected Tropical Diseases*, *10*(8), e0004572. Available from https://doi.org/10.1371/journal.pntd.0004572. Available from 27537413, PMCID: PMC4990199.

Pardi, N., Hogan, M. J., Porter, F. W., & Weissman, D. (2018). mRNA vaccines—A new era in vaccinology. *Nature Reviews. Drug Discovery*, *17*(4), 261–279. Available from https://doi.org/10.1038/nrd.2017.243, Epub 2018 Jan 12. Available from 29326426, PMCID: PMC5906799.

Peng, M., Mo, Y., Wang, Y., Wu, P., Zhang, Y., Xiong, F., Guo, C., Wu, X., Li, Y., Li, X., Li, G., Xiong, W., & Zeng, Z. (2019). Neoantigen vaccine: An emerging tumor immunotherapy. *Molecular Cancer*, *18*(1), 128. Available from https://doi.org/10.1186/s12943-019-1055-6. Available from 31443694, PMCID: PMC6708248.

Petrovsky, N., & Aguilar, J. C. (2004). Vaccine adjuvants: Current state and future trends. *Immunology and Cell Biology*, *82*(5), 488–496. Available from https://doi.org/10.1111/j.0818-9641.2004.01272.x. Available from 15479434.

Petsch, B., Schnee, M., Vogel, A. B., Lange, E., Hoffmann, B., Voss, D., Schlake, T., Thess, A., Kallen, K. J., Stitz, L., & Kramps, T. (2012). Protective efficacy of in vitro synthesized, specific mRNA vaccines against influenza A virus infection. *Nature Biotechnology*, *30*(12), 1210–1216. Available from https://doi.org/10.1038/nbt.2436, Epub 2012 Nov 25. Available from 23159882.

Pichichero, M. E. (2013). Protein carriers of conjugate vaccines: Characteristics, development, and clinical trials. *Human Vaccines & Immunotherapeutics.*, *9*(12), 2505–2523. Available from https://doi.org/10.4161/hv.26109, Epub 2013 Aug 16. Available from 23955057.

Pimenova, T., Nazabal, A., Roschitzki, B., Seebacher, J., Rinner, O., & Zenobi, R. (2008). Epitope mapping on bovine prion protein using chemical cross-linking and mass spectrometry. *Journal of Mass Spectrometry: JMS*, *43*(2), 185–195. Available from https://doi.org/10.1002/jms.1280. Available from 17924399.

Pobre, K., Tashani, M., Ridda, I., Rashid, H., Wong, M., & Booy, R. (2014). Carrier priming or suppression: Understanding carrier priming enhancement of anti-polysaccharide antibody response to conjugate vaccines. *Vaccine*, *32*(13), 1423–1430. Available from https://doi.org/10.1016/j.vaccine.2014.01.047, Epub 2014 Jan 31. Available from 24492014.

Pollard, C., Rejman, J., De Haes, W., Verrier, B., Van Gulck, E., Naessens, T., De Smedt, S., Bogaert, P., Grooten, J., Vanham, G., & De Koker, S. (2013). Type I IFN counteracts the induction of antigen-specific immune responses by lipid-based delivery of mRNA vaccines. *Molecular Therapy: The Journal of the American Society of Gene Therapy*, *21*(1), 251–259. Available from https://doi.org/10.1038/mt.2012.202, Epub 2012 Sep 25. Available from 23011030.

Porter, K. R., & Raviprakash, K. (2017). DNA vaccine delivery and improved immunogenicity. *Current Issues in Molecular Biology*, *22*, 129–138. Available from https://doi.org/10.21775/cimb.022.129, Epub 2016 Nov 10. Available from 27831541.

Rappuoli, R., & Aderem, A. (2011). A 2020 vision for vaccines against HIV, tuberculosis and malaria. *Nature*, *473*(7348), 463–469. Available from https://doi.org/10.1038/nature10124. Available from 21614073.

Rappuoli, R., Bottomley, M. J., D'Oro, U., Finco, O., & De Gregorio, E. (2016). Reverse vaccinology 2.0: Human immunology instructs vaccine antigen design. *The Journal of Experimental Medicine*, *213*(4), 469–481. Available from https://doi.org/10.1084/jem.20151960, Epub 2016 Mar 28. Available from 27022144.

Rappuoli, R. (2001). Reverse vaccinology, a genome-based approach to vaccine development. *Vaccine*, *19*(17-19), 2688–2691. Available from https://doi.org/10.1016/s0264-410x(00)00554-5. Available from 11257410.

Reche, P. A., Glutting, J. P., & Reinherz, E. L. (2002). Prediction of MHC class I binding peptides using profile motifs. *Human Immunology*, *63*(9), 701–709. Available from https://doi.org/10.1016/s0198-8859(02)00432-9. Available from 12175724.

Redding, L., & Weiner, D. B. (2009). DNA vaccines in veterinary use. *Expert Review of Vaccines*, *8*(9), 1251–1276. Available from https://doi.org/10.1586/erv.09.77. Available from 19722897, PMCID: PMC4425421.

Reglinski, M., Ercoli, G., Plumptre, C., Kay, E., Petersen, F. C., Paton, J. C., Wren, B. W., & Brown, J. S. (2018). A recombinant conjugated pneumococcal vaccine that protects against murine infections with a similar efficacy to Prevnar-13. *NPJ Vaccines.*, *3*, 53. Available from https://doi.org/10.1038/s41541-018-0090-4. Available from 30393571, PMCID: PMC6208403.

Richner, J. M., Jagger, B. W., Shan, C., Fontes, C. R., Dowd, K. A., Cao, B., Himansu, S., Caine, E. A., Nunes, B. T. D., Medeiros, D. B. A., Muruato, A. E., Foreman, B. M., Luo, H., Wang, T., Barrett, A. D., Weaver, S. C., Vasconcelos, P. F. C., Rossi, S. L., Ciaramella, G., . . . Diamond, M. S. (2017). Vaccine mediated protection against Zika virus-induced congenital disease. *Cell.*, *170*(2), 273–283.e12. Available from https://doi.org/10.1016/j.cell.2017.06.040. Available from 28708997, PMCID: PMC5546158.

Robbins, J. B., Schneerson, R., Keith, J. M., Miller, M. A., Kubler-Kielb, J., & Trollfors, B. (2009). Pertussis vaccine: A critique. *The Pediatric Infectious Disease Journal*, *28*(3), 237–241. Available from https://doi.org/10.1097/INF.0b013e31818a8958. Available from 19165133, PMCID: PMC3128455.

Rowaiye, A. B., Asala, T., Oli, A. N., Uzochukwu, I. C., Akpa, A., & Esimone, C. O. (2020). The activating receptors of natural killer cells and their inter-switching potentials. *Current Drug Targets*, *21*(16), 1733–1751. Available from https://doi.org/10.2174/1389450121666200910160929. Available from 32914713.

Sanders, B., Koldijk, M., & Schuitemaker, H. (2014). Inactivated viral vaccines. *Vaccine Analysis: Strategies, Principles, and Control.*, 45–80. Available from https://doi.org/10.1007/978-3-662-45024-6_2, PMCID: PMC7189890.

Saylor, K., Gillam, F., Lohneis, T., & Zhang, C. (2020). Designs of antigen structure and composition for improved protein-based vaccine efficacy. *Frontiers in Immunology*, *11*, 283. Available from https://doi.org/10.3389/fimmu.2020.00283. Available from 32153587, PMCID: PMC7050619.

Schnee, M., Vogel, A. B., Voss, D., Petsch, B., Baumhof, P., Kramps, T., & Stitz, L. (2016). An mRNA vaccine encoding rabies virus glycoprotein induces protection against lethal infection in mice and correlates of protection in adult and newborn pigs. *PLoS Neglected Tropical Diseases*, *10*(6), e0004746. Available from https://doi.org/10.1371/journal.pntd.0004746. Available from 27336830, PMCID: PMC4918980.

Sebastian, M., Papachristofilou, A., Weiss, C., Früh, M., Cathomas, R., Hilbe, W., Wehler, T., Rippin, G., Koch, S. D., Scheel, B., Fotin-Mleczek, M., Heidenreich, R., Kallen, K. J., Gnad-Vogt, U., & Zippelius, A. (2014). Phase Ib study evaluating a self-adjuvanted mRNA cancer vaccine (RNActive®) combined with local radiation as consolidation and maintenance treatment for patients with stage IV non-small cell lung cancer. *BMC Cancer*, *14*, 748. Available from https://doi.org/10.1186/1471-2407-14-748. Available from 25288198, PMCID: PMC4195907.

Seib, K. L., Zhao, X., & Rappuoli, R. (2012). Developing vaccines in the era of genomics: A decade of reverse vaccinology. *Clinical Microbiology and Infection: The Official Publication of the European Society of Clinical Microbiology and Infectious Diseases*, *18* (Suppl. 5), 109–116. Available from https://doi.org/10.1111/j.1469-0691.2012.03939.x, Epub 2012 Aug 6. Available from 22882709.

Seo, H. S. (2015). Application of radiation technology in vaccines development. *Clinical and Experimental Vaccine Research.*, *4*(2), 145–158. Available from https://doi.org/10.7774/cevr.2015.4.2.145, Epub 2015 Jul 29. Available from 26273573.

Sharma, R., & Sharma, C. L. (2007). Quadrivalent human papillomavirus recombinant vaccine: The first vaccine for cervical cancers. *Journal of Cancer Research and Therapeutics*, *3*(2), 92–95. Available from https://doi.org/10.4103/0973-1482.34686. Available from 17998730.

Slifka, M. K., & Amanna, I. J. (2018). Passive immunization. *Plotkin's Vaccines*, 84–95.e10. Available from https://doi.org/10.1016/B978-0-323-35761-6.00008-0, Epub 2017 Jul 17. PMCID: PMC7151993.

Stauffer, F., El-Bacha, T., & Da Poian, A. T. (2006). Advances in the development of inactivated virus vaccines. *Recent Patents on Anti-Infective Drug Discovery.*, *1*(3), 291–296. Available from https://doi.org/10.2174/157489106778777673. Available from 18221154.

Straub, T. M., Höner zu Bentrup, K., Orosz-Coghlan, P., Dohnalkova, A., Mayer, B. K., Bartholomew, R. A., Valdez, C. O., Bruckner-Lea, C. J., Gerba, C. P., Abbaszadegan, M., & Nickerson, C. A. (2007). In vitro cell culture infectivity assay for human noroviruses. *Emerging Infectious Diseases*, *13*(3), 396–403. Available from https://doi.org/10.3201/eid1303.060549. Available from 17552092, PMCID: PMC2725917.

Sun, B., Zhao, X., Wu, Y., Cao, P., Movahedi, F., Liu, J., Wang, J., Xu, Z. P., & Gu, W. (2021). Mannose-functionalized biodegradable nanoparticles efficiently deliver DNA vaccine and promote anti-tumor immunity. *ACS Applied Materials & Interfaces.*, *13*(12), 14015–14027. Available from https://doi.org/10.1021/acsami.1c01401, Epub 2021 Mar 22. Available from 33751882.

Tetanus vaccines: WHO position paper – February 2017. (2017). *Weekly Epidemiological Record*, *92*(6), 53–76. English, French. PMID: 28185446.

Tomar, N., & De, R. K. (2014). Immunoinformatics: A brief review. *Methods in Molecular Biology*, *1184*, 23–55. Available from https://doi.org/10.1007/978-1-4939-1115-8_3. Available from 25048118.

Tomar, N., & De, R. K. (2010). Immunoinformatics: An integrated scenario. *Immunology*, *131* (2), 153–168. Available from https://doi.org/10.1111/j.1365-2567.2010.03330.x, Epub 2010 Aug 16. Available from 20722763.

Tonomura, N., McLaughlin, K., Grimm, L., Goldsby, R. A., & Osborne, B. A. (2003). Glucocorticoid-induced apoptosis of thymocytes: Requirement of proteasome-dependent mitochondrial activity. *Journal of Immunology, 170*(5), 2469–2478. Available from https://doi.org/10.4049/jimmunol.170.5.2469. Available from 12594272.

van der Heiden, M., Duizendstra, A., Berbers, G. A. M., Boots, A. M. H., & Buisman, A. M. (2017). Tetanus Toxoid carrier protein induced T-helper cell responses upon vaccination of middle-aged adults. *Vaccine, 35*(42), 5581–5588. Available from https://doi.org/10.1016/j.vaccine.2017.08.056, Epub 2017 Sep 4. Available from 28882444.

Van Regenmortel, M. H. (2016). Structure-based reverse vaccinology failed in the case of HIV because it disregarded accepted immunological theory. *International Journal of Molecular Sciences., 17*(9), 1591. Available from https://doi.org/10.3390/ijms17091591. Available from 27657055, PMCID: PMC5037856.

Vartak, A., & Sucheck, S. J. (2016). Recent advances in subunit vaccine carriers. *Vaccines (Basel), 4*(2), 12. Available from https://doi.org/10.3390/vaccines4020012. Available from 27104575, PMCID: PMC4931629.

Verbeke, R., Lentacker, I., De Smedt, S. C., & Dewitte, H. (2019). Three decades of messenger RNA vaccine development. *Nano Today., 28*, 100766. Available from https://doi.org/10.1016/j.nantod.2019.100766.

Wadman, M. (2020). Public needs to prep for vaccine side effects. *Science (New York, N.Y.), 370*(6520), 1022. Available from https://doi.org/10.1126/science.370.6520.1022. Available from 33243869.

Wang, H., Zhang, Y., Huang, B., Deng, W., Quan, Y., Wang, W., Xu, W., Zhao, Y., Li, N., Zhang, J., Liang, H., Bao, L., Xu, Y., Ding, L., Zhou, W., Gao, H., Liu, J., Niu, P., Zhao, L., ... Yang, X. (2020). Development of an inactivated vaccine candidate, BBIBP-CorV, with potent protection against SARS-CoV-2. *Cell., 182*(3), 713–721.e9. Available from https://doi.org/10.1016/j.cell.2020.06.008, Epub 2020 Jun 6. Available from 32778225.

Weide, B., Pascolo, S., Scheel, B., Derhovanessian, E., Pflugfelder, A., Eigentler, T. K., Pawelec, G., Hoerr, I., Rammensee, H. G., & Garbe, C. (2009). Direct injection of protamine-protected mRNA: Results of a phase 1/2 vaccination trial in metastatic melanoma patients. *Journal of Immunotherapy (Hagerstown, MD: 1997), 32*(5), 498–507. Available from https://doi.org/10.1097/CJI.0b013e3181a00068. Available from 19609242.

Weiner, D. B. (2008). DNA vaccines: Crossing a line in the sand. Introduction to special issue. *Vaccine, 26*(40), 5073–5074. Available from https://doi.org/10.1016/j.vaccine.2008.07.024, Epub 2008 Jul 22. Available from 18652867.

Wraith, D. C., Goldman, M., & Lambert, P. H. (2003). Vaccination and autoimmune disease: What is the evidence? *Lancet, 362*(9396), 1659–1666. Available from https://doi.org/10.1016/S0140-6736(03)14802-7. Available from 14630450.

Zhang, Y., Yang, J., Li, M., Cui, M., Fu, Z. F., Zhao, L., & Zhou, M. (2019). A recombinant rabies virus expressing Fms-like Tyrosine kinase 3 ligand (Flt3L) induces enhanced immunogenicity in mice. *Virologica Sinica., 34*(6), 662–672. Available from https://doi.org/10.1007/s12250-019-00144-x, Epub 2019 Jun 28. Available from 31254272.

Zheng, S. Y., Yu, B., Zhang, K., Chen, M., Hua, Y. H., Yuan, S., Watt, R. M., Zheng, B. J., Yuen, K. Y., & Huang, J. D. (2012). Comparative immunological evaluation of recombinant *Salmonella typhimurium* strains expressing model antigens as live oral vaccines. *BMC Immunology, 13*, 54. Available from https://doi.org/10.1186/1471-2172-13-54. Available from 23013063, PMCID: PMC3503649.

Chapter 3

Dendritic cells as vaccine targets

Frances Iseghohi[1], Aondona Priscilla Yahemba[1], Adekunle Babajide Rowaiye[1] and Angus Nnamdi Oli[2]

[1]Department of Medical Biotechnology, National Biotechnology Development Agency, Abuja, Nigeria, [2]Department of Pharmaceutical Microbiology and Biotechnology, Faculty of Pharmaceutical Sciences, Nnamdi Azikiwe University, Awka, Nigeria

Introduction to dendritic cells

Dendritic cells (DCs) are amongst the earliest identified accessory/nonlymphoid immune cells; they were described in 1973 by Ralph Steinman and Zanvil Cohn as the immune system's most powerful antigen-presenting cells (APCs) (Sabado et al., 2017). As APCs, their main role is to locate and process antigen, and display antigenic peptides on human leukocyte antigen (HLA) proteins [now referred to as major histocompatibility complex (MHC) molecules], which are present on cellular surfaces. The displayed antigenic peptides are recognized by T cells, which culminates in diverse immunological responses (Steinman, 2012), and thus DCs serve as a link between the adaptive and innate immune systems. They also secrete cytokines and growth factors that control the function of diverse immune cells (O'Keeffe et al., 2015), and in particular maintain a balance between inflammatory and tolerogenic T cell responses (Simones et al., 2010). Originating in the bone marrow, DCs differentiate into cells predominantly found in mucosa, skin, and lymphoid tissues.

DCs play a crucial role in a variety of diseases including cancer, hypersensitivity responses, respiratory diseases, infection, and autoimmunity. Antigen recognition of MHC-II peptide epitopes by naïve $CD4^+$ cells results in differentiation into Th1, Th2, and Th17 effector cells (Chapter 1: Immunology and the Concept of Vaccination; (Nakano et al., 2009; Yamane & Paul, 2013). In skin hypersensitivity, $CD4^+$ and $CD8^+$ T lymphocytes are activated, leading to secretion of cytokines, and inflammation mediated in part by chemotaxis, resulting in infiltration of more immune cells (Kaplan et al., 2012; Yazdi et al., 2016). In respiratory diseases such as asthma, DCs capture allergens within the airways, leading to a Th2 response, resulting in inflammation of the airway. DCs are also involved in other lung-related inflammatory diseases (Upham & Hughes, 2006).

Vaccinology and Methods in Vaccine Research. DOI: https://doi.org/10.1016/B978-0-323-91146-7.00010-X

DCs can induce and modulate immunological tolerance in addition to stimulating the immune system. For example, an in vivo animal study demonstrated that depletion of DCs (by targeting a DC transcription factor engineered to express diphtheria toxin receptor) leads to lympho-angiogenesis and myeloproliferative disorders, emphasizing the role of DCs in maintaining immune tolerance (Meredith et al., 2012). Central tolerance depends predominantly on thymic DCs displaying self-antigens to T cells. Thymic DCs delete auto-reactive lymphocytes by negative selection, and positively select regulatory T cells (Treg) for the maintenance of immunological tolerance toward self-antigens. Thymic DCs that play major roles in central tolerance include migratory DCs ($CD8\alpha - CD11b^{+}SIRP\alpha^{+}$), resident DCs ($CD8\alpha^{+}SIRP\alpha^{-}$), and plasmacytoid DCs (pDCs; $CD11c^{int}CD45RA^{int}$) (Audiger et al., 2017; Cools et al., 2007; Hilkens et al., 2010).

DCs have been employed to induce cytotoxic T lymphocyte (CTL)-mediated immune responses against tumor cells in cancer immunotherapy (Lee & Radford, 2019), and DC cancer vaccines have been developed as personalized cancer therapy (Perez, & De Palma, 2019). DCs not only stimulate CD8 + T lymphocytes to produce apoptosis-inducing responses, but also trigger NK cells to release perforins and granzymes capable of killing tumor cells (Castell-Rodríguez et al., 2017).

DCs are found in two forms: mature and immature. Immature DCs (iDCs) recognize disease-causing organisms using pattern recognition receptors (PRRs); these PRRs recognize damage-associated molecular patterns (DAMPs) or pathogen-associated molecular patterns (PAMPs) on pathogen surfaces (Hemmi & Akira, 2005), leading to phagocytosis and lysosomal digestion. Maturation of DCs is then initiated, and DCs move into lymph nodes to deliver antigen to naive T cells (Mbongue et al., 2017). Mature (mDCs) and iDCs upregulate costimulatory proteins (CD40, CD80, CD86, and CD83) that help to stimulate T cells, and CCR7, a chemotactic receptor that encourages homing of lymphocytes and DCs to lymph nodes. The DC maturation process is an important process in vaccine development, as mDCs are capable of eliciting strong immune responses.

In addition to antigen presentation on MHC-II molecules, which is necessary for the activation of CD4 T cells, DCs can activate CD8 cytotoxic T cells (CTL) via a process known as cross-presentation or cross-priming. In this process, antigens or pathogens that have undergone phagocytosis are processed in endosomes and resulting peptide fragments are loaded onto Class I MHC rather than Class II. This is an important mechanism in vaccine-induced immunity for cancer, and intracellular pathogens, which are targeted by CTL.

Origin and types of dendritic cells

Hematopoietic stem cells form the common lymphoid and myeloid ancestral cells (Fig. 3.1), which are progenitors of lymphoid and myeloid families

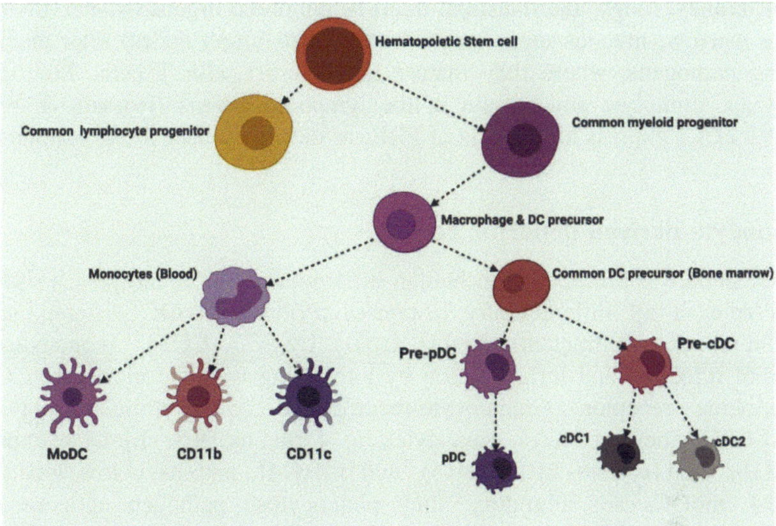

FIGURE 3.1 Origin of human dendritic cells. *Created with BioRender.com. Saxena, M., & Yeretssian, G. (2014). NOD-Like Receptors: Master Regulators of Inflammation and Cancer. Frontiers in immunology, 5, 327. https://doi.org/10.3389/fimmu.2014.00327.*

(Ishikawa et al., 2007). The common myeloid progenitor cells evolve into macrophages and DC precursors, which further develop into monocytes and common DC precursors (CDP) (Belz & Nutt, 2012). The monocytes are disseminated into the blood stream, and during inflammation they develop into MHC-II^{+} Monocyte-derived dendritic cells (moDCs). In the bone marrow, CDP differentiate into preplasmacytoid DCs (pre-pDC) and preconventional DCs (pre-cDC). The pre-cDC differentiate into two subsets, conventional DCs 1 (cDC1) and conventional DCs 2 (cDC2), whereas pre-pDC differentiate into pDC (Geissmann et al., 2010).

Conventional dendritic cells

Conventional DCs (cDCs) (referred to as classical or myeloid DCs) are a group of specialized DCs that are very efficient in antigen processing and presentation. These can be either immature DCs (iDCs) or mature DCs (mDCs). The capability of mDCs to trigger antigen-specific Th0 cells in peripheral lymphoid organs differentiates them from iDCs (Worbs et al., 2017). mDCs also express very significant levels of fascin-1 and costimulatory molecules. cDCs have two classes: cDC1s and cDC2s; both phenotypes are found in humans and mice. cDC populations are distinguished by their molecular markers: cDC1s expresses Klf4, IRF4, Rbpj, and Notch2, while cDC2s express batf3, Flt3, IRF8, and Bc16. cDCs can be migratory or stationary. Migratory cDCs are capable of moving from peripheral organs (skin, liver, kidney, lungs, and intestinal tract) to lymphoid organs (spleen, thymus, bone marrow, mucous membranes, tonsils, and lymph nodes) after encountering pathogens, where they mature and interact with T cells. Stationary DCs are immobile and remain in the lymphoid organs (Hossain & Wall, 2019). cDCs express high levels of PRR, as described later in the chapter.

Monocyte-derived dendritic cells

moDCs share similarities with human inflammatory DCs and play a significant role in all inflammatory responses. Prior to moDC differentiation, monocyte chemoattractant proteins deploy Ly-6C$^+$, CD14$^+$ monocytes to sites of infection and inflammation by interacting with the monocyte CCR2 chemokine receptor. Granulocyte-macrophage colony-stimulating factor (GM-CSF) then activates monocytes to form moDCs. MoDCs express CD11b, CD11c, Fcγ RIII (CD16), and MHC-II, and have low levels of CD14. moDCs are migratory: they phagocytose pathogen antigens and ascend to regional lymph nodes where they help to stimulate and polarize Th cells (Cheong, Choi, et al., 2010; Cheong, Matos, et al., 2010; Cintolo et al., 2012; Lotfi et al., 2019).

Plasmacytoid dendritic cells

pDCs are found within blood, bone marrow, Peyer's patches, thymus, spleen, lymph nodes, and in some peripheral tissues. They produce type I interferons (IFNs) and other cytokines [e.g., tumor-necrosis factor-α (TNF-α) and interleukin-6 (IL-6)] to stimulate innate immunity, antiviral immunity, and systemic autoimmunity (Musumeci et al., 2019). pDCs derive from blood-forming primogenitor cells found in the bone marrow, regulated by the cytokine FLT-3L (FLT-3 ligand) (Hoeffel et al., 2007; McKenna et al., 2005; Ochando et al., 2006; Pulendran et al., 2000; Swiecki & Colonna, 2015). pDCs are efficient captors and presenters of antigen from immune complexes, apoptotic cells, and viruses, and upregulate Toll-like receptors (TLRs) in response to viral infections (Swiecki & Colonna, 2015).

pDCs are stimulated predominantly in response to single-stranded RNA (ssRNA) and single-stranded DNA (ssDNA), through TLR7 and TLR9 receptors. pDCs also regulate immunity through induction of Treg, partly as a result of IL-10 production, and upregulation of receptors including PD-L1 (Gehrie et al., 2011; Matta et al., 2010). Murine pDCs express surface markers BST2 (CD317), CD45R (B220), CD45RA, Siglec-H, and Ly-6C (Musumeci et al., 2019), whereas human pDCs express immunoglobulin-like transcript receptor (ILT 7), IL-3 receptor (CD123), blood DC Ag (BDCA)-2, Gr-1, and B220; they are CD11c low/CD11b negative (Cella et al., 1999; Matta et al., 2010).

Cytokines/growth factors in differentiation of dendritic cells

The differentiation of DCs into cDCs and pDCs is dependent on cytokines including FMS-like tyrosine kinase-3 ligand (Flt3L), an endogenous small molecule that functions as a cytokine and growth factor, and also GM-CSF and macrophage colony-stimulating factor (M-CSF). Flt3L is a major regulator of DC differentiation. Its receptor Flt3 (Fms-like tyrosine kinase-3, or CD135) is expressed on primogenitor cells located inside the bone marrow (Castell-Rodríguez et al., 2017), and on DCs in mucosal and lymphoid organs where it regulates their proliferation (Waskow et al., 2008). DC differentiation from precursor cells is regulated by GM-CSF (Bhattacharya et al., 2015). M-CSF, in addition to its role in macrophage development, can also drive development of both cDCs and pDCs in bone marrow (Onai et al., 2007).

Location of dendritic cells

DCs are situated in diverse locations (Fig. 3.2) in the human body, predominantly in the lymph nodes and epithelia; they express different receptors and molecular markers (Fig. 3.3) based on their location.

Dendritic cell in lymph nodes

DCs resident in lymph nodes are classified into CD8α^+ and CD8α^- subsets. In mice, DCs in the lymph nodes include CD8α^{lo}CD205^{hi} and mature

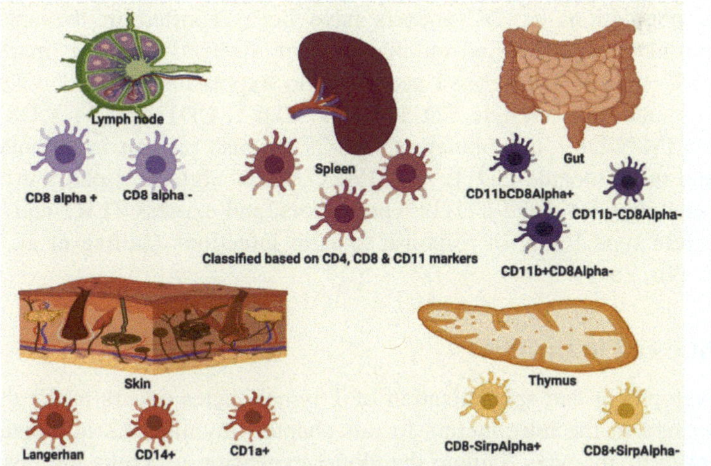

FIGURE 3.2 Location of dendritic cells and their phenotypes. *Created with BioRender.com. Saxena, M., & Yeretssian, G. (2014). NOD-Like Receptors: Master Regulators of Inflammation and Cancer. Frontiers in immunology, 5, 327. https://doi.org/10.3389/fimmu.2014.00327.*

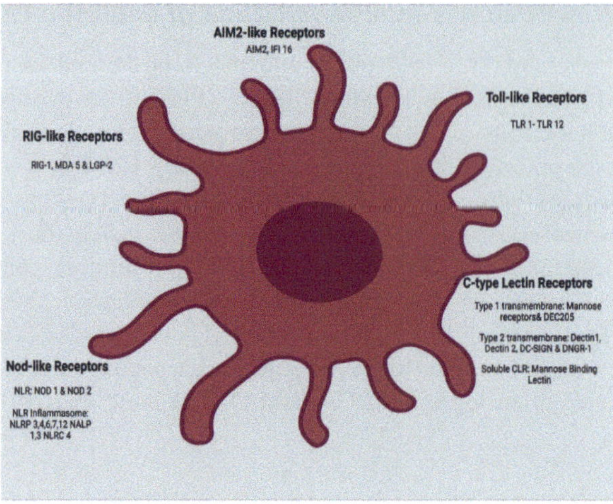

FIGURE 3.3 Dendritic cell receptors. *Created with BioRender.com. Saxena, M., & Yeretssian, G. (2014). NOD-Like Receptors: Master Regulators of Inflammation and Cancer. Frontiers in immunology, 5, 327. https://doi.org/10.3389/fimmu.2014.00327.*

$CD8\alpha^{lo}CD205^{int}$cDC, which relocate from the skin layers. DCs in human lymph nodes upregulate MHC-II and costimulatory molecules on maturation, and function as resident and migratory DCs capable of inducing Th1 and Th2 polarization in naive T lymphocytes.

Dendritic cell in spleen

Various populations of DC subsets have been reported in the spleen of humans and mice, classified on the basis of the CD11 surface marker. In mice, cDC subsets are defined according to expression of CD4, CD8, and CD11. Splenic cDC include $CD11c^{hi}CD4^{-}CD8^{+}$, $CD11c^{hi}CD4^{-}CD8^{-}$, and $CD11c^{hi}CD4^{+}CD8^{-}$ subpopulations. pDCs are also present in the spleen of mice, and these include $CD11c^{int}CD45RA^{+}B220^{+}SiglecH^{+}$. pDCs in human spleen include $CD4^{+}$ and $CD11c^{-}$ precursors, and express TLR7 and TLR9; they secrete type I IFN in response to viral infections (Mittag et al., 2011; Zhou & Wu, 2017).

Dendritic cell in thymus

The development and specialization of T lymphocytes occurs in the thymus. As described in the introduction to this chapter, thymic DCs eliminate self-Ag-specific thymocytes, initiate the clonal expansion of Treg, and positively select T cells with functional TCRs, which have some affinity for self-MHC alleles. Thymic DCs express the signal regulatory protein-α (Sirp-α) and CD8; $CD8^{-}Sirp\alpha^{+}$ and $CD8^{+}Sirp\alpha^{-}$ cDC subsets are also present (Proietto et al., 2008). The CD8α^+Sirp-α^- subset originates inside the thymus from intra-thymic precursors, while CD8α^-Sirpα^+ cells are generated from peripheral DCs. The murine thymus contains two types of cDC (Sirpα^-CD8$^+$ TcDCs and Sirpα^+CD8$^-$ TcDCs) and one pDC subset, sharing similarities with human thymic DCs (Donskoy & Goldschneider, 2003).

Dendritic cell within the skin

Many pDCs, cDCs, and moDCs are located within the epidermal and dermal layers of the skin. The epidermis also contains APCs called Langerhans cells (LCs) that express a C-type lectin receptor (CLR) called langerin (CD207) (Poulin et al., 2007), and also CD45 and MHC-II. The role of LCs is to identify antigens located in the outer layer of the skin and to carry them to the lymph nodes located along the colon and major arteries where they are presented to T cells, initiating immunological responses. Skin DCs express high levels of CD11b, CD11c, and MHC-II surface markers and other costimulatory molecules. Other DCs identified in the skin are CD1c$^+$ DCs, CD14$^+$ DCs, and dermal DCs (DDCs), which are homologous to CD8α^+ DCs present in lymphoid tissues (Haniffa et al., 2012, 2015).

Human skin DC subsets share the same structural features and patterns of gene expression with murine skin DCs. DCs in mice generally express CD11c and CD11b, homologous to CD1c$^+$CD1a + DDCs in humans. The mouse homologs of human CD14$^+$ DDCs are CD11b$^+$CD64$^+$ macrophages. Functionally, murine DC express CD11c, CD80, CD86, DEC-205, MHC-II, monocyte-associated markers (CD11b$^+$Ly6C$^-$CD64$^{lo\text{-}hi}$) and macrophage-associated antigens (Mac-3, F4/80) (Crozat et al., 2010).

Dendritic cell within the gut

DCs in the intestine are important for stimulating and controlling immune responses toward intestinal pathogens, as well as maintaining tolerance toward commensal (protective, gut inhabiting) microbes (Coombes & Powrie, 2008). On antigen detection, intestinal DCs migrate to Peyer's patches and mesenteric lymph nodes (MLN) to activate naive T cells. Gut DCs possess homing properties and influence the homing abilities of T cells that they activate, thus permitting T cells to return to the gut (Butcher et al., 1999).

DCs within the Peyer's patches include cDC subsets: CD11b$^+$CD8α^-, CD11b$^-$CD8α^+ and CD11b$^-$CD8α^-. The DCs in Peyer's patches also express chemokine receptors CX3CR1 and CCR6. CCR6$^+$ cells are deployed to follicle-associated epithelium in response to invasion by enteric pathogens, while CX3CR1 regulates immune tolerance and inflammation within the intestinal mucosa (Niess et al., 2005; Salazar-Gonzalez et al., 2006).

The DCs in the MLN comprise CD11b$^+$CD8α^-, CD11b$^-$CD8α^+, and CD11b$^-$CD8α^- subsets. Migratory DCs in the MLN are in cross-talk with T and B lymphocytes, and evoke immune responses directed at ensuring that there is no inflammation in the intestine, including induction of Tregs. DCs in the MLN highly express integrin CD103, and can initiate very high rates of $CD8^+$ T cell division, as well as IFN-γ secretion (Iwasaki & Kelsall, 2001). DCs found in human lamina propria and small intestine express the CX3CR1 chemokine receptor, CD103 (weakly), and CD11b (Johansson-Lindbom et al., 2005).

Receptors of dendritic cells

Activation of pathogen recognition receptors (PRRs) is crucial in initiating nonspecific (innate) immunity. PRRs differ in their signaling cascades, the type of host response they trigger, and their distribution in tissues. PRRs are mainly expressed by APCs, which include DCs, macrophages, and epithelial cells. PRRs bind to PAMPs found in microorganisms. These PRRs include mannose, lipopolysaccharide, peptidoglycans, lipoproteins, or other molecules depending on the microorganism (Apostolopoulos et al., 2013).

DCs possess several PRRs (Table 3.1) that promote the recognition of pathogens, resulting in the generation of cytokines, capture and degradation of pathogens into antigenic peptides, presentation by MHC to T cells, and eventual stimulation of acquired immune responses. Five types of PRRs have been recognized, namely CLRs, AIM2-like receptor (ALR), nucleotide-binding oligomerization domain (NOD)-like receptors (NLRs), TLRs, and retinoic acid–inducible gene (RIG)-I-like receptors (RLRs) (Akira et al., 2006). Many vaccine adjuvants work by promoting inflammatory responses in macrophage and DCs by activating TLRs and other PRRs: more detail is provided in Chapter 9, on Adjuvants.

C-type lectin receptors

CLRs are also referred to as calcium-dependent (C-type) lectins, and are a part of the lectin superfamily. They comprise carbohydrate-binding C-type lectin domains. CLRs are predominantly transmembrane, and are primarily expressed by DCs, macrophages, and monocytes. Carbohydrate recognition domains (CRDs) enable the CLRs to identify glycan structures on the external surface of pathogens (van Kooyk & Rabinovich, 2008). CLRs are capable of identifying a variety of PAMPs, and stimulating a broad-spectrum of signals that lead to immune responses tailored to specific pathogen attacks. They also recognize diverse endogenous (self-) ligands that regulate homeostasis, including glycans on the surface of tumor cells (Dambuza & Brown, 2015).

CLRs can be categorized into two groups based on their signaling potential. One group contains immunological-receptor tyrosine-based activating motifs (ITAMs) in their integrin cytoplasmic tails, including mannose receptor (MR) and DC-specific intercellular adhesion molecule-3-grabbing nonintegrin (DC-SIGN), described in the next subsection. The ITAM domains

TABLE 3.1 Receptors of DCs.

Receptor	Classes	Sub classes	Examples	Pathways	Cells that express receptor
C-type Lectin Receptors (CLRs)	Type I transmembrane	Mannose Receptor	MRC1	CLR-signaling pathway	DCs, macrophages & monocytes
			MRC2		
		DEC-205	DEC-205		
	Type II transmembrane	Dectin-1	DCAR		
			Clecsf8		
			LOX-1		
			CLEC-1		
			CLEC-2		
			MICL		
			CLEC9A		
		Dectin 2	Dectin-2		
			DCIR		
			BDCA-2		
			Mincle		
		DC-SIGN	DC-SIGN		
		DNGR-1	DNGR-1		
	Soluble CLRs	Mannose-binding lectin	Mannose-binding lectin		

(*Continued*)

TABLE 3.1 (Continued)

Receptor	Classes	Sub classes	Examples	Pathways	Cells that express receptor
Toll-like Receptors	TLR 1–TLR12			TLR signaling pathway	DCs, macrophages, fibroblast cells, epithelial cells
				(MyD88-dependent)	
				(MyD88-independent)	
Nod Like Receptors	NLR	NOD 1		NLR signaling pathway	Epithelial cells, macrophages, lymphocytes, & DCs
		NOD2			
	NLR Inflammasomes	NLRP3/NALP3			
		NLRP1/ NALP1			
		NLRC4 /IPAF			
		NLRP4,6,7, & 12			
RIG-Like Receptors (RLRs)	RIG-I			RLR signaling with MAVS	Macrophages & DCs
	MDA5				
	LGP-2				
AIM2-like receptors (ALR)	AIM2			cGAS-STING	B cells, macrophages & APCs
	IFI16				

APCs, Antigen-presenting cells; *BDCA-2*, blood dendritic cell antigen 2; *DCAR*, dendritic cell immune-activating receptor; *DCIR*, dendritic cell immune receptor; *DC*, dendritic cell; *DNGR-1*, DC NK lectin group receptor-1; *DC-SIGN*, dendritic cell-specific ICAM3-grabbing nonintegrin; *LGP-2*, laboratory of genetics and physiology-2; *MAVS*, mitochondrial antiviral signaling; *MICL*, myeloid inhibitory C-type lectin-like receptor; *RIG-I*, retinoic acid–inducible gene I.

engage with Syk family kinases, leading to the creation of the Card9/Bcl10/Malt1 complex that activates the NF-κB pathway, and other cellular immune responses. Some activating CLRs interact with an ITAM-containing adaptor protein, Fc Receptor gamma chain (FcRγ). The second group possesses immune-receptor tyrosine-based inhibitory motifs (ITIMs) that impede the activation of immune cells responses, thereby regulating immune responses (Chiffoleau, 2018).

On the basis of their molecular structure, CLRs may be subgrouped into type I transmembrane proteins, type II transmembrane proteins, and soluble CLRs. Macrophage MR and DEC-205 are type I transmembrane CLRs, which have multiple CRDs or CRD-like domains. Type II transmembrane CLRs are Dectin-1, Dectin-2, DC-SIGN, DC NK lectin group receptor-1 (DNGR-1), and macrophage-inducible C-type lectin (Mincle). Mannose-binding lectin is a soluble CLR.

CLRs have been exploited for their diverse therapeutic potential. Signals from CLRs initiate the development of adaptive immunity, especially Th1 and Th17 responses. DEC-205 and CLEC9A have the potential to be used as targets in cancer vaccine design, and synthesized ligands of Mincle have been developed to preferentially augment immunological responses. CLRs are vital in immunological synapse formation, DC trafficking, inducing humoral and cellular immunity, and linking both natural and adaptive immune responses (Apostolopoulos et al., 2013; Jang et al., 2015).

Mannose receptor or CD206

MR belongs to the class of CLR expressed by iDCs and macrophages (Szolnoky et al., 2001). In its extracellular region, MR has one fibronectin-type II domain, eight CRDs, and one cysteine-rich N-terminal domain. MR has been targeted by DNA vaccines, for example, incorporating OVA DNA into mannosylated liposomes induced strong cytotoxic T lymphocyte responses in mice compared to nonmannosylated complexes (Hattori et al., 2006; Minigo et al., 2007).

DC-SIGN

DC-SIGN is expressed by DCs and macrophages. It binds to high mannose *N*-glycans on the external surface of many pathogenic organisms, and to ICAM3, a receptor found on the surface of granulocytes. DC-SIGN has been demonstrated to enhance viral infection of cells, including SARS-CoV-2 and HIV, via binding to gp120.

DEC-205

DEC-205 (also referred to as lymphocyte antigen Ly75 or CD205) is a membrane protein found on the surface of endosomes, and is expressed by human DCs, monocytes, and lymphocytes (Kato et al., 2006). In mice, it is expressed on DCs present in the medullary and cortical compartments of thymus (Shrimpton et al., 2009). DEC-205 binds to carbohydrates on microorganisms and mediates their endocytosis (Butler et al., 2007; Jiang et al., 1995).

Dectin-1 family

Dectin-1

Dectin-1 (or DC-associated C-type lectin-1), a β-glucan receptor of the CLR family, is a G-protein coupled receptor found predominantly in macrophages, monocytes, DCs, DDCs, $CD8^-CD4^-$ DCs, mast cells, microglia, neutrophils, T cell subsets, B cells, and eosinophils (Ariizumi et al., 2000). Binding to β-glucan induces cellular responses such as the maturation of DCs, antigen uptake, and production of chemokines and cytokines such as IL-1β, IL-6, TNF-α, CDCL1, and GM-CSF (Brown, 2006). Dectin-1 recognizes β-1,3 and β-1,6-linked glucans present in fungal, bacterial, and plant cell walls, including *Candida*, *Saccharomyces*, *Coccidioides*, *Pneumocystis*, *Penicillium*, and *Aspergillus*, excluding *Cryptococcus* species (Ryan et al., 2002). Several authors have also reported inhibitory functions of Dectin-1, in sterile inflammation and peritumoral immune tolerance (Daley et al., 2017; Seifert et al., 2015).

The dendritic cell immune-activating receptor

DC immune-activating receptor (DCAR) is expressed by DCs localized within the lungs, bone-marrow, and spleen, with lower levels in lymph node and skin DCs, and in selected NK cells. DCAR is characterized by a short cytoplasmic domain and a lack of a specific signaling motif. However, it has been shown that the receptor is associated with the FcRγ chain, which provides the activating ITAM domains (Kanazawa et al., 2003).

Lectin-like oxidized low-density lipoprotein receptor-1

Also called OLR1 or CLEC8A, low-density lipoprotein receptor-1 (LOX-1) belongs to the Dectin-1 cluster of CLRs. LOX-1 is present in endothelial cells, platelets, DCs, human macrophages, and B cells, and can stimulate humoral immunity. LOX-1 stimulates secretion of CCR10, APRIL, and BAFF to enhance plasma cell differentiation and migration to mucosal sites. Fusion of influenza hemagglutinin 1 subunit to LOX-1 resulted in generation of a protective antibody response in rhesus macaques. Several studies have shown LOX-1 to be a good biomarker for cancer diagnosis and a potential therapeutic target, being found in high levels in colorectal, prostate, breast, and gastric cancers (Joo et al., 2014; Liang et al., 2007; Murdocca et al., 2016).

CLEC-1

C-type lectin-like receptor 1 (CLEC-1 or CLEC1A) is found in monocytes, macrophages, DCs, and endothelial cells of humans and mice. CLEC-1 expression is enhanced by TGFβ and decreased by pro-inflammatory stimuli. CLEC-1 is a promising therapeutic target for controlling T cell-mediated adverse immune responses (Robles et al., 2017).

CLEC-2

C-type lectin-like receptor 2 (CLEC-2) is found on platelets, granulocytes, monocytes, DCs, platelets, NK cells, megakaryocytes, and liver sinusoidal endothelial cells. It is also known as CLEC2B and CLEC1B. CLEC-2 binds to podoplanin, a mucin-like sialoglycoprotein expressed on a range of normal tissues and some cancers. CLEC-2 is needed in the formation and preservation of lymph node micro-architecture, and maintenance of the reticular network. CLEC-2 also functions in regulating inflammatory response by promoting DC movement on stromal surfaces, by stimulating secretion of IL-10 and modulating TLR signaling (Acton et al., 2012, 2014; Mourão-Sá et al., 2011).

Myeloid inhibitory C-type lectin-like receptor

Myeloid inhibitory C-type lectin-like receptor (MICL) [or CLEC12A, DC-associated C-type lectin 2 (DCAL-2), CLL1, C-type lectin-like molecule-1 (CLL-1)] is found on DCs, macrophages, $CD8^+$ T cells in peripheral blood, B cells, monocytes and granulocytes, where it regulates the activation of myeloid cells during inflammation (Marshall et al., 2004).

CLEC9A

C-type lectin-like receptor 2 (CLEC9A), also referred to as DNGR1, DNGR-1, or CD370, is found in human $BDCA3^+$ DCs, where it serves as a potential target for the delivery of antigens in vivo for improving vaccine efficacy and stimulating antitumor immunity (Yan et al., 2016). It is also expressed on subsets of mouse $CD8\alpha^+$ and $CD4^+$ DCs (Schreibelt et al., 2012).

Dectin-2 family

Structurally, Dectin-2 receptors have six exons and a common structure comprising a single extracellular CRD, a stalk region, a transmembrane domain, and a short cytoplasmic domain (Graham & Brown, 2009). All but one (DC immune receptor, DCIR) associate with the FcRγ chain to provide the intracellular ITAM signaling domain.

Dectin-2

Dectin-2 (a type II transmembrane receptor) is a receptor for high mannose structures, and is found on DCs, monocytes, macrophages, and neutrophils. Dectin-2 recognizes pathogenic microbes including *Saccharomyces cerevisiae*, *Histoplasma capsulatum*, and *Cryptococcus neoformans* (Fernandes et al., 1999). Dectin-2 also plays a role in eliciting allergic immune responses: Dectin-2 on murine bone marrow–derived DCs can bind *Aspergillus fumigatus* antigen and house dust mite extract in a mannose-dependent fashion, leading to release of pro-inflammatory cysteinyl leukotrienes (Barrett et al., 2009).

Dectin-2 on DCs shows potential to be a target protein for vaccination approaches: for example, in mice, an anti-Dectin-2 monoclonal antibody fused to antigen elicits a $CD8^+$ T cell response (Carter et al., 2006). In another study, a lentivector which encodes the human cancer antigen NY-ESO-1 expressed from the Dectin-2 promoter was observed to activate $CD4^+$ and $CD8^+$ T cells (Lopes et al., 2008). An immune complex formed by anti-Dectin-2 monoclonal antibodies and ovalbumin was reported to increase antigen display to cytotoxic T cells (Carter et al., 2006).

Clecsf8

Clecsf8 is a poorly characterized member of the Dectin-2 family, mostly expressed in macrophages, monocytes, and DCs. Clecsf8 was first reported in mice, through a differential display of gene products with macrophage restricted expression. Transcripts were mainly found in macrophages present in the peritoneum, and at reduced levels in lymph nodes, lung, bone marrow, and spleen (Arce et al., 2004; Balch et al., 1998).

Dendritic cell immune receptor

DCIR possesses a long cytoplasmic tail with an ITIM domain, and therefore acts as an inhibitory receptor. It is the first ever-documented DC-expressed ITIM-bearing C-type lectin, and differs in this respect from other Dectin-2 family members, which associate with the ITAM containing (activating) FcRγ chain. DCIR is mostly found on macrophages, moDCs, pDCs, monocytes, neutrophils, and B cells (Bates et al., 1999). DCIR mRNA is expressed in peripheral blood leukocytes of the humans, and at reduced levels in lymphoid tissues. Conversely, in mice DCIR mRNA is expressed maximally in lymph node and spleen.

DCIR has several observed functions:

- Firstly, to inhibit cytokine release following TLR engagement, for example inhibition of IFN-α release in response to TLR9 engagement, and inhibition of TLR-8 induced IL-12 and TNF-α release (Meyer-Wentrup et al., 2008, 2009);

- Secondly, to bind and internalize antigens, targeting them to endosomes for degradation and antigen presentation, including cross-priming of CD8 T cells. Anti-DCIR monoclonal antibody fused to influenza matrix protein, melanoma antigen MART-1, or HIV gag led to internalization by DCs and stimulation of $CD8^+$ T cells in vitro (Apostolopoulos et al., 2013);
- Lastly, DCIR appears to bind HIV-1 and promotes uptake into nondegrading endosomes, leading to infection of CD4 T cells (Lambert et al., 2008).

The blood dendritic cell antigen 2, or Clec4C

Blood DC antigen 2 (BDCA-2) is a type II C-type lectin found in human pDCs. As with most Dectin-2 family members, BDCA-2 signaling relies on the ITAM sequence in the FcRγ chain leading to signaling through Syk, phospholipase C-gamma (PLCγ2), Vav1, and Erk1/2. BDCA-2 signaling inhibits the secretion of type I IFNs and other cytokines in response to TLR engagement (Cao et al., 2007; Dzionek et al., 2001; Wu et al., 2008).

Macrophage-inducible C-type lectin

Mincle is highly expressed in macrophages following activation by IFN-γ, IL-6, TNF-α, or LPS (Matsumoto et al., 1999). Signaling through Mincle requires association with FcRγ, leading to Syk activation, formation of the CARD9–Bcl-10–MALT1 complex, NF-κB activation, and proinflammatory cytokine release. Mincle recognizes a wide range of glycans and glycolipids found on the surface of bacteria and fungi, for example on *C. albicans* (Wells et al., 2008). Mincle can also induce an inflammatory response to necrotic cells, by responding to spliceosome-associated protein 130, which is released during cellular necrosis (Yamasaki et al., 2008, 2009).

Toll-like receptors

TLRs are vital in innate (nonspecific) immune responses. They are found in DCs and macrophages, and are able to detect both invading microbes and endogenous harmful products from dead/damaged tissues (Fig. 3.4). The TLR family is made up of 10 members, TLR1–TLR10 in humans, and 12 (TLR1–TLR9, TLR11–TLR13) in mice (Kawai & Akira, 2011).

Individual TLRs have external leucine-rich repeats (LRRs) and intracellular Toll-IL-1 receptor (TIR) signaling domains. TLRs recognize PAMPs and DAMPs; after binding to ligand they recruit adapter proteins to trigger the MyD88-dependent IL-1 receptor–TLR signaling pathway with consequent NF-κB stimulation (Kawasaki & Kawai, 2014).

Endogenous ligands of TLRs include plasma membrane constituents, cytoplasmic proteins, hyaluronan and fibrinogen, and organelle components.

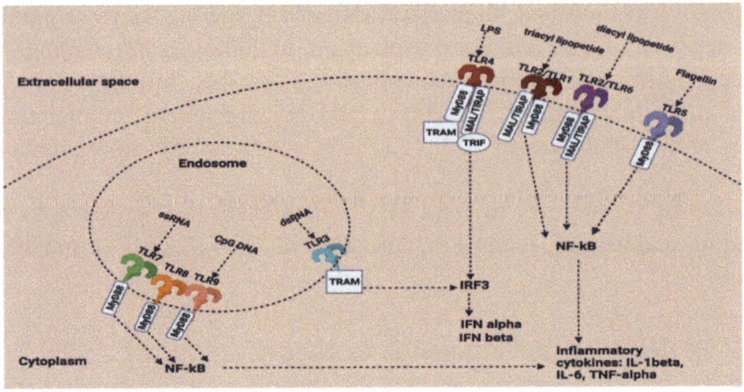

FIGURE 3.4 Toll-like receptor signaling. *Created with BioRender.com. Saxena, M., & Yeretssian, G. (2014). NOD-Like Receptors: Master Regulators of Inflammation and Cancer. Frontiers in immunology, 5, 327. https://doi.org/10.3389/fimmu.2014.00327.*

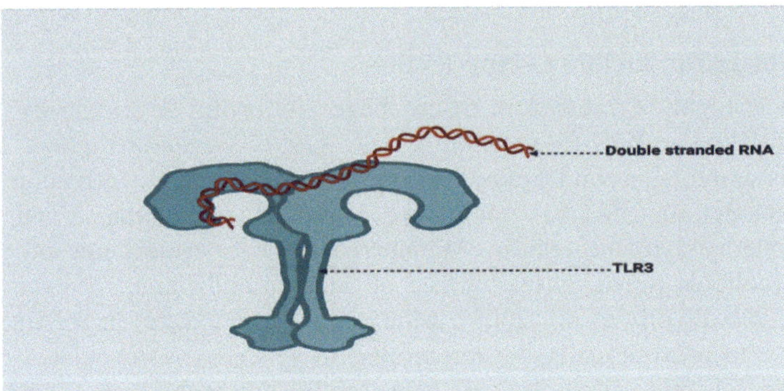

FIGURE 3.5 TLR3 attached to double-stranded RNA. *Created with BioRender.com. Saxena, M., & Yeretssian, G. (2014). NOD-Like Receptors: Master Regulators of Inflammation and Cancer. Frontiers in immunology, 5, 327. https://doi.org/10.3389/fimmu.2014.00327.*

Exogenous TLR ligands include bacteria flagellin, lipopolysaccharide (LPS), lipoarabinomannan (LAM), zymosan, peptidoglycan (PGN), lipopeptides, lipomannans, lipoteichoic acid (LTA), lipoglycans, ssRNA and double-stranded RNA (dsRNA) from bacteria and viruses (El-Zayat et al., 2019), as shown in Fig. 3.5.

As described below, TLR stimulation ultimately leads to NF-κB activation and cytokine release. In addition, MHC-II and costimulatory moieties CD80/CD86 and CD40 are upregulated on surfaces of DCs. These molecules interact with CD28 and CD40 ligand on naive $CD4^+$ T cells to trigger T cell priming (Jin et al, 2012).

PAMP recognition by toll-like receptors

TLRs are grouped into cell-surface TLRs and intracellular TLRs. The former predominantly detect components of microbial cell membranes including proteins, lipids, and lipoproteins; they comprise TLR1, TLR2, TLR4−6, and TLR10. TLR3, TLR7−9, and TLR11−13 are intracellular TLRs present in intracellular vesicles, which recognize mainly pathogen nucleic acids (Kawai & Akira, 2010).

A good knowledge of how ligands interact with TLRs, and the outcome of the resulting signaling pathways, will help in the design/discovery of therapeutic compounds that can target TLRs for the treatment of inflammation and vaccine development (Dowling & Mansell, 2016).

TLR1, TLR2, and TLR6

TLR2 can form dimers with TLR1, TLR6, and possibly TLR10 to produce physiologically active heterodimers (Dowling & Dellacasagrande, 2016). TLR6/TLR2 complexes (found on several immune cells) recognize diacyl lipopeptides including *Mycoplasma fermentans* macrophage-activating lipopeptide (MALP-2) (Da Silva et al., 2012; Okusawa et al., 2004; Takeuchi et al., 2001, 2002).

TLR3

TLR3 receptors are able to identify dsRNA of viral origin, as well as self-RNA. Viral dsRNA (created during replication) binds TLR3, inducing type I IFN (IFN-α/β) generation and NF-κB stimulation. TLR3 is expressed by cDCs and epithelial cells found in the airway, uterus, cornea, vagina, cervix, and intestine. TLR3 can promote cross-presentation to CD8 T cells specific for viral antigens following phagocytosis of virally infected cells by DCs, and TLR3 recognition of viral RNA (Schulz et al., 2005).

TLR4

DCs, granulocytes, B and T cells, and macrophages express cell surface TLR4, which binds to lipopolysaccharide (LPS) on Gram-negative bacteria. TLR4 can also recognize fungal cell surface PAMPs, and the use of TLR knock-out mice has suggested different roles for TLR2 and TLR4 in identifying pathogenic fungi (Netea et al., 2004).

TLR5

TLR5 is located within epithelial cells, iDCs, and monocytes. It is found in the lungs where it helps to combat respiratory pathogens. Flagellin, the main protein component of the bacterial flagellum, is a TLR5 ligand (Bambou et al., 2004; Smith & Ozinsky, 2002). The flagellin produced by *Campylobacter jejuni* and *Helicobacter pylori* lack pro-inflammatory properties, hence enabling them to evade flagellin-specific host immunological responses (Andersen-Nissen et al., 2005).

TLR7 and TLR8

TLR7 and TLR8 share homology and both bind to viral RNA; they are located in endosomes. Murine TLR7 and human TLR8 are able to identify guanosine- or uridine-rich single-stranded viral RNA (ssRNA). TLR7 and TLR8 are both present in mice, although only TLR7 is active (Akira et al., 2006; Heil et al., 2004; Jurk et al., 2002).

TLR9

TLR9 is predominantly located in the endosomes, and is a receptor for A/D-type and B/K-type CpG DNA. B/K-type CpG DNA is a powerful stimulator of inflammatory cytokines including IL-12 and TNF-α. A/D-type CpG DNA strongly induces IFN-α but has weakened capability to induce IL-12 from pDCs (Krug et al., 2004). TLR9 can distinguish host from bacterial DNA because of the high frequency of unmethylated CpG dinucleotides in bacteria (Boulé et al., 2004). TLR9 has been demonstrated to respond to murine cytomegalovirus and herpes simplex virus 1 and 2 (Krug et al., 2001, 2004).

TLR11

TLR11 is located within bladder epithelial cells where it inhibits infection by uropathogenic bacteria, in mice (not humans). TLR11 recognizes flagellin, for example in *Salmonella typhi*, and profilin, for example in *Toxoplasma gondii* (Mathur et al., 2012; Zhang et al., 2013).

TLR12

TLR12 is located within murine myeloid cells, where it forms a heterodimer with TLR11 and can bind to profilin (Koblansky et al., 2013).

Toll-like receptor signaling pathways

TLR dimerization stimulates different signaling pathways that originate in the TIR domain. TLR signaling comprises two distinctive pathways, both of which activate activation protein-1 (AP-1) and NF-κB, a transcription factor that triggers the transcription of cytokines, chemokines, and immunoreceptors contributing to inflammatory responses (Kabelitz et al., 2006; Kawai & Akira, 2011):

1. The myeloid differentiation factor 88 (MyD88) dependent pathway, and
2. The TRIF-dependent pathway.

TRIF stands for TIR domain-containing adapter-inducing IFN-β. TRIF is also commonly referred to as the TIR domain containing adapter molecule (TICAM).

MyD88-dependent pathway

All TLRs except TLR3 interact with MyD88 (Broad et al., 2007). For TLR2, TLR4, and TLR9, MyD88 is recruited by TIR domain-containing adaptor protein (TIRAP). TIRAP has a lipid-binding domain that binds to phosphatidylinositol 4,5-bisphosphate (PIP_2) on the cell membrane, and with phosphatidylinositol 3-phosphate (PI(3)P) in endosomes, resulting in the creation of TLR4 and TLR9 signaling complexes (Bonham et al., 2014; Kawasaki & Kawai, 2014).

MyD88 connects to the TIR domain of TLRs via a C-terminal TIR domain and an N-terminal "death domain" (Akira & Takeda, 2004). On activation, MyD88 directs IRAK-4 to TLRs via direct communication of the death domains of the two proteins, and expedites IRAK-4-mediated stimulation of IRAK-1. Activated IRAK-1 interacts with TRAF6, resulting in the stimulation of two separate signal transduction pathways. One pathway stimulates AP-1 by triggering MAP kinases. The other pathway stimulates the TAK1 complex, which activates IκB kinase (IKK) (Takeda & Akira, 2005). The IKK complex comprises IKKα and IKKβ catalytic subunits, and the regulatory subunit IKKγ. The IKK complex phosphorylates the NF-κB inhibitory protein IκBα, degrading it and inducing nuclear translocation of NF-κB.

TRIF-dependent pathway

The TRIF pathway is commonly used by TLR3 and TLR4 receptors, which recognize dsRNA and LPS (TLR4 also signals through MyD88, as described above). TRAM links TRIF to the TLR4 signaling pathway, whereas TLR3 does not require this adapter protein.

TRIF interacts with TRAF6 and TRAF3. The TRAF3 pathway activates the transcription factor IRF-3, leading to IFN-β production, and upregulation of a number of IFN-inducible genes (Akira & Takeda, 2004; Doyle et al., 2002; Toshchakov et al., 2002). TRAF6 recruits receptor interacting protein 1, leading to MAP kinase and then NF-κB activation. TRIF also activates TANK (TRAF-associated NF-κB activator kinase or I-TRAF) via TRAF3 (Kawai & Akira, 2011).

T cell responses stimulated by toll-like receptors

TLR-driven T cell polarization to Th1, Th2, Th17, and Treg (Fig. 3.6) is dependent on the type of DC expressing the TLR, the cytokines present, and also on antigen dose (Aragane et al., 2003; Jin et al, 2012). TLRs drive Th1 responses via the release of IL-12p70, IL-6, and IFN-α from DCs. Th17

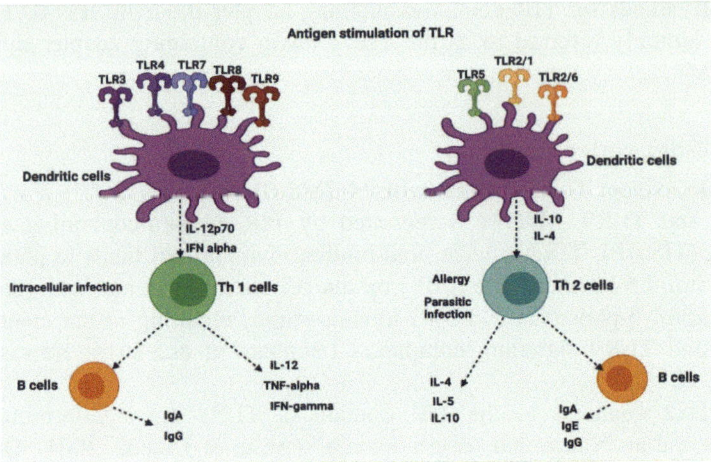

FIGURE 3.6 TLRs and T helper cell induction. *Created with BioRender.com. Saxena, M., & Yeretssian, G. (2014). NOD-Like Receptors: Master Regulators of Inflammation and Cancer. Frontiers in immunology, 5, 327. https://doi.org/10.3389/fimmu.2014.00327.*

responses are driven by IL-21 and IL-23, and Tregs by IL-10 (Dowling & Mansell, 2016; Jin et al., 2012).

NOD-like receptors

NOD-like receptors, also known as nucleotide-binding oligomerization domain-like receptors or NLR, are PRRs principally engaged in detecting intracellular PAMPs that have, for example, entered the cell by phagocytosis. They are located in macrophages, lymphocytes, and DCs, and also in epithelial cells. They can recognize a variety of ligands including: flagellin, peptidoglycan, fungal hyphae, viral RNA, cholesterol crystals, ATPs, uric acid from host cells, asbestos, alloy particles, alum, silica, and skin irritants from environmental origins (Ting et al., 2008).

There are about 22 human NLRs. The domain structure consists of (1) a modifiable N-terminal protein interaction domain, for example a caspase recruitment (caspase activation and recruitment domain, CARD) or pyrin (PYD) domain, (2) a central nucleotide-binding domain (NACHT), and (3) C-terminal LRRs. The C-terminal LRR domain functions as a cofactor during ligand binding and activator sensing, while the N-terminal domain serves as an effector domain (Schroder & Tschopp, 2010). NLRs are classified according to their N-terminal domain: (1) NLRA (acidic transactivation domain), (2) NLRB (baculovirus inhibitor of apoptosis protein repeat, or BIR), (3) NLRC (CARD domain), (4) NLRP (Pyrin domain), and (5) NLRX (an unidentified domain) (Saxena & Yeretssian, 2014).

NOD1 and NOD2

NOD1 recognizes meso-diaminopimelic acid (meso-DAP), commonly present in Gram-negative bacteria, while NOD2 binds the muramyl dipeptide peptidoglycan layer of degraded Gram-negative bacterial cell walls (Inohara et al., 2001; Jang et al., 2015). NOD1 and NOD2 engage receptor interacting protein 2 after ligand activation, which activates the MAPK and NF-κB pathways, resulting in enhanced transcription of pro-inflammatory cytokines (Girardin et al., 2003; Kim et al., 2016; Yamamoto et al., 2003). NOD1 and NOD2 initiate autophagy (lysosomal degradation of cellular components and also of invading bacteria) by recruiting ATG16L1 (Travassos et al., 2010).

NOD-like receptor inflammasomes

Inflammasomes are multiprotein complexes that stimulate caspase proteins, mainly Caspase-1. Inflammasomes are activated either by PAMPs from bacteria, fungi, and viruses or by sterile activators, which include DAMPs and stimulants from the environment (Davis et al., 2011). There are several kinds of inflammasomes containing one or two NLR proteins, the adapter molecule apoptosis-associated speck-like protein containing a CARD (ASC) and a pro-Caspase (Kufer & Sansonetti, 2011). Caspases are either pro-inflammatory or pro-apoptotic; Caspase-1 is pro-inflammatory and activates IL-1β and IL-18 by cleaving the pro-cytokines (Lamkanfi et al., 2007).

NLRP3/NALP3 inflammasome

NLRP3 consists of the ASC (PYCARD) adaptor, NLRP3, and Caspase-1. It is activated by LPS, dsRNA, PGN, and LTA, and in particular responds to fungi: *Porphyromonas gingivalis*, *A. fumigatus*, *C. albicans*, and *S. cerevisiae*; bacteria: *Klebsiella pneumoniae*, *Listeria monocytogenes*, *Chlamydia pneumoniae*, *Mycobacterium tuberculosis*, Legionella *pneumophila*, *Staphylococcus aureus*, and viruses including influenza (Carlsson et al., 2010; Case et al., 2013; Gross et al., 2009; Kanneganti et al., 2006; Mariathasan et al., 2004, 2006; Park et al., 2014). The NLRP3 inflammasome is also triggered by silica, alum, and asbestos via reactive oxygen species (Cassel et al., 2008; Hornung et al., 2008), and by small molecules including trinitrophenylchloride, dinitrofluorobenzene, and trinitrochlorobenzene (Sutterwala et al., 2006).

NLRP1/NALP1 inflammasome

NLRP1 is found in the respiratory system's epithelial cells, lymphocytes, and myeloid cells in large numbers. The NLRP1 inflammasome is triggered in humans by the lethal toxin (LT) from *Bacillus anthracis* (Boyden & Dietrich, 2006), whereas in mice it is activated by *T. gondii* (Ewald et al., 2014). The LT anthrax toxin antigen activates NLRP1 inflammasomes in the cytoplasm resulting in rapid cell death, a process known as pyroptosis (Bauernfeind & Hornung, 2013). The inflammasome can also be activated by viral proteases (Tsu et al., 2021).

NLRC4 /NAIP (IPAF) inflammasome

NLRC4 is found in immune and epithelial cells. Forming oligomeric structures, it interacts with NAIPs (NLR family apoptosis inhibitory proteins), which are receptors for microbial molecules. The NLRC4 inflammasome can be triggered by several stimuli including flagellin secreted from Gram-negative bacteria (Amer et al., 2006), and is important for the detection of commensal (nonpathogenic organisms including gut bacteria). Activation of the inflammasome can lead to activation of pro-cytokines, induction of pyroptosis, and release of leukotrienes and pro-inflammatory prostaglandins including LTB4 and PGE2 (Duncan & Canna, 2018).

RIG-like receptors

The immune system uses RLRs, located in the cytosol, to identify patterns in invading viral pathogens (Goubau et al., 2013). RLRs trigger signal transduction events to activate downstream transcription factors, that stimulate type I IFN production and antiviral gene expression, ultimately controlling viral infection.

Three RLRs are known, namely retinoic acid–inducible protein I (RIG-I), melanoma differentiation factor-5 (MDA5), and laboratory of genetics and physiology-2 (LGP-2). They have a core DExD/H box RNA helicase domain and a repressor domain located inside the C-terminal domain. MDA5 and RIG-I are similar in structure, with N-terminal regions comprising tandem CARDs that ultimately induce type I IFN expression (Loo & Gale, 2007; Yoneyama et al., 2004). LGP-2 binds to dsRNA but lacks CARD domains; it is thought to regulate the other RLRs in both a positive and negative fashion (Yoneyama et al., 2005).

Viral RNA activates RLRs, which undergo conformational changes that enhance homotypic CARD–CARD interactions with mitochondrial antiviral signaling (MAVS)—a vital adaptor protein required for RLR signal transduction, also known as IPS-1. MAVS signals to IκB kinase-ε (IKKε) and TANK-binding kinase 1 (TBK1), which will then stimulate IRFs (IRF3 and IRF7) and NF-κB, resulting in transcription of type 1 IFN, and other immuno-modulatory genes (Rehwinkel & Gack, 2020).

AIM2-like receptors

The IFN-inducible protein AIM2 (not found in melanoma 2) belongs to the HIN-200 family. It is an inflammasome component, and recognizes bacterial and viral cytoplasmic DNA. It interacts with cytoplasmic nucleic acids via its C-terminal oligonucleotide/oligosaccharide-binding domain, and an ASC adapter via its N-terminal DAPIN/pyrin domain, to trigger pro-caspase-1 (Caneparo et al., 2018; Ludlow et al., 2005). As with other inflammasomes, CARD leads to caspase activation followed by pyroptosis, and activation of IL-1 and IL-18 (Lugrin & Martinon, 2018).

Strategies for targeting vaccines to dendritic cell receptors

DCs have been utilized as treatments for cancer, autoimmunity, and infectious diseases, leading to regulation of T cell immunity. DC-based vaccines are of two types: antigen-expressing DC based vaccines generated ex vivo, and in vivo DC directed vaccines (Chen et al., 2016). The transfer of ex vivo-cultured DCs containing target antigens was the first attempt to utilize DCs to evoke protective immunity in humans. This approach has been widely explored for the treatment for cancer, as described in Chapter 7: Approaches to Cancer Vaccination.

In vivo dendritic cell targeted vaccines

DC-targeting strategies couple protein antigens, DNA, or drug molecules to ligands or antibodies which recognize specific receptors expressed by DCs, thus stimulating an immune response against the targeted antigen (Caminschi & Shortman, 2012).

Ligand-based dendritic cell–targeted vaccines

Linking proteins to PRR ligands (PRRLs) makes it possible to produce antigen–PRR-ligand conjugate vaccines (Chen et al., 2016).

TLR9 is the most commonly targeted PRR in antigen–PRR-ligand conjugate vaccines; it recognizes CpG, which has adjuvant activity. For example, conjugating the model antigen OVA to CpG using chemical crosslinking produced a strong adaptive immune response in mice, inducing potent Th1 cells and IgG2a antibody responses as well as cross-primed CTLs (Chen et al., 2016; Horner et al., 2001). A second proof of concept (POC) experiment cross-linked HIV gp120 protein to CpG, which elicited antibody, and $CD4^{+}$ and $CD8^{+}$ T cell responses (Tighe et al., 2000). Several other TLRs including DC-SIGN, and CD91 (a heat shock protein receptor), have been targeted by ligand-antigen fusions in POC experiments (Chen et al., 2016).

Antibody-based dendritic cell–targeted vaccines

Monoclonal antibodies have been developed against several DC receptor proteins including CLRs, in order to target antigens to DCs and elicit immune responses. Antigens can be attached to antibodies via chemical conjugation or, more commonly, genetic fusion. Targeting of specific DC subsets can be achieved by selecting specific receptors, for example langerin/CD207 is mainly expressed by subsets of LCs and DDCs, while BDCA-2 (CLEC4C) is expressed mainly by pDCs (Chen et al., 2016). DEC-205 is expressed by $CD8^+$ DCs, LCs, DDCs, and B cells. Antigens fused to an anti-DEC-205 monoclonal antibody to achieve DC targeting include Survivin, HIV gag, ovalbumin (Ova), and HER2/neu. The stimulation of antigen-specific immunological responses requires an adjuvant, for example CpG; in the absence of adjuvant, DEC-205 targeting can induce tolerance (Hawiger et al., 2001; Lahoud et al., 2012). One study showed that anti-DEC-205 antibody-Ova injected into mice via the skin experienced very rapid uptake by LCs which promoted T cell immunity (Flacher et al., 2010). Ova-conjugated anti-DEC-205 antibodies produce strong MHC-I cross-presentation to CD8 T cells if $CD8^+$ DCs are targeted, and elevated Ova presentation to MHC Class II when targeting $CD8^-$ DCs (Bonifaz et al., 2002, 2004).

DNGR1 is highly expressed by $CD8^+$ DCs and minimally by pDCs. Antigens targeted to DCs with an anti-DNGR1 antibody stimulate antigen-specific CTL responses in mice, via cross-presentation (De Souza Apostólico et al., 2019; Sancho et al., 2009).

Murine CD11c has also been explored as a vaccine target in POC experiments. It is found in every DC subset (including $CD8^+$ DC and $CD8^-$ DC) at high levels. It is, therefore, a good target for triggering both the humoral and cellular immune responses. An antibody specific for murine CD11c combined with antigens and targeted to DCs will generate antigen-specific B-cell immune responses after a single immunization.

Fc receptors, which recognize IgG, IgA, and IgE ($Fc\gamma R$, $Fc\alpha R$, and $Fc\varepsilon R$), participate in many immune-related processes including ingestion of cells by phagocytosis, antigen presentation, and immune complex–mediated DC maturation. Prostate-specific antigen conjugated to monoclonal antibodies recognizing $Fc\gamma R$ was internalized and cross-presented by humanized THP-1 cells (Sancho et al., 2009).

References

Acton, S. E., Astarita, J. L., Malhotra, D., Lukacs-Kornek, V., Franz, B., Hess, P. R., Jakus, Z., Kuligowski, M., Fletcher, A. L., Elpek, K. G., Bellemare-Pelletier, A., Sceats, L., Reynoso, E. D., Gonzalez, S. F., Graham, D. B., Chang, J., Peters, A., Woodruff, M., Kim, Y. A., ... Turley, S. J. (2012). Podoplanin-rich stromal networks induce dendritic cell motility via activation of the C-type lectin receptor CLEC-2. *Immunity*, *37*(2), 276–289. Available from https://doi.org/10.1016/j.immuni.2012.05.022.

Acton, S. E., Farrugia, A. J., Astarita, J. L., Mourão-Sá, D., Jenkins, R. P., Nye, E., Hooper, S., Van Blijswijk, J., Rogers, N. C., Snelgrove, K. J., Rosewell, I., Moita, L. F., Stamp, G., Turley, S. J., Sahai, E., & Sousa, C. R. (2014). Dendritic cells control fibroblastic reticular network tension and lymph node expansion. *Nature*, *514*(7253), 498–502. Available from https://doi.org/10.1038/nature13814.

Akira, S., & Takeda, K. (2004). Toll-like receptor signalling. *Nature Reviews Immunology*, *4*(7), 499–511. Available from https://doi.org/10.1038/nri1391.

Akira, S., Uematsu, S., & Takeuchi, O. (2006). Pathogen recognition and innate immunity. *Cell, 124*(4), 783–801. Available from https://doi.org/10.1016/j.cell.2006.02.015.

Amer, A., Franchi, L., Kanneganti, T. D., Body-Malapel, M., Özören, N., Brady, G., Meshinchi, S., Jagirdar, R., Gewirtz, A., Akira, S., & Núñez, G. (2006). Regulation of *Legionella* phagosome maturation and infection through flagellin and host Ipaf. *Journal of Biological Chemistry, 281*(46), 35217–35223. Available from https://doi.org/10.1074/jbc.M604933200.

Andersen-Nissen, E., Smith, K. D., Strobe, K. L., Rassoulian Barrett, S. L., Cookson, B. T., Logan, S. M., & Aderem, A. (2005). Evasion of Toll-like receptor 5 by flagellated bacteria. *Proceedings of the National Academy of Sciences of the United States of America, 102*(26), 9247–9252. Available from https://doi.org/10.1073/pnas.0502040102.

Apostolopoulos, V., Thalhammer, T., Tzakos, A. G., & Stojanovska, L. (2013). Targeting antigens to dendritic cell receptors for vaccine development. *Journal of Drug Delivery*, 1–22. Available from https://doi.org/10.1155/2013/869718.

Aragane, Y., Maeda, A., Schwarz, A., Tezuka, T., Ariizumi, K., & Schwarz, T. (2003). Involvement of dectin-2 in ultraviolet radiation-induced tolerance. *Journal of Immunology, 171*(7), 3801–3807. Available from https://doi.org/10.4049/jimmunol.171.7.3801.

Arce, I., Martínez-Muñoz, L., Roda-Navarro, P., & Fernández-Ruiz, E. (2004). The human C-type lectin CLECSF8 is a novel monocyte/macrophage endocytic receptor. *European Journal of Immunology, 34*(1), 210–220. Available from https://doi.org/10.1002/eji.200324230.

Ariizumi, K., Shen, G. L., Shikano, S., Xu, S., Ritter, R., Kumamoto, T., Edelbaum, D., Morita, A., Bergstresser, P. R., & Takashima, A. (2000). Identification of a novel, dendritic cell-associated molecule, dectin-1, by subtractive cDNA cloning. *Journal of Biological Chemistry, 275*(26), 20157–20167. Available from https://doi.org/10.1074/jbc.M909512199.

Audiger, C., Rahman, M. J., Yun, T. J., Tarbell, K. V., & Lesage, S. (2017). The importance of dendritic cells in maintaining immune tolerance. *Journal of Immunology, 198*(6), 2223–2231. Available from https://doi.org/10.4049/jimmunol.1601629.

Balch, S. G., McKnight, A. J., Seldin, M. F., & Gordon, S. (1998). Cloning of a novel C-type lectin expressed by murine macrophages. *Journal of Biological Chemistry, 273*(29), 18656–18664. Available from https://doi.org/10.1074/jbc.273.29.18656.

Bambou, J. C., Giraud, A., Menard, S., Begue, B., Rakotobe, S., Heyman, M., Taddei, F., Cerf-Bensussan, N., & Gaboriau-Routhiau, V. (2004). In vitro and ex vivo activation of the TLR5 signaling pathway in intestinal epithelial cells by a commensal *Escherichia coli* strain. *Journal of Biological Chemistry, 279*(41), 42984–42992. Available from https://doi.org/10.1074/jbc.M405410200.

Barrett, N. A., Maekawa, A., Rahman, O. M., Austen, K. F., & Kanaoka, Y. (2009). Dectin-2 recognition of house dust mite triggers cysteinyl leukotriene generation by dendritic cells. *Journal of Immunology, 182*(2), 1119–1128. Available from https://doi.org/10.4049/jimmunol.182.2.1119.

Bates, E. E. M., Fournier, N., Garcia, E., Valladeau, J., Durand, I., Pin, J. J., Zurawski, S. M., Patel, S., Abrams, J. S., Lebecque, S., Garrone, P., & Saeland, S. (1999). APCs express DCIR, a novel C-type lectin surface receptor containing an immunoreceptor tyrosine-based inhibitory motif. *Journal of Immunology, 163*(4), 1973–1983.

Bauernfeind, F., & Hornung, V. (2013). Of inflammasomes and pathogens – Sensing of microbes by the inflammasome. *EMBO Molecular Medicine, 5*(6), 814–826. Available from https://doi.org/10.1002/emmm.201201771.

Belz, G. T., & Nutt, S. L. (2012). Transcriptional programming of the dendritic cell network. *Nature Reviews Immunology, 12*(2), 101–113. Available from https://doi.org/10.1038/nri3149.

Bhattacharya, P., Thiruppathi, M., Elshabrawy, H. A., Alharshawi, K., Kumar, P., & Prabhakar, B. S. (2015). GM-CSF: An immune modulatory cytokine that can suppress autoimmunity. *Cytokine*, *75*(2), 261–271. Available from https://doi.org/10.1016/j.cyto.2015.05.030.

Bonham, K. S., Orzalli, M. H., Hayashi, K., Wolf, A. I., Glanemann, C., Weninger, W., Iwasaki, A., Knipe, D. M., & Kagan, J. C. (2014). A promiscuous lipid-binding protein diversifies the subcellular sites of toll-like receptor signal transduction. *Cell*, *156*(4), 705–716. Available from https://doi.org/10.1016/j.cell.2014.01.019.

Bonifaz, L. C., Bonnyay, D. P., Charalambous, A., Darguste, D. I., Fujii, S. I., Soares, H., Brimnes, M. K., Moltedo, B., Moran, T. M., & Steinman, R. M. (2004). In vivo targeting of antigens to maturing dendritic cells via the DEC-205 receptor improves T cell vaccination. *Journal of Experimental Medicine*, *199*(6), 815–824. Available from https://doi.org/10.1084/jem.20032220.

Bonifaz, L., Bonnyay, D., Mahnke, K., Rivera, M., Nussenzweig, M. C., & Steinman, R. M. (2002). Efficient targeting of protein antigen to the dendritic cell receptor DEC-205 in the steady state leads to antigen presentation on major histocompatibility complex class I products and peripheral CD8 + T cell tolerance. *Journal of Experimental Medicine*, *196*(12), 1627–1638. Available from https://doi.org/10.1084/jem.20021598.

Boulé, M. W., Broughton, C., Mackay, F., Akira, S., Marshak-Rothstein, A., & Rifkin, I. R. (2004). Toll-like receptor 9-dependent and -independent dendritic cell activation by chromatin-immunoglobulin G complexes. *Journal of Experimental Medicine*, *199*(12), 1631–1640. Available from https://doi.org/10.1084/jem.20031942.

Boyden, E. D., & Dietrich, W. F. (2006). Nalp1b controls mouse macrophage susceptibility to anthrax lethal toxin. *Nature Genetics*, *38*(2), 240–244. Available from https://doi.org/10.1038/ng1724.

Broad, A., Kirby, J. A., & Jones, D. E. J. (2007). Toll-like receptor interactions: Tolerance of MyD88-dependent cytokines but enhancement of MyD88-independent interferon-β production. *Immunology*, *120*(1), 103–111. Available from https://doi.org/10.1111/j.1365-2567.2006.02485.x.

Brown, G. D. (2006). Dectin-1: A signalling non-TLR pattern-recognition receptor. *Nature Reviews Immunology*, *6*(1), 33–43. Available from https://doi.org/10.1038/nri1745.

Butcher, E. C., Williams, M., Youngman, K., Rott, L., & Briskin, M. (1999). Lymphocyte trafficking and regional immunity. *Advances in Immunology*, *72*(72), 209–253. Available from https://doi.org/10.1016/s0065-2776(08)60022-x.

Butler, M., Morel, A. S., Jordan, W. J., Eren, E., Hue, S., Shrimpton, R. E., & Ritter, M. A. (2007). Altered expression and endocytic function of CD205 in human dendritic cells, and detection of a CD205-DCL-1 fusion protein upon dendritic cell maturation. *Immunology*, *120*(3), 362–371. Available from https://doi.org/10.1111/j.1365-2567.2006.02512.x.

Caminschi, I., & Shortman, K. (2012). Boosting antibody responses by targeting antigens to dendritic cells. *Trends in Immunology*, *33*(2), 71–77. Available from https://doi.org/10.1016/j.it.2011.10.007.

Caneparo, V., Landolfo, S., Gariglio, M., & De Andrea, M. (2018). *The absent in melanoma 2-like receptor IFN-inducible protein 16 as an inflammasome regulator in systemic lupus erythematosus: The dark side of sensing microbes. Frontiers in Immunology* (9). Available from https://doi.org/10.3389/fimmu.2018.01180.

Cao, W., Zhang, L., Rosen, D. B., Bover, L., Watanabe, G., Bao, M., Lanier, L. L., & Liu, Y. J. (2007). BDCA2/FcεRIγ complex signals through a novel BCR-like pathway in human plasmacytoid dendritic cells. *PLoS Biology*, *5*(10), 2190–2200. Available from https://doi.org/10.1371/journal.pbio.0050248.

Carlsson, F., Kim, J., Dumitru, C., Barck, K. H., Carano, R. A. D., Sun, M., Diehl, L., & Brown, E. J. (2010). Host-detrimental role of Esx-1-mediated inflammasome activation in mycobacterial infection. *PLoS Pathogens*, *6*(5), 1–12. Available from https://doi.org/10.1371/journal.ppat.1000895.

Carter, R. W., Thompson, C., Reid, D. M., Wong, S. Y. C., & Tough, D. F. (2006). Induction of CD8 + T cell responses through targeting of antigen to Dectin-2. *Cellular Immunology*, *239* (2), 87–91. Available from https://doi.org/10.1016/j.cellimm.2006.05.001.

Case, C. L., Kohler, L. J., Lima, J. B., Strowig, T., Zoete, M. R. D., Flavell, R. A., Zamboni, D. S., & Roy, C. R. (2013). Caspase-11 stimulates rapid flagellin-independent pyroptosis in response to *Legionella pneumophila*. *Proceedings of the National Academy of Sciences of the United States of America*, *110*(5), 1851–1856. Available from https://doi.org/10.1073/pnas.1211521110.

Cassel, S. L., Eisenbarth, S. C., Iyer, S. S., Sadler, J. J., Colegio, O. R., Tephly, L. A., Carter, A. B., Rothman, P. B., Flavell, R. A., & Sutterwala, F. S. (2008). The Nalp3 inflammasome is essential for the development of silicosis. *Proceedings of the National Academy of Sciences of the United States of America*, *105*(26), 9035–9040. Available from https://doi.org/10.1073/pnas.0803933105.

Castell-Rodríguez, A., Piñón-Zárate, G., Herrera-Enríquez, M., Jarquín-Yáñez, K., & Medina-Solares, I. (2017). *Dendritic cells: Location, function, and clinical implications. Biology of myelomonocytic cells*. London: IntechOpen. Available from https://doi.org/10.5772/intechopen.68352.

Cella, M., Jarrossay, D., Faccheth, F., Alebardi, O., Nakajima, H., Lanzavecchia, A., & Colonna, M. (1999). Plasmacytoid monocytes migrate to inflamed lymph nodes and produce large amounts of type I interferon. *Nature Medicine*, *5*(8), 919–923. Available from https://doi.org/10.1038/11360.

Chen, P., Liu, X., Sun, Y., Zhou, P., Wang, Y., & Zhang, Y. (2016). Dendritic cell targeted vaccines: Recent progresses and challenges. *Human Vaccines & Immunotherapeutics*, *12*(3), 612–622. Available from https://doi.org/10.1080/21645515.2015.1105415.

Cheong, C., Choi, J. H., Vitale, L., He, L. Z., Trumpfheller, C., Bozzacco, L., Do, Y., Nchinda, G., Park, S. H., Dandamudi, D. B., Shrestha, E., Pack, M., Lee, H. W., Keler, T., Steinman, R. M., & Park, C. G. (2010). Improved cellular and humoral immune responses in vivo following targeting of HIV Gag to dendritic cells within human anti-human DEC205 monoclonal antibody. *Blood*, *116*(19), 3828–3838. Available from https://doi.org/10.1182/blood-2010-06-288068.

Cheong, C., Matos, I., Choi, J. H., Dandamudi, D. B., Shrestha, E., Longhi, M. P., Jeffrey, K. L., Anthony, R. M., Kluger, C., Nchinda, G., Koh, H., Rodriguez, A., Idoyaga, J., Pack, M., Velinzon, K., Park, C. G., & Steinman, R. M. (2010). Microbial stimulation fully differentiates monocytes to DC-SIGN/CD209 + dendritic cells for immune T cell areas. *Cell*, *143*(3), 416–429. Available from https://doi.org/10.1016/j.cell.2010.09.039.

Chiffoleau, E. (2018). C-type lectin-like receptors as emerging orchestrators of sterile inflammation represent potential therapeutic targets. *Frontiers in Immunology*, *9*. Available from https://doi.org/10.3389/fimmu.2018.00227.

Cintolo, J. A., Datta, J., Mathew, S. J., & Czerniecki, B. J. (2012). Dendritic cell-based vaccines: Barriers and opportunities. *Future Oncology*, *8*(10), 1273–1299. Available from https://doi.org/10.2217/fon.12.125.

Cools, N., Ponsaerts, P., Van Tendeloo, V. F. I., & Berneman, Z. N. (2007). Regulatory T cells and human disease. *Clinical and Developmental Immunology*, *2007*. Available from https://doi.org/10.1155/2007/89195.

Coombes, J. L., & Powrie, F. (2008). Dendritic cells in intestinal immune regulation. *Nature Reviews Immunology*, *8*(6), 435–446. Available from https://doi.org/10.1038/nri2335.

Crozat, K., Guiton, R., Guilliams, M., Henri, S., Baranek, T., Schwartz-Cornil, I., Malissen, B., & Dalod, M. (2010). Comparative genomics as a tool to reveal functional equivalences between human and mouse dendritic cell subsets. *Immunological Reviews*, *234*(1), 177–198. Available from https://doi.org/10.1111/j.0105-2896.2009.00868.x.

Da Silva, R. A. B., Ferreira, P. D. F., De Rossi, A., Nelson-Filho, P., & Silva, L. A. B. (2012). Toll-like receptor 2 knockout mice showed increased periapical lesion size and osteoclast number. *Journal of Endodontics*, *38*(6), 803–813. Available from https://doi.org/10.1016/j.joen.2012.03.017.

Daley, D., Mani, V. R., Mohan, N., Akkad, N., Ochi, A., Heindel, D. W., Lee, K. B., Zambirinis, C. P., Pandian, G. S. D. B., Savadkar, S., Torres-Hernandez, A., Nayak, S., Wang, D., Hundeyin, M., Diskin, B., Aykut, B., Werba, G., Barilla, R. M., Rodriguez, R., & Miller, G. (2017). Dectin 1 activation on macrophages by galectin 9 promotes pancreatic carcinoma and peritumoral immune tolerance. *Nature Medicine*, *23*(5), 556–567. Available from https://doi.org/10.1038/nm.4314.

Dambuza, I. M., & Brown, G. D. (2015). C-type lectins in immunity: Recent developments. *Current Opinion in Immunology*, *32*, 21–27. Available from https://doi.org/10.1016/j.coi.2014.12.002.

Davis, B. K., Wen, H., & Ting, J. P. Y. (2011). The inflammasome NLRs in immunity, inflammation, and associated diseases. *Annual Review of Immunology*, *29*, 707–735. Available from https://doi.org/10.1146/annurev-immunol-031210-101405.

De Souza Apostólico, J., Lunardelli, V. A. S., Yamamoto, M. M., Cunha-Neto, E., Boscardin, S. B., & Rosa, D. S. (2019). Poly(I:C) potentiates T cell immunity to a dendritic cell targeted HIV-multiepitope vaccine. *Frontiers in Immunology*, *10*. Available from https://doi.org/10.3389/fimmu.2019.00843.

Donskoy, E., & Goldschneider, I. (2003). Two developmentally distinct populations of dendritic cells inhabit the adult mouse thymus: Demonstration by differential importation of hematogenous precursors under steady state conditions. *Journal of Immunology*, *170*(7), 3514–3521. Available from https://doi.org/10.4049/jimmunol.170.7.3514.

Dowling, J. K., & Dellacasagrande, J. (2016). Toll-like receptors: Ligands, cell-based models, and readouts for receptor action. *Methods in Molecular Biology*, *1390*, 3–27. Available from https://doi.org/10.1007/978-1-4939-3335-8_1.

Dowling, J. K., & Mansell, A. (2016). Toll-like receptors: The Swiss army knife of immunity and vaccine development. *Clinical and Translational Immunology*, *5*(5). Available from https://doi.org/10.1038/cti.2016.22.

Doyle, S. E., Vaidya, S. A., O'Connell, R., Dadgostar, H., Dempsey, P. W., Wu, T. T., Rao, G., Sun, R., Haberland, M. E., Modlin, R. L., & Cheng, G. (2002). IRF3 mediates a TLR3/TLR4-specific antiviral gene program. *Immunity*, *17*(3), 251–263. Available from https://doi.org/10.1016/S1074-7613(02)00390-4.

Duncan, J. A., & Canna, S. W. (2018). The NLRC4 inflammasome. *Immunological Reviews*, *281*(1), 115–123. Available from https://doi.org/10.1111/imr.12607.

Dzionek, A., Sohma, Y., Nagafune, J., Cella, M., Colonna, M., Facchetti, F., Günther, G., Johnston, I., Lanzavecchia, A., Nagasaka, T., Okada, T., Vermi, W., Winkels, G., Yamamoto, T., Zysk, M., Yamaguchi, Y., & Schmitz, J. (2001). BDCA-2, a novel plasmacytoid dendritic cell-specific type II C-type lectin, mediates antigen capture and is a potent inhibitor of interferon α/β induction. *Journal of Experimental Medicine*, *194*(12), 1823–1834. Available from https://doi.org/10.1084/jem.194.12.1823.

El-Zayat, S. R., Sibaii, H., & Mannaa, F. A. (2019). Toll-like receptors activation, signaling, and targeting: An overview. *Bulletin of the National Research Centre*. Available from https://doi.org/10.1186/s42269-019-0227-2.

Ewald, S. E., Chavarria-Smith, J., & Boothroyd, J. C. (2014). NLRP1 is an inflammasome sensor for *Toxoplasma gondii*. *Infection and Immunity*, *82*(1), 460–468. Available from https://doi.org/10.1128/IAI.01170-13.

Fernandes, M. J., Finnegan, A. A., Siracusa, L. D., Brenner, C., Iscove, N. N., & Calabretta, B. (1999). Characterization of a novel receptor that maps near the natural killer gene complex: Demonstration of carbohydrate binding and expression in hematopoietic cells. *Cancer Research*, *59*(11), 2709–2717.

Flacher, V., Tripp, C. H., Stoitzner, P., Haid, B., Ebner, S., Del Frari, B., Koch, F., Park, C. G., Steinman, R. M., Idoyaga, J., & Romani, N. (2010). Epidermal Langerhans cells rapidly capture and present antigens from C-type lectin-targeting antibodies deposited in the dermis. *Journal of Investigative Dermatology*, *130*(3), 755–762. Available from https://doi.org/10.1038/jid.2009.343.

Gehrie, E., Van der Touw, W., Bromberg, J. S., & Ochando, J. C. (2011). Plasmacytoid dendritic cells in tolerance. *Methods in Molecular Biology (Clifton, N.J.)*, *677*, 127–147.

Geissmann, F., Manz, M. G., Jung, S., Sieweke, M. H., Merad, M., & Ley, K. (2010). Development of monocytes, macrophages, and dendritic cells. *Science (New York, N.Y.)*, *327*(5966), 656–661. Available from https://doi.org/10.1126/science.1178331.

Girardin, S. E., Boneca, I. G., Carneiro, L. A. M., Antignac, A., Jéhanno, M., Viala, J., Tedin, K., Taha, M. K., Labigne, A., Zähringer, U., Coyle, A. J., DiStefano, P. S., Bertin, J., Sansonetti, P. J., & Philpott, D. J. (2003). Nod1 detects a unique muropeptide from gram-negative bacterial peptidoglycan. *Science (New York, N.Y.)*, *300*(5625), 1584–1587. Available from https://doi.org/10.1126/science.1084677.

Goubau, D., Deddouche, S., & Reis e Sousa, C. (2013). Cytosolic sensing of viruses. *Immunity*, *38*(5), 855–869. Available from https://doi.org/10.1016/j.immuni.2013.05.007.

Graham, L. M., & Brown, G. D. (2009). The Dectin-2 family of C-type lectins in immunity and homeostasis. *Cytokine*, *48*(1–2), 148–155. Available from https://doi.org/10.1016/j.cyto.2009.07.010.

Gross, O., Poeck, H., Bscheider, M., Dostert, C., Hannesschläger, N., Endres, S., Hartmann, G., Tardivel, A., Schweighoffer, E., Tybulewicz, V., Mocsai, A., Tschopp, J., & Ruland, J. (2009). Syk kinase signalling couples to the Nlrp3 inflammasome for anti-fungal host defence. *Nature*, *459*(7245), 433–436. Available from https://doi.org/10.1038/nature07965.

Haniffa, M., Gunawan, M., & Jardine, L. (2015). Human skin dendritic cells in health and disease. *Journal of Dermatological Science*, *77*(2), 85–92. Available from https://doi.org/10.1016/j.jdermsci.2014.08.012.

Haniffa, M., Shin, A., Bigley, V., McGovern, N., Teo, P., See, P., Wasan, P. S., Wang, X. N., Malinarich, F., Malleret, B., Larbi, A., Tan, P., Zhao, H., Poidinger, M., Pagan, S., Cookson, S., Dickinson, R., Dimmick, I., Jarrett, R. F., ... Ginhoux, F. (2012). Human tissues contain CD141 hi cross-presenting dendritic cells with functional homology to mouse CD103 + nonlymphoid dendritic cells. *Immunity*, *37*(1), 60–73. Available from https://doi.org/10.1016/j.immuni.2012.04.012.

Hattori, Y., Kawakami, S., Lu, Y., Nakamura, K., Yamashita, F., & Hashida, M. (2006). Enhanced DNA vaccine potency by mannosylated lipoplex after intraperitoneal administration. *Journal of Gene Medicine*, *8*(7), 824–834. Available from https://doi.org/10.1002/jgm.910.

Hawiger, D., Inaba, K., Dorsett, Y., Guo, M., Mahnke, K., Rivera, M., Ravetch, J. V., Steinman, R. M., & Nussenzweig, M. C. (2001). Dendritic cells induce peripheral T cell unresponsiveness under steady state conditions in vivo. *Journal of Experimental Medicine, 194*(6), 769–779. Available from https://doi.org/10.1084/jem.194.6.769.

Heil, F., Hemmi, H., Hochrein, H., Ampenberger, F., Kirschning, C., Akira, S., Lipford, G., Wagner, H., & Bauer, S. (2004). Species-specific recognition of single-stranded RNA via toll-like receptor 7 and 8. *Science (New York, N.Y.), 303*(5663), 1526–1529. Available from https://doi.org/10.1126/science.1093620.

Hemmi, H., & Akira, S. (2005). TLR signalling and the function of dendritic cells. *Chemical Immunology and Allergy, 86*, 120–135. Available from https://doi.org/10.1159/000086657.

Hilkens, C. M. U., Isaacs, J. D., & Thomson, A. W. (2010). Development of dendritic cell-based immunotherapy for autoimmunity. *International Reviews of Immunology, 29*(2), 156–183. Available from https://doi.org/10.3109/08830180903281193.

Hoeffel, G., Ripoche, A. C., Matheoud, D., Nascimbeni, M., Escriou, N., Lebon, P., Heshmati, F., Guillet, J. G., Gannagé, M., Caillat-Zucman, S., Casartelli, N., Schwartz, O., De la Salle, H., Hanau, D., Hosmalin, A., & Marañón, C. (2007). Antigen crosspresentation by human plasmacytoid dendritic cells. *Immunity, 27*(3), 481–492. Available from https://doi.org/10.1016/j.immuni.2007.07.021.

Horner, A. A., Datta, S. K., Takabayashi, K., Belyakov, I. M., Hayashi, T., Cinman, N., Nguyen, M. D., Van Uden, J. H., Berzofsky, J. A., Richman, D. D., & Raz, E. (2001). Immunostimulatory DNA-based vaccines elicit multifaceted immune responses against HIV at systemic and mucosal sites. *Journal of Immunology, 167*(3), 1584–1591. Available from https://doi.org/10.4049/jimmunol.167.3.1584.

Hornung, V., Bauernfeind, F., Halle, A., Samstad, E. O., Kono, H., Rock, K. L., Fitzgerald, K. A., & Latz, E. (2008). Silica crystals and aluminum salts activate the NALP3 inflammasome through phagosomal destabilization. *Nature Immunology, 9*(8), 847–856. Available from https://doi.org/10.1038/ni.1631.

Hossain, M. K., & Wall, K. A. (2019). Use of dendritic cell receptors as targets for enhancing anti-cancer immune responses. *Cancers, 11*(3). Available from https://doi.org/10.3390/cancers11030418.

Inohara, N., Ogura, Y., Chen, F. F., Muto, A., & Nuñez, G. (2001). Human Nod1 confers responsiveness to bacterial lipopolysaccharides. *Journal of Biological Chemistry, 276*(4), 2551–2554. Available from https://doi.org/10.1074/jbc.M009728200.

Ishikawa, F., Niiro, H., Iino, T., Yoshida, S., Saito, N., Onohara, S., Miyamoto, T., Minagawa, H., Fujii, S. I., Shultz, L. D., Harada, M., & Akashi, K. (2007). The developmental program of human dendritic cells is operated independently of conventional myeloid and lymphoid pathways. *Blood, 110*(10), 3591–3600. Available from https://doi.org/10.1182/blood-2007-02-071613.

Iwasaki, A., & Kelsall, B. L. (2001). Unique functions of CD11b +, CD8α +, and double-negative Peyer's patch dendritic cells. *Journal of Immunology, 166*((8), 4884–4890. Available from https://doi.org/10.4049/jimmunol.166.8.4884.

Jang, J. H., Shin, H. W., Lee, J. M., Lee, H. W., Kim, E. C., & Park, S. H. (2015). An overview of pathogen recognition receptors for innate immunity in dental pulp. *Mediators of Inflammation, 2015*. Available from https://doi.org/10.1155/2015/794143.

Jiang, W., Swiggard, W. J., Heufler, C., Peng, M., Mirza, A., Steinman, R. M., & Nussenzweig, M. C. (1995). The receptor DEC-205 expressed by dendritic cells and thymic epithelial cells is involved in antigen processing. *Nature, 375*(6527), 151–155. Available from https://doi.org/10.1038/375151a0.

Jin, B., Sun, T., Yu, X.-H., Yang, Y.-X., & Yeo, A. E. T. (2012). *The effects of TLR activation on T-cell development and differentiation. Clinical and Developmental Immunology* (2012, p. 836485).

Johansson-Lindbom, B., Svensson, M., Pabst, O., Palmqvist, C., Marquez, G., Förster, R., & Agace, W. W. (2005). Functional specialization of gut CD103 + dendritic cells in the regulation of tissue-selective T cell homing. *Journal of Experimental Medicine*, *202*(8), 1063–1073. Available from https://doi.org/10.1084/jem.20051100.

Joo, H. M., Li, D., Dullaers, M., Kim, T. W., Duluc, D., Upchurch, K., Xue, Y., Zurawski, S., LeGrand, R., Liu, Y. J., Kuroda, M., Zurawski, G., & Oh, S. K. (2014). C-type lectin-like receptor LOX-1 promotes dendritic cell-mediated class-switched B cell responses. *Immunity*, *41*(4), 592–604. Available from https://doi.org/10.1016/j.immuni.2014.09.009.

Jurk, M., Heil, F., Vollmer, J., Schetter, C., Krieg, A. M., Wagner, H., Lipford, G., & Bauer, S. (2002). Human TLR7 or TLR8 independently confer responsiveness to the antiviral compound R-848 [1]. *Nature Immunology*, *3*(6), 499. Available from https://doi.org/10.1038/ni0602-499.

Kabelitz, D., Wesch, D., & Oberg, H. H. (2006). Regulation of regulatory T cells: Role of dendritic cells and toll-like receptors. *Critical Reviews in Immunology*, *26*(4), 291–305.

Kanazawa, N., Tashiro, K., Inaba, K., & Miyachi, Y. (2003). Dendritic cell immunoactivating receptor, a novel C-type lectin immunoreceptor, acts as an activating receptor through association with Fc receptor γ chain. *Journal of Biological Chemistry*, *278*(35), 32645–32652. Available from https://doi.org/10.1074/jbc.M304226200.

Kanneganti, T. D., Body-Malapel, M., Amer, A., Park, J. H., Whitfield, J., Franchi, L., Taraporewala, Z. F., Miller, D., Patton, J. T., Inohara, N., & Núñez, G. (2006). Critical role for Cryopyrin/Nalp3 in activation of caspase-1 in response to viral infection and double-stranded RNA. *Journal of Biological Chemistry*, *281*(48), 36560–36568. Available from https://doi.org/10.1074/jbc.M607594200.

Kaplan, D. H., Igyártó, B. Z., & Gaspari, A. A. (2012). Early immune events in the induction of allergic contact dermatitis. *Nature reviews. Immunology*, *12*(2), 114–124. Available from https://doi.org/10.1038/nri3150.

Kato, M., McDonald, K. J., Khan, S., Ross, I. L., Vuckovic, S., Chen, K., Munster, D., MacDonald, K. P. A., & Hart, D. N. J. (2006). Expression of human DEC-205 (CD205) multilectin receptor on leukocytes. *International Immunology*, *18*(6), 857–869. Available from https://doi.org/10.1093/intimm/dxl022.

Kawai, T., & Akira, S. (2010). The role of pattern-recognition receptors in innate immunity: Update on Toll-like receptors. *Nature Immunology*, *11*(5), 373–384. Available from https://doi.org/10.1038/ni.1863.

Kawai, T., & Akira, S. (2011). Regulation of innate immune signalling pathways by the tripartite motif (TRIM) family proteins. *EMBO Molecular Medicine*, *3*(9), 513–527. Available from https://doi.org/10.1002/emmm.201100160.

Kawasaki, T., & Kawai, T. (2014). Toll-like receptor signaling pathways. *Frontiers in Immunology*, 5. Available from https://doi.org/10.3389/fimmu.2014.00461.

Kim, Y. K., Shin, J. S., & Nahm, M. H. (2016). NOD-like receptors in infection, immunity, and diseases. *Yonsei Medical Journal*, *57*(1), 5–14. Available from https://doi.org/10.3349/ymj.2016.57.1.5.

Koblansky, A. A., Jankovic, D., Oh, H., Hieny, S., Sungnak, W., Mathur, R., Hayden, M. S., Akira, S., Sher, A., & Ghosh, S. (2013). Recognition of profilin by toll-like receptor 12 is critical for host resistance to *Toxoplasma gondii*. *Immunity*, *38*(1), 119–130. Available from https://doi.org/10.1016/j.immuni.2012.09.016.

Krug, A., French, A. R., Barchet, W., Fischer, J. A. A., Dzionek, A., Pingel, J. T., Orihuela, M. M., Akira, S., Yokoyama, W. M., & Colonna, M. (2004). TLR9-dependent recognition of MCMV by IPC and DC generates coordinated cytokine responses that activate antiviral

NK cell function. *Immunity, 21*(1), 107–119. Available from https://doi.org/10.1016/j.immuni.2004.06.007.

Krug, A., Rothenfusser, S., Hornung, V., Jahrsdrfer, B., Blackwell, S., Ballas, Z. K., Endres, S., Krieg, A. M., & Hartmann, G. (2001). Identification of CpG oligonucleotide sequences with high induction of IFN-α/β in plasmacytoid dendritic cells. *European Journal of Immunology, 31*(7), 2154–2163. Available from https://doi.org/10.1002/1521-4141(200107)31:72154:AID-IMMU21543.0.CO;2-U.

Kufer, T. A., & Sansonetti, P. J. (2011). NLR functions beyond pathogen recognition. *Nature Immunology, 12*(2), 121–128. Available from https://doi.org/10.1038/ni.1985.

Lahoud, M. H., Ahmet, F., Zhang, J. G., Meuter, S., Policheni, A. N., Kitsoulis, S., Lee, C. N., O'Keeffe, M., Sullivan, L. C., Brooks, A. G., Berry, R., Rossjohn, J., Mintern, J. D., Vega-Ramos, J., Villadangos, J. A., Nicola, N. A., Nussenzweig, M. C., Stacey, K. J., Shortman, K., & Caminschi, I. (2012). DEC-205 is a cell surface receptor for CpG oligonucleotides. *Proceedings of the National Academy of Sciences of the United States of America, 109*(40), 16270–16275. Available from https://doi.org/10.1073/pnas.1208796109.

Lambert, A. A., Gilbert, C., Richard, M., Beaulieu, A. D., & Tremblay, M. J. (2008). The C-type lectin surface receptor DCIR acts as a new attachment factor for HIV-1 in dendritic cells and contributes to trans- and cis-infection pathways. *Blood, 112*(4), 1299–1307. Available from https://doi.org/10.1182/blood-2008-01-136473.

Lamkanfi, M., Kanneganti, T. D., Franchi, L., & Núñez, G. (2007). Caspase-1 inflammasomes in infection and inflammation. *Journal of Leukocyte Biology, 82*(2), 220–225. Available from https://doi.org/10.1189/jlb.1206756.

Lee, Y. S., & Radford, K. J. (2019). The role of dendritic cells in cancer. *International review of cell and molecular biology, 348*, 123–178. Available from https://doi.org/10.1016/bs.ircmb.2019.07.006.

Liang, M., Zhang, P., & Fu, J. (2007). Up-regulation of LOX-1 expression by TNF-α promotes trans-endothelial migration of MDA-MB-231 breast cancer cells. *Cancer Letters, 258*(1), 31–37. Available from https://doi.org/10.1016/j.canlet.2007.08.003.

Loo, Y. M., & Gale, M. (2007). Viral regulation and evasion of the host response. *Current Topics in Microbiology and Immunology, 316*, 295–313. Available from https://doi.org/10.1007/978-3-540-71329-6_14.

Lopes, L., Dewannieux, M., Gileadi, U., Bailey, R., Ikeda, Y., Whittaker, C., Collin, M. P., Cerundolo, V., Tomihari, M., Ariizumi, K., & Collins, M. K. (2008). Immunization with a lentivector that targets tumor antigen expression to dendritic cells induces potent CD8 + and CD4 + T-cell responses. *Journal of Virology, 82*(1), 86–95. Available from https://doi.org/10.1128/JVI.01289-07.

Lotfi, N., Thome, R., Rezaei, N., Zhang, G. X., Rezaei, A., Rostami, A., & Esmaeil, N. (2019). Roles of GM-CSF in the pathogenesis of autoimmune diseases: An update. *Frontiers in Immunology*, 10. Available from https://doi.org/10.3389/fimmu.2019.01265.

Ludlow, L. E. A., Johnstone, R. W., & Clarke, C. J. P. (2005). The HIN-200 family: More than interferon-inducible genes? *Experimental Cell Research, 308*(1), 1–17. Available from https://doi.org/10.1016/j.yexcr.2005.03.032.

Lugrin, J., & Martinon, F. (2018). The AIM2 inflammasome: Sensor of pathogens and cellular perturbations. *Immunological Reviews, 281*(1), 99–114. Available from https://doi.org/10.1111/imr.12618.

Mariathasan, S., Hewton, K., Monack, D. M., Vucic, D., French, D. M., Lee, W. P., Roose-Girma, M., Erickson, S., & Dixit, V. M. (2004). Differential activation of the inflammasome by caspase-1 adaptors ASC and Ipaf. *Nature*, *430*(6996), 213–218. Available from https://doi.org/10.1038/nature02664.

Mariathasan, S., Weiss, D. S., Newton, K., McBride, J., O'Rourke, K., Roose-Girma, M., Lee, W. P., Weinrauch, Y., Monack, D. M., & Dixit, V. M. (2006). Cryopyrin activates the inflammasome in response to toxins and ATP. *Nature*, *440*(7081), 228–232. Available from https://doi.org/10.1038/nature04515.

Marshall, A. S. J., Willmen, J. A., Lin, H. H., Williams, D. L., Gordon, S., & Brown, G. D. (2004). Identification and characterization of a novel human myeloid inhibitory C-type lectin-like receptor (MICL) that is predominantly expressed on granulocytes and monocytes. *Journal of Biological Chemistry*, *279*(15), 14792–14802. Available from https://doi.org/10.1074/jbc.M313127200.

Mathur, R., Oh, H., Zhang, D., Park, S. G., Seo, J., Koblansky, A., Hayden, M. S., & Ghosh, S. (2012). A mouse model of salmonella typhi infection. *Cell*, *151*(3), 590–602. Available from https://doi.org/10.1016/j.cell.2012.08.042.

Matsumoto, M., Tanaka, T., Kaisho, T., Sanjo, H., Copeland, N. G., Gilbert, D. J., Jenkins, N. A., & Akira, S. (1999). A novel LPS-inducible C-type lectin is a transcriptional target of NF-IL6 in macrophages. *Journal of Immunology*, *163*(9), 5039–5048.

Matta, B. M., Castellaneta, A., & Thomson, A. W. (2010). Tolerogenic plasmacytoid DC. *European Journal of Immunology*, *40*(10), 2667–2676. Available from https://doi.org/10.1002/eji.201040839.

Mbongue, J. C., Nieves, H. A., Torrez, T. W., & Langridge, W. H. R. (2017). The role of dendritic cell maturation in the induction of insulin-dependent diabetes mellitus. *Frontiers in Immunology*, 8. Available from https://doi.org/10.3389/fimmu.2017.00327.

McKenna, K., Beignon, A. S., & Bhardwaj, N. (2005). Plasmacytoid dendritic cells: Linking innate and adaptive immunity. *Journal of Virology*, *79*(1), 17–27. Available from https://doi.org/10.1128/JVI.79.1.17-27.2005.

Meredith, M. M., Liu, K., Darrasse-Jeze, G., Kamphorst, A. O., Schreiber, H. A., Guermonprez, P., Idoyaga, J., Cheong, C., Yao, K. H., Niec, R. E., & Nussenzweig, M. C. (2012). Expression of the zinc finger transcription factor zDC (Zbtb46, Btbd4) defines the classical dendritic cell lineage. *Journal of Experimental Medicine*, *209*(6), 1153–1165. Available from https://doi.org/10.1084/jem.20112675.

Meyer-Wentrup, F., Benitez-Ribas, D., Tacken, P. J., Punt, C. J. A., Figdor, C. G., De Vries, I. J. M., & Adema, G. J. (2008). Targeting DCIR on human plasmacytoid dendritic cells results in antigen presentation and inhibits IFN-α production. *Blood*, *111*(8), 4245–4253. Available from https://doi.org/10.1182/blood-2007-03-081398.

Meyer-Wentrup, F., Cambi, A., Joosten, B. J., Looman, M. W., De Vries, I. J. M., Figdor, C. G., & Adema, G. J. (2009). DCIR is endocytosed into human dendritic cells and inhibits TLR8-mediated cytokine production. *Journal of Leukocyte Biology*, *85*(3), 518–525. Available from https://doi.org/10.1189/jlb.0608352.

Minigo, G., Scholzen, A., Tang, C. K., Hanley, J. C., Kalkanidis, M., Pietersz, G. A., Apostolopoulos, V., & Plebanski, M. (2007). Poly-L-lysine-coated nanoparticles: A potent delivery system to enhance DNA vaccine efficacy. *Vaccine*, *25*(7), 1316–1327. Available from https://doi.org/10.1016/j.vaccine.2006.09.086.

Mittag, D., Proietto, A. I., Loudovaris, T., Mannering, S. I., Vremec, D., Shortman, K., Wu, L., & Harrison, L. C. (2011). Human dendritic cell subsets from spleen and blood are similar in phenotype and function but modified by donor health status. *Journal of Immunology*, *186*(11), 6207–6217. Available from https://doi.org/10.4049/jimmunol.1002632.

Mourão-Sá, D., Robinson, M. J., Zelenay, S., Sancho, D., Chakravarty, P., Larsen, R., Plantinga, M., van Rooijen, N., Soares, M. P., Lambrecht, B., & Reis e Sousa, C. (2011). CLEC-2 signaling via Syk in myeloid cells can regulate inflammatory responses. *European Journal of Immunology*, *41*(10), 3040–3053. Available from https://doi.org/10.1002/eji.201141641.

Murdocca, M., Mango, R., Pucci, S., Biocca, S., Testa, B., Capuano, R., Paolesse, R., Sanchez, M., Orlandi, A., Di Natale, C., Novelli, G., & Sangiuolo, F. (2016). The lectin-like oxidized LDL receptor-1: A new potential molecular target in colorectal cancer. *Oncotarget*, *7*(12), 14765–14780. Available from https://doi.org/10.18632/oncotarget.7430.

Musumeci, A., Lutz, K., Winheim, E., & Krug, A. B. (2019). What makes a PDC: Recent advances in understanding plasmacytoid DC development and heterogeneity. *Frontiers in Immunology*, *10*. Available from https://doi.org/10.3389/fimmu.2019.01222.

Nakano, H., Lin, K. L., Yanagita, M., Charbonneau, C., Cook, D. N., Kakiuchi, T., & Gunn, M. D (2009). Blood-derived inflammatory dendritic cells in lymph nodes stimulate acute T helper type 1 immune responses. *Nature Immunology*, *10*(4), 394–402. Available from https://doi.org/10.1038/ni.1707.

Netea, M. G., Van Der Graaf, C., Van Der Meer, J. W. M., & Kullberg, B. J. (2004). Recognition of fungal pathogens by toll-like receptors. *European Journal of Clinical Microbiology and Infectious Diseases*, *23*(9), 672–676. Available from https://doi.org/10.1007/s10096-004-1192-7.

Niess, J. H., Brand, S., Gu, X., Landsman, L., Jung, S., McCormick, B. A., Vyas, J. M., Boes, M., Ploegh, H. L., Fox, J. G., Littman, D. R., & Reinecker, H. C. (2005). CX3CR1-mediated dendritic cell access to the intestinal lumen and bacterial clearance. *Science (New York, N.Y.)*, *307*(5707), 254–258. Available from https://doi.org/10.1126/science.1102901.

Ochando, J. C., Homma, C., Yang, Y., Hidalgo, A., Garin, A., Tacke, F., Angeli, V., Li, Y., Boros, P., Ding, Y., Jessberger, R., Trinchieri, G., Lira, S. A., Randolph, G. J., & Bromberg, J. S. (2006). Alloantigen-presenting plasmacytoid dendritic cells mediate tolerance to vascularized grafts. *Nature Immunology*, *7*(6), 652–662. Available from https://doi.org/10.1038/ni1333.

O'Keeffe, M., Mok, W. H., & Radford, K. J. (2015). Human dendritic cell subsets and function in health and disease. *Cellular and Molecular Life Sciences*, *72*(22), 4309–4325. Available from https://doi.org/10.1007/s00018-015-2005-0.

Okusawa, T., Fujita, M., Nakamura, J. I., Into, T., Yasuda, M., Yoshimura, A., Hara, Y., Hasebe, A., Golenbock, D. T., Morita, M., Kuroki, Y., Ogawa, T., & Shibata, K. I. (2004). Relationship between structures and biological activities of mycoplasmal diacylated lipopeptides and their recognition by toll-like receptors 2 and 6. *Infection and Immunity*, *72*(3), 1657–1665. Available from https://doi.org/10.1128/IAI.72.3.1657-1665.2004.

Onai, N., Obata-Onai, A., Schmid, M. A., Ohteki, T., Jarrossay, D., & Manz, M. G. (2007). Identification of clonogenic common Flt3 + M-CSFR + plasmacytoid and conventional dendritic cell progenitors in mouse bone marrow. *Nature Immunology*, *8*(11), 1207–1216. Available from https://doi.org/10.1038/ni1518.

Park, E., Na, H. S., Song, Y. R., Shin, S. Y., Kim, Y. M., & Chung, J. (2014). Activation of NLRP3 and AIM2 inflammasomes by *Porphyromonas gingivalis* infection. *Infection and Immunity*, *82*(1), 112–123. Available from https://doi.org/10.1128/IAI.00862-13.

Perez, C. R., & De Palma, M (2019). Engineering dendritic cell vaccines to improve cancer immunotherapy. *Nature communications*, *10*(1), 1–10. Available from https://doi.org/10.1038/s41467-019-13368-y.

Poulin, L. F., Henri, S., De Bovis, B., Devilard, E., Kissenpfennig, A., & Malissen, B. (2007). The dermis contains langerin + dendritic cells that develop and function independently of epidermal Langerhans cells. *Journal of Experimental Medicine, 204*(13), 3119–3131. Available from https://doi.org/10.1084/jem.20071724.

Proietto, A. I., Van Dommelen, S., Zhou, P., Rizzitelli, A., D'Amico, A., Steptoe, R. J., Naik, S. H., Lahoud, M. H., Liu, Y., Zheng, P., Shortman, K., & Wu, L. (2008). Dendritic cells in the thymus contribute to T-regulatory cell induction. *Proceedings of the National Academy of Sciences of the United States of America, 105*(50), 19869–19874. Available from https://doi.org/10.1073/pnas.0810268105.

Pulendran, B., Banchereau, J., Burkeholder, S., Kraus, E., Guinet, E., Chalouni, C., Caron, D., Maliszewski, C., Davoust, J., Fay, J., & Palucka, K. (2000). Flt3-ligand and granulocyte colony-stimulating factor mobilize distinct human dendritic cell subsets in vivo. *Journal of Immunology, 165*(1), 566–572. Available from https://doi.org/10.4049/jimmunol.165.1.566.

Rehwinkel, J., & Gack, M. U. (2020). RIG-I-like receptors: Their regulation and roles in RNA sensing. *Nature Reviews Immunology, 20*(9), 537–551. Available from https://doi.org/10.1038/s41577-020-0288-3.

Robles, M. D. L., Pallier, A., Huchet, V., Le Texier, L., Remy, S., Braudeau, C., Delbos, L., Moreau, A., Louvet, C., Brosseau, C., Royer, P. J., Magnan, A., Halary, F., Josien, R., Cuturi, M. C., Anegon, I., & Chiffoleau, E. (2017). Cell-surface C-type lectin-like receptor CLEC-1 dampens dendritic cell activation and downstream Th17 responses. *Blood Advances, 1*(9), 557–568. Available from https://doi.org/10.1182/bloodadvances.2016002360.

Ryan, E. J., Marshall, A. J., Magaletti, D., Floyd, H., Draves, K. E., Olson, N. E., & Clark, E. A. (2002). Dendritic cell-associated lectin-1: A novel dendritic cell-associated, C-type lectin-like molecule enhances T cell secretion of IL-4. *Journal of Immunology, 169*(10), 5638–5648. Available from https://doi.org/10.4049/jimmunol.169.10.5638.

Sabado, R. L., Balan, S., & Bhardwaj, N. (2017). Dendritic cell-based immunotherapy. *Cell Research, 27*(1), 74–95. Available from https://doi.org/10.1038/cr.2016.157.

Salazar-Gonzalez, R. M., Niess, J. H., Zammit, D. J., Ravindran, R., Srinivasan, A., Maxwell, J. R., Stoklasek, T., Yadav, R., Williams, I. R., Gu, X., McCormick, B. A., Pazos, M. A., Vella, A. T., Lefrancois, L., Reinecker, H. C., & McSorley, S. J. (2006). CCR6-mediated dendritic cell activation of pathogen-specific T cells in Peyer's patches. *Immunity, 24*(5), 623–632. Available from https://doi.org/10.1016/j.immuni.2006.02.015.

Sancho, D., Joffre, O. P., Keller, A. M., Rogers, N. C., Martínez, D., Hernanz-Falcón, P., Rosewell, I., & Sousa, C. R. E. (2009). Identification of a dendritic cell receptor that couples sensing of necrosis to immunity. *Nature, 458*(7240), 899–903. Available from https://doi.org/10.1038/nature07750.

Saxena, M, & Yeretssian, G (2014). NOD-like receptors: Master regulators of inflammation and cancer. *Frontiers in Immunology, 5*, 327. Available from https://doi.org/10.3389/fimmu.2014.00327.

Schreibelt, G., Klinkenberg, L. J. J., Cruz, L. J., Tacken, P. J., Tel, J., Kreutz, M., Adema, G. J., Brown, G. D., Figdor, C. G., & De Vries, I. J. M. (2012). The C-type lectin receptor CLEC9A mediates antigen uptake and (cross-)presentation by human blood BDCA3 + myeloid dendritic cells. *Blood, 119*(10), 2284–2292. Available from https://doi.org/10.1182/blood-2011-08-373944.

Schroder, K., & Tschopp, J. (2010). The inflammasomes. *Cell, 140*(6), 821–832. Available from https://doi.org/10.1016/j.cell.2010.01.040.

Schulz, O., Diebold, S. S., Chen, M., Näslund, T. I., Nolte, M. A., Alexopoulou, L., Azuma, Y. T., Flavell, R. A., Liljeström, P., & Reis e Sousa, C. (2005). Toll-like receptor 3 promotes cross-priming to virus-infected cells. *Nature, 433*(7028), 887–892. Available from https://doi.org/10.1038/nature03326.

Seifert, L., Deutsch, M., Alothman, S., Alqunaibit, D., Werba, G., Pansari, M., Pergamo, M., Ochi, A., Torres-Hernandez, A., Levie, E., Tippens, D., Greco, S. H., Tiwari, S., Ly, N. N. G., Eisenthal, A., van Heerden, E., Avanzi, A., Barilla, R., Zambirinis, C. P., & Miller, G. (2015). Dectin-1 Regulates hepatic fibrosis and hepatocarcinogenesis by suppressing TLR4 signaling pathways. *Cell Reports*, *13*(9), 1909–1921. Available from https://doi.org/10.1016/j.celrep.2015.10.058.

Shrimpton, R. E., Butler, M., Morel, A. S., Eren, E., Hue, S. S., & Ritter, M. A. (2009).). CD205 (DEC-205): A recognition receptor for apoptotic and necrotic self. *Molecular Immunology*, *46*(6), 1229–1239. Available from https://doi.org/10.1016/j.molimm.2008.11.016.

Simones, T., Shepherd, D. M., & Moser, M. (2010). *Dendritic cells*. (2nd ed., pp. 155–170). *Comprehensive toxicology*, (Vol. 5, pp. 155–170). Elsevier Inc. Available from https://doi.org/10.1016/B978-0-08-046884-6.00608-4.

Smith, K. D., & Ozinsky, A. (2002). Toll-like receptor-5 and the innate immune response to bacterial flagellin. *Current Topics in Microbiology and Immunology*, *270*, 93–108. Available from https://doi.org/10.1007/978-3-642-59430-4_6.

Steinman, R. M. (2012). Decisions about dendritic cells: Past, present, and future. *Annual Review of Immunology*, *30*, 1–22. Available from https://doi.org/10.1146/annurev-immunol-100311-102839.

Sutterwala, F. S., Ogura, Y., Szczepanik, M., Lara-Tejero, M., Lichtenberger, G. S., Grant, E. P., Bertin, J., Coyle, A. J., Galán, J. E., Askenase, P. W., & Flavell, R. A. (2006). Critical role for NALP3/CIAS1/cryopyrin in innate and adaptive immunity through its regulation of caspase-1. *Immunity*, *24*(3), 317–327. Available from https://doi.org/10.1016/j.immuni.2006.02.004.

Swiecki, M., & Colonna, M. (2015). The multifaceted biology of plasmacytoid dendritic cells. *Nature Reviews Immunology*, *15*(8), 471–485. Available from https://doi.org/10.1038/nri3865.

Szolnoky, G., Bata-Csörgö, Z., Kenderessy, A. S., Kiss, M., Pivarcsi, A., Novák, Z., Newman, K. N., Michel, G., Ruzicka, T., Maródi, L., Dobozy, A., & Kemény, L. (2001). A mannose-binding receptor is expressed on human keratinocytes and mediates killing of *Candida albicans*. *Journal of Investigative Dermatology*, *117*(2), 205–213. Available from https://doi.org/10.1046/j.1523-1747.2001.14071.x.

Takeda, K, & Akira, S (2005). Toll-like receptors in innate immunity. *International Immunology*, *17*(1), 1–14. Available from https://doi.org/10.1093/intimm/dxh186.

Takeuchi, O., Kawai, T., Mühlradt, P. F., Morr, M., Radolf, J. D., Zychlinsky, A., Takeda, K., & Akira, S. (2001). Discrimination of bacterial lipoproteins by Toll-like recepttor 6. *International Immunology*, *13*(7), 933–940. Available from https://doi.org/10.1093/intimm/13.7.933.

Takeuchi, O., Sato, S., Horiuchi, T., Hoshino, K., Takeda, K., Dong, Z., Modlin, R. L., & Akira, S. (2002). Cutting edge: Role of Toll-like receptor 1 in mediating immune response to microbial lipoproteins. *Journal of Immunology*, *169*(1), 10–14. Available from https://doi.org/10.4049/jimmunol.169.1.10.

Tezuka, H., Abe, Y., Iwata, M., Takeuchi, H., Ishikawa, H., Matsushita, M., Shiohara, T., Akira, S., & Ohteki, T. (2007). Regulation of IgA production by naturally occurring TNF/iNOS-producing dendritic cells. *Nature*, *448*(7156), 929–933. Available from https://doi.org/10.1038/nature06033.

Tighe, H., Takabayashi, K., Schwartz, D., Marsden, R., Beck, L., Corbeil, J., Richman, D. D., Eiden, J. J., Jr, Spiegelberg, H. L., & Raz, E. (2000). Conjugation of protein to immunostimulatory DNA results in a rapid, long-lasting and potent induction of cell-mediated and humoral immunity. European journal of immunology, *30*(7), 1939–1947. https://doi.org/10.1002/1521-4141(200007)30:7 < 1939::AID-IMMU1939 > 3.0.CO;2-#

Ting, J. P. Y., Lovering, R. C., Alnemri, E. S., Bertin, J., Boss, J. M., Davis, B. K., Flavell, R. A., Girardin, S. E., Godzik, A., Harton, J. A., Hoffman, H. M., Hugot, J. P., Inohara, N., MacKenzie, A., Maltais, L. J., Nunez, G., Ogura, Y., Otten, L. A., Philpott, D., ... Ward, P. A. (2008). The NLR gene family: A standard nomenclature. *Immunity*, *28*(3), 285–287. Available from https://doi.org/10.1016/j.immuni.2008.02.005.

Toshchakov, V., Jones, B. W., Perera, P. Y., Thomas, K., Cody, M. J., Zhang, S., Williams, B. R. G., Major, J., Hamilton, T. A., Fenton, M. J., & Vogel, S. N. (2002). TLR4, but not TLR2, mediates IFN-β-induced STATIα/β-dependent gene expression in macrophages. *Nature Immunology*, *3*(4), 392–398. Available from https://doi.org/10.1038/ni774.

Travassos, L. H., Carneiro, L. A. M., Ramjeet, M., Hussey, S., Kim, Y. G., Magalhes, J. G., Yuan, L., Soares, F., Chea, E., Le Bourhis, L., Boneca, I. G., Allaoui, A., Jones, N. L., Nũez, G., Girardin, S. E., & Philpott, D. J. (2010). Nod1 and Nod2 direct autophagy by recruiting ATG16L1 to the plasma membrane at the site of bacterial entry. *Nature Immunology*, *11*(1), 55–62. Available from https://doi.org/10.1038/ni.1823.

Tsu, B. V., Beierschmitt, C., Ryan, A. P., Agarwal, R., Mitchell, P. S., & Daugherty, M. D. (2021). Diverse viral proteases activate the NLRP1 inflammasome. *Elife*, *10*, e60609. Available from https://doi.org/10.7554/eLife.60609.

Upham, J. W, & Hughes, T. L (2006). Dendritic Cells. In G. J. Laurent, & S. D Shapiro (Eds.), *Encyclopedia of Respiratory Medicine*. (Vol. 3). Academic Press Elsevier.

van Kooyk, Y., & Rabinovich, G. A. (2008). Protein-glycan interactions in the control of innate and adaptive immune responses. *Nature Immunology*, *9*(6), 593–601. Available from https://doi.org/10.1038/ni.f.203.

Waskow, C., Liu, K., Darrasse-Jèze, G., Guermonprez, P., Ginhoux, F., Merad, M., Shengelia, T., Yao, K., & Nussenzweig, M. (2008). The receptor tyrosine kinase Flt3 is required for dendritic cell development in peripheral lymphoid tissues. *Nature Immunology*, *9*(6), 676–683. Available from https://doi.org/10.1038/ni.1615.

Wells, C. A., Salvage, J. A. J., Li, X., Hitchens, K., Butcher, S., Murray, R. Z., Beckhouse, A. G., Lo, Y. L. S., Manzanero, S., Cobbold, C., Schroder, K., Ma, B., Orr, S., Stewart, L., Lebus, D., Sobieszczuk, P., Hume, D. A., Stow, J., Blanchard, H., & Ashman, R. B. (2008). The macrophage-inducible C-type lectin, Mincle, is an essential component of the innate immune response to *Candida albicans*. *Journal of Immunology*, *180*(11), 7404–7413. Available from https://doi.org/10.4049/jimmunol.180.11.7404.

Worbs, T., Hammerschmidt, S. I., & Förster, R. (2017). Dendritic cell migration in health and disease. *Nature Reviews Immunology*, *17*(1), 30–48. Available from https://doi.org/10.1038/nri.2016.116.

Wu, P., Wu, J., Liu, S., Han, X., Lu, J., Shi, Y., Wang, J., Lu, L., & Cao, X. (2008). TLR9/TLR7-triggered downregulation of BDCA2 expression on human plasmacytoid dendritic cells from healthy individuals and lupus patients. *Clinical Immunology*, *129*(1), 40–48. Available from https://doi.org/10.1016/j.clim.2008.06.004.

Yamamoto, M., Sato, S., Hemmi, H., Hoshino, K., Kaisho, T., Sanjo, H., Takeuchi, O., Sugiyama, M., Okabe, M., Takeda, K., & Akira, S. (2003). Role of adaptor TRIF in the MyD88-independent toll-like receptor signaling pathway. *Science (New York, N.Y.)*, *301*(5633), 640–643. Available from https://doi.org/10.1126/science.1087262.

Yamane, H, & Paul, W. E. (2013). Early signaling events that underlie fate decisions of naive CD4(+) T cells toward distinct T-helper cell subsets. *Immunological reviews*, *252*(1), 12–23. Available from https://doi.org/10.1111/imr.12032.

Yamasaki, S., Ishikawa, E., Sakuma, M., Hara, H., Ogata, K., & Saito, T. (2008). Mincle is an ITAM-coupled activating receptor that senses damaged cells. *Nature Immunology*, *9*(10), 1179–1188. Available from https://doi.org/10.1038/ni.1651.

Yamasaki, S., Matsumoto, M., Takeuchi, O., Matsuzawa, T., Ishikawa, E., Sakuma, M., Tateno, H., Uno, J., Hirabayashi, J., Mikami, Y., Takeda, K., Akira, S., & Saito, T. (2009). C-type lectin Mincle is an activating receptor for pathogenic fungus, *Malassezia*. *Proceedings of the National Academy of Sciences of the United States of America*, *106*(6), 1897–1902. Available from https://doi.org/10.1073/pnas.0805177106.

Yan, Z., Wu, Y., Du, J., Li, G., Wang, S., Cao, W., Zhou, X., Wu, C., Zhang, D., Jing, X., Li, Y., Wang, H., Gao, Y., & Qi, Y. (2016). A novel peptide targeting Clec9a on dendritic cell for cancer immunotherapy. *Oncotarget*, *7*(26), 40437–40450. Available from https://doi.org/10.18632/oncotarget.9624.

Yazdi, A. S, Röcken, M, & Ghoreschi, K (2016). Cutaneous immunology: basics and new concepts. *Seminars in immunopathology*, *38*(1), 3–10. Available from https://doi.org/10.1007/s00281-015-0545-x.

Yoneyama, M., Kikuchi, M., Matsumoto, K., Imaizumi, T., Miyagishi, M., Taira, K., Foy, E., Loo, Y. M., Gale, M., Akira, S., Yonehara, S., Kato, A., & Fujita, T. (2005). Shared and unique functions of the DExD/H-box helicases RIG-I, MDA5, and LGP2 in antiviral innate immunity. *Journal of Immunology*, *175*(5), 2851–2858. Available from https://doi.org/10.4049/jimmunol.175.5.2851.

Yoneyama, M., Kikuchi, M., Natsukawa, T., Shinobu, N., Imaizumi, T., Miyagishi, M., Taira, K., Akira, S., & Fujita, T. (2004). The RNA helicase RIG-I has an essential function in double-stranded RNA-induced innate antiviral responses. *Nature Immunology*, *5*(7), 730–737. Available from https://doi.org/10.1038/ni1087.

Zhang, S., Kodys, K., Li, K., & Szabo, G. (2013). Human type 2 myeloid dendritic cells produce interferon-λ and amplify interferon-α in response to hepatitis C virus infection. *Gastroenterology*, *144*(2), 414–e27. Available from https://doi.org/10.1053/j.gastro.2012.10.034.

Zhou, H., & Wu, L. (2017). The development and function of dendritic cell populations and their regulation by miRNAs. *Protein & Cell*, *8*(7), 501–513. Available from https://doi.org/10.1007/s13238-017-0398-2.

Further reading

Diebold, S. S., Kaisho, T., Hemmi, H., Akira, S., & Reis e Sousa, C. (2004). Innate antiviral responses by means of TLR7-mediated recognition of single-stranded RNA. *Science (New York, N.Y.)*, *303*(5663), 1529–1531. Available from https://doi.org/10.1126/science.1093616.

Kawai, T., & Akira, S. (2006). TLR signaling. *Cell Death and Differentiation*, *13*(5), 816–825. Available from https://doi.org/10.1038/sj.cdd.4401850.

McGovern, N., Schlitzer, A., Gunawan, M., Jardine, L., Shin, A., Poyner, E., Green, K., Dickinson, R., Wang, X. N., Low, D., Best, K., Covins, S., Milne, P., Pagan, S., Aljefri, K., Windebank, M., Saavedra, D. M., Larbi, A., Wasan, P. S., ... Haniffa, M. (2014). Human dermal CD14+ cells are a transient population of monocyte-derived macrophages. *Immunity*, *41*(3), 465–477. Available from https://doi.org/10.1016/j.immuni.2014.08.006.

Chapter 4

Vaccine development: from the laboratory to the field

Linda Onyeka Anagu[1] and Nana Efua Andoh[2]
[1]Department of Pharmaceutical Microbiology and Biotechnology, Nnamdi Azikiwe University, Awka, Nigeria, [2]Department of Parasitology, Noguchi Memorial Institute for Medical Research, University of Ghana, Accra, Ghana

Introduction

This chapter focuses on the evaluation of potential antigenic targets and candidate vaccines including the use of genetic studies, protein studies, and immunogenicity assessment. These can be used to determine the efficacy of candidate vaccines and approaches to inducing and maintaining protective immune responses. Established immune parameters enable the prediction of the level of protection provided by the candidate vaccine against disease or infection, known as correlates of protection. The methods described here can also be used during clinical testing and immuno-monitoring. Some of the methods, for example those described under the genetic and protein studies, can be applied to any antigen, while those described in section "Other methods of in vitro evaluation of immunogenicity" are mainly applicable to *Plasmodium*-specific antigens.

Antigen selection for new vaccines

The first vaccines were whole viruses or bacteria (killed or attenuated), and contained antigens vital in inducing a protective immune response, similar to that produced by exposure to the target pathogen (Cunningham et al., 2016). However, these vaccines were sometimes highly immunogenic, caused adverse reactions, and could generate pathology that was close to that produced by the infection targeted by the vaccine (Cunningham et al., 2016). As a result of this, new approaches such as subunit vaccines that only contain key pathogen—derived components are mainly being used for modern vaccine design. Subunit vaccines consist of one or more specific antigens of the pathogen and are produced through genetic engineering or purification

Vaccinology and Methods in Vaccine Research. DOI: https://doi.org/10.1016/B978-0-323-91146-7.00011-1

methods (Wang et al., 2020). Indeed, a large number of severe acute respiratory syndrome coronavirus 2 (SARS-CoV-2) vaccines expressing only the spike protein have been developed. The spike protein (S protein) that covers the surface of SARS-CoV-2 binds to angiotensin-converting enzyme 2 receptors on the surface of human cells and mediates viral entry into the cell (Letko et al., 2020). The S protein induces neutralizing antibodies (Bisht et al., 2004; Yang et al., 2004) that are protective, making it an appropriate antigen candidate for SARS-CoV-2 vaccines (see Chapter 8).

Several factors have to be considered when selecting the appropriate candidate antigens. Antigens need to be able to stimulate innate and T-cell-dependent responses, and produce a strong B-cell (antibody) response that leads to the neutralization and elimination of pathogens (Tan et al., 2016). Also, candidate antigens should be able to protect against different strains/mutations of the same pathogen (Tan et al., 2016), so should possess sequences common to the majority of pathogen strains. Antigenic drift (minute antigenic changes caused by mutations that lead to new strains) and antigenic shift (major antigen changes that lead to a new subtype of a virus) have to be considered when selecting candidate antigens, as this enables the selection of strain-transcending vaccines (Carrat & Flahault, 2007; Tan et al., 2016).

Antigens have to be evaluated for immunogenicity in vivo, ideally using animal models of infection that mimic human disease, which can elucidate whether the vaccine-induced immune response can confer protection against infection (Brinks et al., 2011). Cell-based in vitro assays can be used to evaluate antibody function, for example, to test whether antibodies generated in vivo (e.g., in mice, rats, or rabbits) can neutralize pathogen in a cellular infection model. After suitable antigens have been selected for vaccine development, appropriate vaccine delivery systems such as viral vectors have to be considered and tested (see Chapter 2). The resulting candidate vaccines are then formulated with a variety of additional components such as adjuvants (see Chapter 9), stabilizers, and/or preservatives. Formulation of candidate vaccines, the use of adjuvants, and the choice of an optimal vaccine delivery system can enhance the immunogenicity and potency of the vaccine. New vaccine formulations are manufactured to good manufacturing practice (GMP) (see Chapter 10) and tested in clinical trials as described in Chapter 11.

Genetic studies

Candidate antigens can be evaluated as recombinant proteins, expressed in a variety of genetically engineered cell lines including bacterial, yeast, or mammalian. They can also be expressed from RNA, DNA, or viral vectors. Genetic engineering is used in making tools for the evaluation of candidate antigens; for example, it may sometimes be helpful to make a modified form of a pathogen that infects mice (e.g., *Plasmodium berghei*) engineered to express an antigen from a similar pathogen that infects humans (e.g.,

Plasmodium falciparum) to evaluate human vaccines in mice. These pathogens may also be engineered to express fluorescent or luciferase reporters under the control of a promoter that allows for evaluation of immune responses to antigens, for example by measuring the reduction in pathogen fluorescense.

Expression of the antigenic target gene of interest

Candidate antigens can be cloned and expressed to evaluate immunogenicity directly in vivo. In the laboratory, various steps are involved in cloning the gene that encodes the protein of interest. After the expression of the protein has been induced in a genetically modified host system, it is purified and characterized. The two broad steps involved in expressing an antigen are discussed below.

Cloning of the target antigen of interest

The process begins with the construction of plasmids expressing the gene that encodes the target antigen. Genetic sequences encoding a part or the whole of an antigenic target can be cloned into a plasmid to express the corresponding amino acid sequence in an artificial expression system. In the design and construction of such plasmids, there are several considerations including codon optimization of the antigen for expression in the chosen system (e.g., mammalian cells, yeast, or bacteria), the inclusion of sequences for initiation/termination of transcription, and translation, a promoter appropriate for the expression system, a gene encoding a selectable marker for antibiotic resistance, and restriction enzyme sites for subcloning procedures (Rosano & Ceccarelli, 2014). After a gene has been cloned, it is then expressed. It is to be noted that cloning is technically challenging and requires careful training.

Numerous candidate subunit vaccines under development today were first cloned, then expressed and analyzed. For example, the malaria vaccine candidate MSP-1_{42} was isolated using restriction cloning by Christian and colleagues. MSP-1_{42} is the C-terminal subunit of the *Plasmodium* merozoite surface protein 1 (MSP-1) and it was demonstrated that polyclonal serum raised against this antigen recognized native MSP-1 in *Plasmodium*-infected red blood cells (Epp et al., 2003). The method described below is restriction cloning, but other newer methods of cloning are now popular. The transmission-blocking malaria vaccine candidate *P. falciparum* P48/45 has been expressed using a "gene insertion/marker out" (GIMO) transfection and selection method (Othman et al., 2018). For an overview of other cloning methods, please see Basic Science Methods for clinical researchers (Bertero et al., 2017).

Protocol

1. Select/design the target antigen to be expressed, as an amino acid sequence.
2. Choose an expression system. The expression system consists of the cell or microorganism that synthesizes the antigen, the expression cassette containing the gene to be cloned, and the expression vector. The type of expression system chosen will determine if codon optimization is necessary. Codon optimization is a process that improves translation efficiency; for example, when expressing a eukaryotic gene in a prokaryotic system, some codons may need to be replaced as the prokaryotic organism is biased toward using specific codons for certain amino acids. Several in silico codon optimization tools exist, including IDT and Genewiz, and gene synthesis companies utilize their own proprietary software.
3. Design the nucleotide sequence that encodes the target antigen. This can be obtained from any nucleic acid database including GenBank. If codon optimization is needed, the entire artificial gene encoding the target antigen is synthesized using a specialist gene synthesis company. Alternatively, the amino acid sequence of a target gene can be obtained from a protein database and back-translated to a nucleotide sequence, based on the codon frequency table for the selected expression system (Mazumdar et al., 2010).
4. Design primers complementary to the target gene. Software such as Snapgene can be used, and the specificity and heterodimerization of the primers must be analyzed, for example using NCBIblast or IDT. Sequences encoding a restriction site (with a few additional flanking nucleotides to facilitate efficient enzymatic cleavage) can be included in the primer sequence. Primers are synthesized commercially. This step is not required for de novo synthesis of a codon-optimized version of the antigen.
5. Amplify the gene encoding the target antigen from a DNA template (e.g., the extracted genome of a target pathogen) in a process called polymerase chain reaction (PCR). Please refer to any biotechnology book on how to carry out PCR. This step is not required for the codon-optimized antigen approach.
6. Clone the expression cassette or codon-optimized synthetic gene (containing the necessary restriction site provided by the gene synthesis company) into an expression vector, under a promoter that is appropriate for the selected expression system. The promoter can be inducible, for example by the inclusion of the lac operator, as shown in the plasmid map in Fig. 4.1, which is activated upon the addition of isopropyl-β-D-thiogalactopyranoside (IPTG). The gene can also be fused in frame with a hexahistidine or other affinity tag at its N or C terminus, to enable purification of the expressed protein. This creates the expression construct. There are numerous commercially available expression vectors into which the gene of interest/expression cassette can be inserted.

7. A microorganism, usually the *Escherichia coli* BL21 strain, is transformed with the expression plasmid. Please see the linked protocol below. There are also several cell-free expression systems.

The plasmid modified here is the same plasmid used in the link below. The plasmid map shows the different parts of an expression vector and does not represent an actual plasmid that has been used to express the MTRAP antigen.

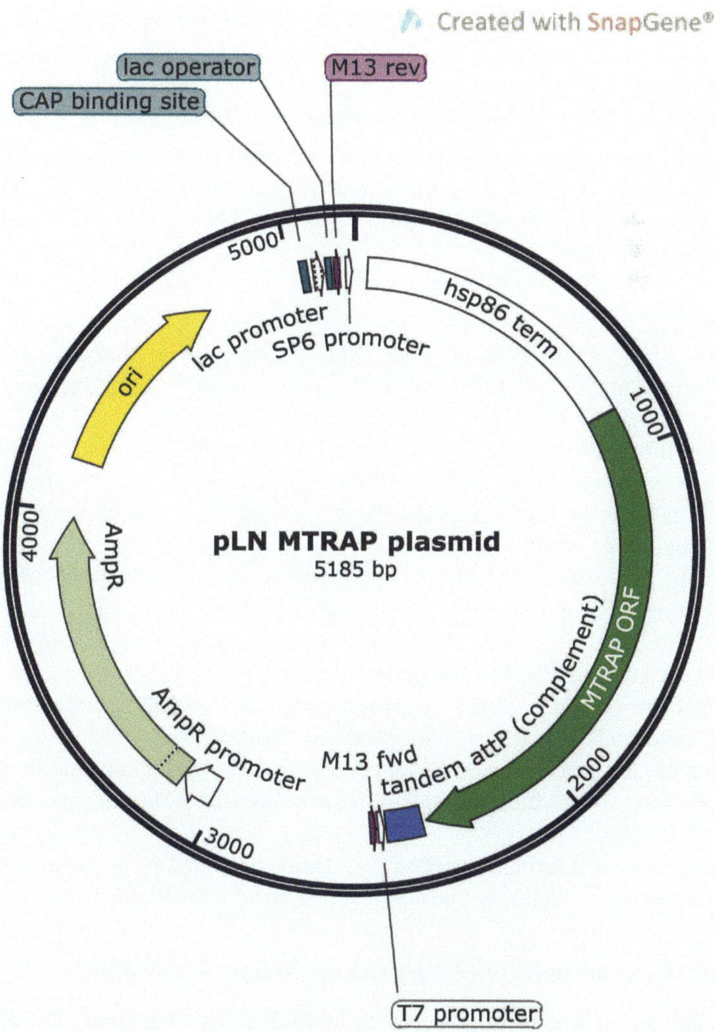

FIGURE 4.1 MTRAP (a malaria antigen) in the multiple cloning site between T7 and SP6 polymerase transcription initiation sequences. There is no affinity tag in this plasmid that can be used to purify the expressed protein.

Restriction cloning can be used to create a luciferase reporter vaccine, i.e. a genetically modified whole-cell vaccine that expresses luciferase. The resulting genetically-modified organism (GMO) can be used to study the bio-distribution of a vaccine in order to determine the best route of administration. A step-by-step process for generating a transgenic line is presented here (https://doi.org/10.17504/protocols.io.btsxnnfn). Whole sporozoite vaccines have been evaluated in this way (Ploemen et al., 2013).

Expression and purification of the recombinant antigen

There are various platforms for the expression of recombinant proteins as presented in Table 10.1 of the GMP chapter, but *E. coli* has become a particularly popular platform (Rosano & Ceccarelli, 2014) and is suitable for small-scale laboratory experiments. We will focus on using this organism transformed with a plasmid containing an IPTG inducible promoter. The following procedure can be used for the expression and purification of a recombinant target antigen.

Protocol

1. Grow the transformed *E. coli* BL21 strain in a shaking incubator (220 rpm) at 37°C in Luria Broth (LB) containing an antibiotic destroyed by an enzyme encoded by the expression construct.
2. When the optical density of the broth culture at 600 nm reaches 0.6–0.7, induce expression by 1 mM IPGT for 3 hours.
3. Harvest cells by centrifugation for 10 minutes at 5000 g. Wash in PBS and centrifuge again.
4. Disrupt the cells by sonication or use French Press Lysis at 1000 p.s.i. Enzymatic disruption of the cell membrane using lysozyme is also feasible. Alternatively, you can use a combination of both. Centrifuge and save the supernatant and the pellet.
5. Detect the presence of the target antigen in the soluble (supernatant) and insoluble (pellet) fractions using western blotting.
6. After the antigen has been detected, then a large-scale expression process is repeated to produce enough protein for purification and use in subsequent studies.
7. Size exclusion chromatography, or affinity purification using glutathione-S-transferase or histidine binding resins, can be used to purify the proteins.

Genetically modified whole organisms in vaccine studies

GMOs can be rationally engineered as attenuated live vaccines. Reasons for such engineering may include:

1. Generation of a mutant strain of the organism that will be unable to cause disease (e.g., by deleting genes that cause pathogenicity).

2. Generation of transgenic pathogens that can infect animals, containing target antigens from human pathogens. This is used to provide a proof of concept. For example, several chimeric rodent malaria parasites expressing human *Plasmodium* proteins have been developed (reviewed in Othman et al., 2017) and a quick protocol for generating these has been published (Salman et al., 2015).

GMOs can also be used to determine if an antigen is essential for the survival of the pathogen, which is a desirable attribute for a vaccine candidate, for example, by genetic knockout of the antigenic target. Immune responses targeting essential proteins are more likely to prevent infection than those targeting redundant pathways.

Designing genetically engineered live-attenuated vaccines

This approach involves attenuating the virulence of a pathogen, through genetic modification of specific gene(s) required by the pathogen to cause disease or to survive long enough to cause disease. The approach has been applied to bacteria and parasites. In the bacteria *Acinetobacter baumannii*, *Pseudomonas aeruginosa*, and *Staphylococcus aureus*, the genes encoding glutamate racemase and D-amino acid transaminase have been inactivated so that the resulting mutant is incapable of converting specific substrates to D-glutamate. D-Glutamate is an essential component of bacteria peptidoglycan, and the live-attenuated vaccines in this study were D-Glu auxotrophic strains, preventing their long-term survival (Cabral et al., 2017).

Among the parasites, *Plasmodium* has been the focus of engineered live-attenuated vaccines. This is exemplified in the design of *P. falciparum* Δ*p52* sporozoites. The sporozoites of this malaria parasite have their *p52* gene disrupted using single cross-over integration, and their growth is arrested after hepatocyte invasion. Immunity is developed against the attenuated organism due to its ability to initiate and establish infection, just like the wild-type parasite, but the infection does not progress to the blood stage, which causes malaria symptoms. These sporozoites were referred to as *P. falciparum* genetically attenuated sporozoites (GAS), developed by Shahid Khan and his colleagues (van Schaijk et al., 2008) and have been also referred to as *P. falciparum* genetically attenuated parasites (*Pf*GAP).

A $p36^{-}/p52^{-}$ double knockout line of *P. falciparum* became the first GAP to be used in a human trial as a whole organism malaria vaccine candidate (Spring et al., 2013), because for *P. falciparum* Δ*p52* GAS reversal to the wild-type parasite can occur at low frequency that could result in breakthrough infections (van Schaijk et al., 2008). Unfortunately, there was one breakthrough infection out of the six volunteers in the $p36^{-}/p52^{-}$ trial (Spring et al., 2013), but the production of antibodies and cytokines was induced. This problem was solved by knocking out an additional gene, *sap1*, in PfGAP3KO, a potential whole organism vaccine candidate

(Kublin et al., 2017), although cryopreservation may reduce its efficiency as demonstrated for *P. berghei* (Prinz et al., 2018). It is noteworthy that gene disruptions in *Plasmodium* can also be achieved using the novel FRT/FLP-recombinase system (van Schaijk et al., 2010) or CRISPR-Cas9 (Cui et al., 2021). Both GAS/GAP and radiation attenuated sporozoites (RAS) induce comparable sterile immunity (Finney et al., 2014), but the advantage of GAS/GAP over RAS is that uniformity of dosage (homogeneity of the parasite genotype/phenotype) is not an issue, and so scaling up parasite production is easier (van Schaijk et al., 2008).

After generation of GAS, the parasites have to be characterized, by demonstrating that correct disruption of the target gene(s) was achieved, assessing the adequacy of the attenuation and also hepatocyte invasion and subsequent growth arrest. Then their ability to provide immunity will be determined by measuring immune parameters, for example, using fluorescence-activated cell sorting (FACS) in a flow cytometer to measure lymphocyte induction (van Schaijk et al., 2008). Animal models provide a good understanding of a potential vaccine candidate, so before *Pf*GAP was designed, p52 and/or its ortholog p36p (which are syntenic on chromosome 10) were deleted in the rodent malaria parasites *P. berghei* (van Dijk et al., 2005) or *P. yoelii* (Labaied et al., 2007) and assessed. In mice, *P. berghei* Δp52 sporozoites (referred to as *P. berghei* p52$^-$GAP) induce a long-lasting effector memory CD8 + T cell response in the liver (Douradinha et al., 2011), and this has been used as a correlate of protection in assessing candidate vaccines comprising GAS. Antibodies have been also found to contribute to the induced sterile immunity (Finney et al., 2014; Spring et al., 2013). The ability of such engineered whole organisms to induce immunity can be evaluated using the methods discussed below.

Evaluation of a genetically attenuated parasite as an engineered live-attenuated vaccine

Successful knockout of the target gene(s) can be assessed by genotyping, using PCR or southern blotting. Characterization also involves testing the ability of the parasite to produce gametocytes, which are needed to produce sporozoites in the mosquito. The ability of sporozoites to move toward hepatocytes and invade them is determined by analyzing their gliding motility and their ability to traverse the hepatocyte and establish an infection. Finally, because reversal to the wild-type parasite is a possibility, the adequacy of the attenuation (is determined by analyzing development in the hepatocyte, using immunofluorescence, or assessed in vivo by infecting model animals and monitoring for any disease manifestation (Annoura et al., 2012).

The ability of the characterized GAS/GAP to produce sterile immunity or at least induce immunity against malaria can be evaluated using various methods that evaluate cellular and humoral immune responses, both of which are required.

Methods used to evaluate cellular responses include the assessment of CD4 and CD8 cytokine production. These cytokines can be detected in leukocytes isolated from the liver and spleen after an experimental vaccination in animals, or peripheral blood mononuclear cells (PBMCs) derived from the whole blood of volunteers in a clinical study. Intracellular cytokine staining (ICS) coupled with flow cytometry is a popular method, which is discussed below. Quantitative PCR (qPCR) has also been used to measure cytokine responses.

In evaluating humoral responses, analysis of the plasma concentration(s) of antibodies to the immunodominant *P. falciparum* sporozoites antigens, such as circumsporozoite protein (CSP) and apical membrane antigen 1 (AMA1), can be carried out by using enzyme-linked immunosorbent assay (ELISA) or western blotting. The details of these techniques are discussed below. These approaches should be followed by evaluating the ability of the antibodies to inhibit sporozoite invasion of hepatocytes. This traditionally involves a microscopy-based assessment of inhibition, however, a medium-throughput, more reproducible flow cytometry–based inhibition of sporozoite invasion assay can also be used (Kaushansky et al., 2012). Alternatively, for more robust analysis, sporozoites expressing a reporter protein [e.g., green fluorescence protein (GFP)] can be used via a flow cytometric readout (Rodríguez-Galán et al., 2017).

Animal models and controlled human malaria infection models can be used to directly measure vaccine efficacy, providing that the proper ethical approval and consent are obtained.

Basic protein studies

Enzyme-linked immunosorbent assay (ELISA)

ELISA is an assay technique used to detect small amounts of antibodies, antigens, peptides, proteins, glycoproteins, and viruses in biological samples (Alhajj & Farhana, 2020; Quintero-Ronderos et al., 2013). ELISAs can be used to assess the humoral immune response, where B cells recognize foreign bodies (antigens) and produce antibodies against these antigens. Due to the sensitivity and specificity of ELISAs, this assay is widely used in vaccine research to evaluate antigenic targets, measure the immune response, and evaluate vaccination-induced herd immunity. There are different types of ELISAs including direct ELISA, indirect ELISA, competitive ELISA, and sandwich ELISA (Fig. 4.2) (Alhajj & Farhana, 2020).

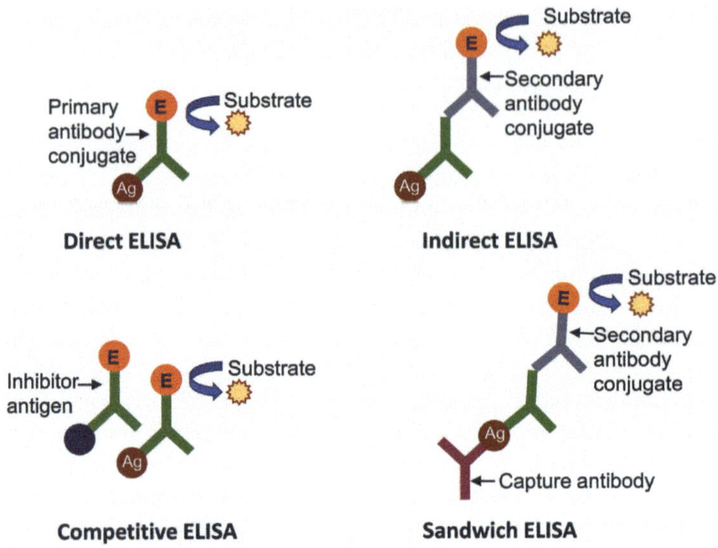

FIGURE 4.2 A schematic representation of the different types of ELISAs. The different types of ELISA include direct ELISA, indirect ELISA, competitive ELISA, and sandwich ELISA. *E*, Enzyme; *Ag*, antigen.

Direct ELISA

In a direct ELISA, an unknown quantity of antigen is immobilized on a microtiter plate. A specific primary antibody conjugated to an enzyme such as horseradish peroxidase (HRP) is then added to a plate to allow it to react with the antigen. Subsequently, a colorless substrate is added to the plate producing a colored product (considered as a positive result) that is measured by a spectrophotometer or a plate reader (Berg et al., 2002; Sakamoto et al., 2018). The colored product is proportional to the amount of antigen bound in the initial reaction. Direct ELISAs are quicker than the other types of ELISAs since this technique requires only a few steps and only one antibody is used. However, direct ELISAs can have a high background since the immobilization of antigens is unspecific.

Indirect ELISA

Indirect ELISAs involve immobilization of an antigen on a microtiter plate, similar to direct ELISAs. An unconjugated primary antibody is added to the plate and this binds the immobilized antigen. A secondary antibody conjugated to an enzyme is then added, which binds the primary antibody. The substrate is then added for color development. Indirect ELISAs are commonly used in vaccinology to measure antibody responses, with serum samples providing the primary antibody. Indirect ELISAs are highly sensitive and cheaper than direct ELISA since fewer labeled antibodies are needed. However, cross-reactivity between secondary antibodies may occur during indirect ELISA (Alhajj & Farhana, 2020).

Competitive ELISA

Competitive ELISAs can detect the presence of antigens or antibodies, although in the context of vaccine development they are more commonly used to detect antibodies in serum. Typically, an enzyme conjugated antibody is used to detect antigen coated onto the ELISA plate. In the presence of test sample (antibodies from serum) there is competition for antigen binding between the unconjugated serum antibodies and the conjugated detection antibody, resulting in signal reduction. A positive test is therefore indicated by the absence or reduction of color, and indicates that antibodies are present in the serum. One of the major advantages of a competitive ELISA is that it is highly sensitive, requires no sample processing, and can detect antigens of small size, if the assay is formatted to detect antigen rather than antibodies (Kohl & Ascoli, 2017). However, competitive ELISAs can be complex and time-consuming to develop (Aydin, 2015).

Sandwich ELISA

Sandwich ELISAs are very effective in detecting the target antigen in an unknown sample. The target protein must have at least two antigenic sites that can bind two different antibodies. In a sandwich ELISA, a capture antibody specific to the target antigen is used to coat a microtiter plate. Hydrophobic interactions allow the antibody to adhere to the plate (Cafruny et al., 1986). The antigen of interest is then added to the plate. In some wells, a known amount of antigen is added to create a standard curve. The detection antibody conjugated to an enzyme is added to the plate. This is known as direct sandwich ELISA. In an indirect sandwich ELISA, an unlabeled detection antibody is added to the plate followed by the addition of a conjugated antibody. The capture antibody must be of different species to the detection antibody to prevent cross-reactivity with the conjugated antibody in indirect ELISAs. Following the addition of the detection antibody, the substrate is added for color development. Sandwich ELISAs are more sensitive than the direct and indirect ELISAs since the antigen of interest is detected by two antibodies. However, optimization of antibodies can be problematic and cross-reactivity between capture and detection antibodies might take place (Alhajj & Farhana, 2020).

Endpoint titer ELISA

Endpoint titer ELISAs can be used to monitor the immunogenicity of vaccine candidates. The endpoint titer is usually the highest or last dilution of a sample that gives a reading above the cut-off absorbance (which is the 3 standard deviations above the negative control) (Hartman et al., 2018; Miura et al., 2008). Endpoint titer ELISAs differ from standard ELISAs in that no standards are used in the assays.

Monitoring the immune response in vaccine research

After vaccination, the immune response to vaccines has to be assessed. The immune response can be analyzed by measuring the total amount of antibodies such as IgM, IgG, or IgA, usually in the serum of vaccinated patients. In the early stages of a primary antibody response, B cells secrete IgM into the blood (Clem, 2011; den Hartog et al., 2020). IgM only lasts a few weeks in the blood and detection of IgM usually indicates recent exposure to an infection or pathogen (den Hartog et al., 2020). Differentiation of B cells results in an Ig class switch recombination from IgM to IgG, IgA, or IgE (Roco et al., 2019; Siegrist, 2008). In the majority of vaccines, particularly vaccines given via the intramuscular route, B cells mainly produce IgG (den Hartog et al., 2020; Siegrist, 2008). The IgG response following vaccination can last for months or years. ELISAs are commonly used in vaccine research to measure the levels of antibodies, particularly IgM or IgG, in samples from immunized patients. Below are examples of ELISA used in malaria and coronavirus-19 (COVID-19) research to determine the antibody response following vaccination.

Malaria, a life-threatening disease caused by the *Plasmodium* parasite, is responsible for an estimated 229 million cases and over 409,000 deaths worldwide in 2019 (World Health Organization, 2020). Due to the severity of this disease, vaccines such as RTS,S/AS01E have been developed to reduce morbidity and mortality from malaria. RTS,S/AS01E is a virus like particle composed of the hepatitis B virus surface antigen protein and a large segment of the circumsporozoite protein (CSP) (Dobaño, Ubillos, et al., 2019). To determine the immunogenicity of RTS,S/AS01E, studies by Dobaño, Sanz, et al. (2019) used ELISAs to analyze CSP IgG plasma/serum concentrations following immunization of volunteers with the vaccine. CSP-specific IgG levels in the samples were detected using rabbit antihuman IgG conjugated to HRP (Dobaño, Ubillos, et al., 2019). Results showed a significant increase in IgG concentrations binding to CSP in the serum of immunized individuals and indicated that the RTS,S/AS01E vaccine produced a strong IgG response (Dobaño, Sanz, et al., 2019). This study is an example of how ELISAs can be used to determine the immunogenicity of a vaccine.

The novel SARS-CoV-2 virus causes coronavirus disease 19 (COVID-19) (Liu et al., 2020). COVID-19 has affected over 220 countries worldwide and was responsible for over 195 million cases and 4.2 million deaths worldwide as of July 28, 2021 (World Health Organization, 2021). SARS-CoV-2 has four structural proteins including spike (S), envelope (E), membrane (M), and nucleocapsid (N) (Liu et al., 2020; Yoshimoto, 2020). The S protein and the N protein are the main immunogens. Liu et al. (2020) used an ELISA based on the recombinant SARS-CoV-2 N protein (rN) and recombinant SARS-CoV-2 S protein (rS) to detect IgM and IgG antibodies in patients who tested positive for COVID-19 (Liu et al., 2020). The rN and rS ELISAs used sandwich and competitive ELISA principles to detect IgM and IgG, respectively, in the serum of patients. Results showed an increase in IgM and IgG levels against the rN and rS proteins were detected by both ELISAs, with the highest sensitivity detected 10 days after disease onset (Liu et al., 2020). The study indicated that rN- and rS-based ELISAs could be used as a diagnostic method for detection of COVID-19 particularly 10 days after disease onset (Liu et al., 2020). ELISAs can be used to evaluate the immunogenicity of COVID-19 vaccines and also to assess the prevalence of COVID-19 in a population after vaccination.

General sandwich ELISA protocol

Sandwich ELISAs are widely used in vaccine research; a detailed protocol of the sandwich ELISA technique is described below. Detailed protocols for direct, indirect, and competitive ELISA can be found in Kohl and Ascoli (2017).

Materials and methods

96-well plate, Eppendorf tubes, multichannel pipettes, pipettes, pipette tips, water bottle, troughs, aspirator, plate reader

Buffers

- Coating buffer, wash buffer—phosphate buffer saline (PBS)/Tween (0.05%), PBS, blocking buffer, diluent buffer—1% BSA (bovine serum albumin) in PBS, capture antibody, detection antibody, HRP or alkaline phosphatase (AP) or β-D-galactosidase
- Substrates—for HRP, substrates used are TMB (3,3′,5,5′-tetramethylbenzidine), o-phenylenediamine, and 5-aminosalicylic acid. For AP, the substrate used is p-nitrophenyl phosphate
- 2M H_2SO_4 stop solution

Preparation of samples

1. Centrifuge samples, for example, serum from patients at approximately 3000 rpm for 10 minutes at 4°C.
2. Aliquot supernatant of serum into Eppendorf tubes and store at −80°C.
3. Remove samples from −80°C to thaw overnight the day before the experiment.
4. After thawing, centrifuge samples at 10,000 rpm for 5 minutes at 4°C.

5. Dilute samples in diluent buffer or blocking buffer before using in experiments to obtain accurate results and to allow the sample absorbance values to fall within the range of the standard curve.

Sandwich ELISA procedure

1. Coat wells of a microtiter plate with a capture antibody diluted in the coating buffer overnight at 4°C (Fig. 4.3).
2. Wash the 96-well plate with wash buffer to remove any unbound antibodies. Remove excess buffer in microtiter plate using an aspirator (Fig. 4.3).
3. Coat wells with blocking buffer for approximately 1 hour to block all unbound sites, reduce background, and prevent false-positive results. Aspirate wells after blocking (Fig. 4.3).
4. Add reconstituted standards and diluted samples to appropriate wells (Fig. 4.3). Add positive and negative controls to the appropriate wells to validate the experiment. Run standards, samples, and controls in duplicate or triplicate for approximately 2 hours (duration depends on samples). A plate layout is needed in ELISA experiments to record standards and samples added to wells.
5. Aspirate each well and wash with washing buffer (Fig. 4.3), 4 times. To wash, fill each well with a wash buffer using a multichannel pipette or a squirt bottle. Remove liquid in wells after each wash step to remove unbound antigens and to reduce background noise. Excess liquid in wells should be aspirated or decanted following the last wash.
6. Add diluted detection antibody to each well (Fig. 4.3), cover the plates with an adhesive plastic, and incubate at room temperature for 2 hours. Repeat the wash procedure in step 5.

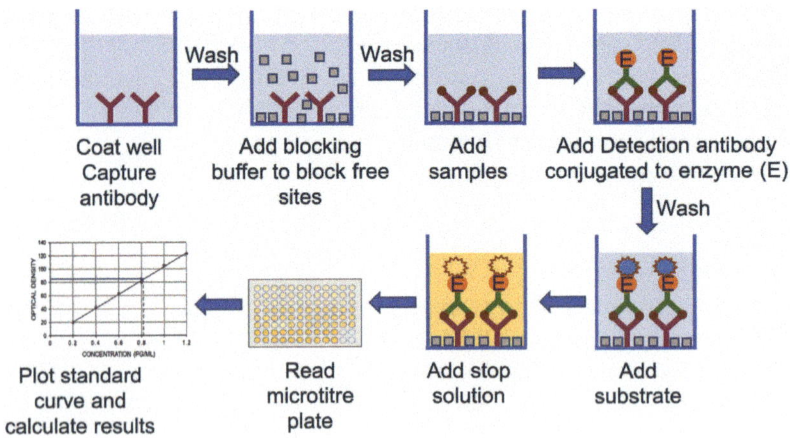

FIGURE 4.3 Schematic of sandwich ELISA procedure.

7. Add HRP (if detection antibody is not conjugated with HRP) for 30 minutes to 2 hours. Repeat the wash procedure in step 5.
8. Add substrate solution such as TMB to each well and incubate for about 30 minutes (exact time depends on the experiment and may require optimization) at room temperature in the dark (Fig. 4.3).
9. Add stop solution to each well (Fig. 4.3).
10. Determine the optical density of each well using a microplate reader at 450 nm (Fig. 4.3).

The microtiter plate is coated with a capture antibody overnight. Blocking buffer is added to each well and then samples are added. A detection antibody conjugated to an enzyme (E) is added and then the substrate is added. Stop solution is added and the microplate is read in a plate reader. The standard curve is plotted and results are calculated.

Data analysis

1. Calculate the average absorbance values of duplicate or triplicate readings.
2. Subtract the average zero standard (blank) from each standard, control, and sample.
3. To create the standard curve, plot the mean absorbance of standards (*Y*-axis) against the protein concentration (*X*-axis) using excel or a suitable computer program.
4. Draw the best fit curve through the points in the graph plotted (Fig. 4.4).
5. To calculate the concentration of target proteins in samples, extend a horizontal line from the absorbance value on the *Y*-axis to the standard curve (Fig. 4.4). At the point of intersection draw a vertical line to the *X*-axis and read the corresponding protein concentration (Fig. 4.4).

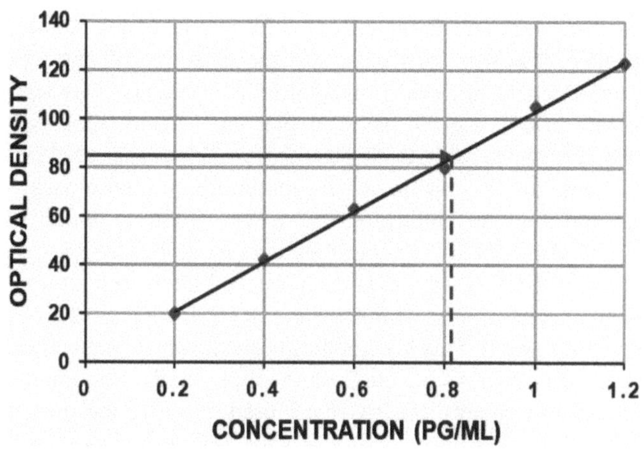

FIGURE 4.4 Schematic of a standard curve.

6. If diluted samples were used in the experiment, multiply the concentration obtained from the standard curve by the dilution factor.

An example of how to generate a standard curve and determine the concentration of the protein of interest in an experimental sample.

Western blotting

Western blotting is a sensitive analytical technique used to detect a specific protein, as well as posttranslational modifications, in a given sample (Mishra et al., 2017). In this technique, sodium dodecyl sulfate-polyacrylamide gel electrophoresis (SDS-PAGE) is used to separate a mixture of proteins based on molecular weight. Subsequently, proteins are transferred from the gel to a membrane [usually nitrocellulose or polyvinylidene fluoride (PVDF)] with a high affinity for proteins. The hydrophobic binding sites on the membrane are then blocked to reduce the nonspecific binding of detection antibodies to the membrane. The membrane is incubated with a primary antibody that is specific for the protein (antigen) of interest, forming an antigen–antibody complex. A secondary antibody that is specific to the host species of the primary antibody (e.g., if a primary antibody is raised in a goat, an antigoat secondary antibody is used) is added to the membrane. This antibody is usually conjugated to a fluorescent or chemiluminescent molecule, which will produce a detectable signal when a substrate is added. The signal intensity correlates with the abundance of the target protein, which can be visually assessed and also quantified using densitometry.

Detection of antibodies postvaccination and antigenic targets in vitro using western blotting

Western blotting is a sensitive assay that can be used to monitor the immune response after vaccination (Bisht et al., 2004; Yang et al., 2004). He et al. (2004) developed a western blot assay for the detection of antibodies against SARS-CoV infection using sera from SARS-CoV patients (Dowlatshahi et al., 2021; He et al., 2004). Results showed that the western blot assay demonstrated 98.3% specificity and 90.9% sensitivity in detecting antibodies against SARS-COV, comparable with that of an immunofluorescence assay (He et al., 2004). Data from the study indicated that western blotting could generally be used in the clinical diagnosis of viral infections including SARS-CoV and also in the detection of the antibody response produced after vaccination.

The leading vaccine candidate for malaria, RTS,S, contains the NANP repeat region of the CSP protein. This vaccine has been shown to induce partial protection against *P. falciparum* infection, and elicit antibodies against CSP in humans. Western blot assays are frequently used to verify the presence of specific antigens such as CSP in recombinant vaccines and are used as QC release tests during GMP manufacture. Kumar et al. (2013) developed a western blot assay for the detection of recombinant PfCSP in *P. falciparum* sporozoites from mosquito glands (Kumar et al., 2013). The assay was based on the immune reaction between anti-PfCSP monoclonal antibodies and the NANP repeat of PfCSP (Kumar et al., 2013). Results showed that the assay was sensitive and could detect very small quantities of recombinant PfCSP. The western blot assay can thus be used in estimating the immunological components present in a vaccine, and also in the quantification of CSP in sporozoite vaccines.

General western blot detailed protocol

Below is a detailed protocol of the western blotting technique.

Materials: Filter paper, membrane (PVDF, nitrocellulose), forceps;
Equipment: Power supply, gel-cutting knife, rocker platform, scanner, tank transfer system;
Buffers (Table 4.1): Lysis buffer, loading buffer, running buffer, 1 × transfer buffer, blocking buffer, washing buffer [1XTris-Buffered Saline with Tween 20 (TBST)], PBS.

Antibodies: Primary antibody, secondary antibody. The primary antibody used in western blot should be able to recognize denatured proteins and thus should not be conformationally dependent.

SDS PAGE gel: Detailed protocols of preparation of SDS PAGE gels can be found in a journal article written by Mahmood and Yang (2012).

Sample preparation

Proteins need to be extracted effectively from different samples such as cells or tissues before western blotting. Lysis buffers break down the cell membranes, releasing the intracellular contents of the cells. Protease and phosphatase inhibitors are commonly added to the lysis buffer to prevent the degradation of the extracted proteins at cold temperatures. The protein concentration of samples may be determined through colorimetric assays such as the Bradford assay (Desjardins et al., 2010) before being loaded onto an SDS-PAGE gel, to ensure equal loading of samples. Samples must be diluted in a loading buffer made up of a dye (such as bromophenol blue) and glycerol. Glycerol is added to the loading buffer to increase the density of samples and the dye is added to aid in the visualization of samples on the gel. Samples are then heated for approximately 5 minutes at 100°C to denature protein structure before loading.

Western blot procedure

The sequential steps needed to carry out a successful western blotting technique are highlighted in Fig. 4.5.

Samples are loaded onto the gel and electrophoresis is used to separate proteins according to their molecular weight. After gel electrophoresis, a

TABLE 4.1 Buffers required for western blotting.

1 × SDS-PAGE running buffer (pH 8.3)		
Chemical	**Final concentration**	**Weight/Volume**
Tris-base	25 mM	3.03 g
Glycine	192 mM	14.4 g
SDS	0.1% (w/v)	1 g
Makeup to 1 L with dH2O		
1X Transfer buffer		
Chemical	**Final concentration**	**Weight/Volume**
Tris-base	25 mM	6.06 g
Glycine	192 mM	28.8 g
Methanol	10% (v/v)	400 mL
Makeup to 2 L with dH2O		
Washing Buffer 1X TBST (pH 7.6)		
Chemical	**Final concentration**	**Weight/Volume**
Tris base	20 mM	2.4 g
NaCl	150 mM	8.0 g
dH2O	–	800 mL
Total volume		1 L
Sample Buffer		
Chemical	**Final concentration**	**Weight/Volume**
Tris HCl	150 mM	15 mL (of a 1 M stock)
Glycerol	150 mM	25 mL
SDS	12%	12 g
Bromophenol	0.05%	0.05 g
Total volume		100 mL

transfer sandwich is prepared and run in a transfer apparatus to enable the transfer of proteins from the gel to the membrane. The membrane is blocked and then incubated with a primary antibody. A secondary antibody is subsequently added. Enhanced Chemiluminescence (ECL) is added to the membrane and bands are visualized using a chemiluminescence imaging system.

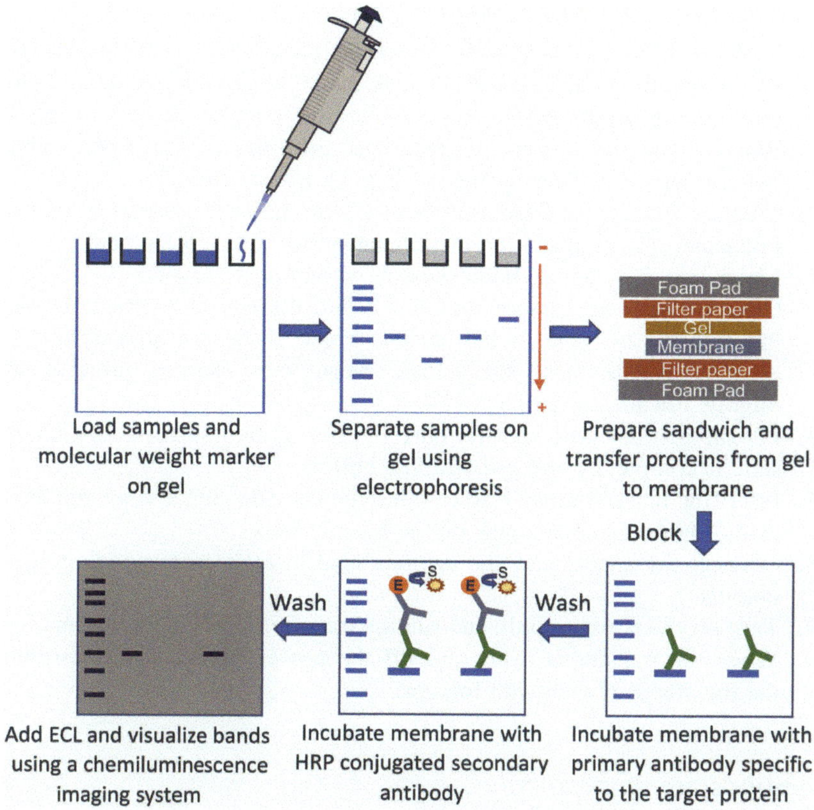

FIGURE 4.5 Schematic showing the sequential process of the western blot procedure.

Gel electrophoresis protocol

1. Place the gel in the electrophoresis apparatus and fill the buffer chambers with 1X running buffer to the indicator line on the chamber. Make sure the running buffer completely covers the gels.
2. Carefully remove combs from the gel and wash wells with running buffer.
3. Load equal amounts of protein into wells alongside a mixture of protein molecular weight markers. The loaded samples should include positive and negative controls. For quantitation, serial dilutions of a protein standard (usually the protein of interest) can be added into wells.
4. Place the lid on the gel tank and turn on the electrophoresis power pack.
5. Run the gel at 150 V for 2 hours or 200 V for 45 minutes or until the bromophenol blue dye is at the bottom of the gel.

Protocol for transfer of proteins

1. Soak membranes, foam pads, and filter papers in transfer buffer. If a PVDF membrane is being used, soak for 30 seconds in methanol before moving to transfer buffer.
2. Remove gel plates from the apparatus and separate the plates using a gel-cutting knife. Transfer the gel into 1X transfer buffer.
3. Create a transfer sandwich by installing the transfer cassette as follows: Filter pad → foam pad → gel membrane → filter pad → foam pad.
4. Make sure there are no bubbles between any of the layers.
5. Close the transfer cassette and slot it into the electrode assembly making sure the gel side of the cassette holder faces the cathode and the membrane side faces the anode. Negatively charged proteins will migrate toward the anode.
6. Fill the gel tank with transfer buffer ensuring the transfer sandwich cassette is completely covered with the buffer.
7. Insert the black cathode lead (−) into the cathode jack and the red anode lead (+) into the anode jack on the transfer tank.
8. Connect the anode lead and cathode lead to their corresponding power outputs.
9. Turn on the power supply and run the transfer at 100 V for 1 hour.
10. When the transfer is over, turn off the power supply and disassemble the transfer sandwich with forceps.

Immunostaining protocol

1. Stain membrane with Ponceau S to check if proteins have been transferred to the membrane.
2. Rinse off Ponceau S staining from the membrane with washing buffer.
3. Block membrane with blocking buffer for 1 hour.
4. Incubate the membrane with the primary antibody in blocking buffer at 4°C overnight or 1–2 hours at room temperature.
5. Wash the membrane three times with washing buffer for 5 minutes to remove unbound antibodies.
6. Incubate the membrane with secondary antibody in blocking buffer for 1–2 hours at room temperature.
7. Wash membrane twice for 5 minutes with washing buffer to remove any unbound antibodies.
8. Prepare ECL substrate following the manufacturer's recommendations. Apply substrate onto the blot for 1–5 minutes and remove excess reagent.
9. Cover membrane in plastic wrap. Remove air bubbles between the blot and the surface of the plastic wrap.
10. Image the blot using film or a cooled charge-coupled device.
11. Use an image analysis software to quantify the protein band intensity.

Flow cytometry

Flow cytometry is a technique used to measure the chemical and physical characteristics of single cells in solution. It is used in different fields such as molecular biology, immunology, virology, parasitology, and microbiology, in research and clinical settings. In flow cytometry, a cell suspension is injected into the central channel of a fluid system that is surrounded by a physiological buffer of fast-flowing fluid known as the sheath fluid (Suthanthiraraj & Graves, 2013).

As each cell goes through a laser beam, physical structures present in the cell may be struck, resulting in the scattering of light (Cossarizza & Chang, 2018). Forward scatter (FS) is measured by a detector in front of the light beam (McKinnon, 2018; Mealey & Long, 2018). Side scatter (SS) is measured by detectors found on the side of the light beam. FS is proportional to the size of the cell or cell surface (McKinnon, 2018; Mealey & Long, 2018). SS is proportional to cell granularity or internal complexity (Reggeti & Bienzle, 2011). FS and SS can be used to differentiate different cell types in a heterogeneous cell population. Although the light scatter signal is influenced by particle size, other factors such as the wavelength of the laser, refractive index, and sheath fluid can contribute to the light scatter signal (Kerker, 1983; Loken & Houck, 1981; Reardon et al., 2014; Salzman et al., 1979). The addition of fluorescent dyes, for example, propidium iodide or fluorochromes (e.g., fluorescein isothiocyanate, phycoerythrin) to cells can give quantitative and qualitative information relating to the structural properties of cells (Cossarizza & Chang, 2018; McKinnon, 2018). Fluorochromes are commonly conjugated to antibodies recognizing specific cell surface markers. In fluorescence flow cytometry, a peak is produced as the fluorescent labeled cells pass through the laser beam (Adan et al., 2017). Each fluorescent dye or fluorochrome has its characteristic excitation and emission spectra. Every cell in a sample that passes through the laser intercept is recorded as a distinct event (McKinnon, 2018). The intensity or amount of scattered light and fluorescent emissions is detected by photodiodes or photomultiplier tubes and converted into a voltage pulse (McKinnon, 2018). Voltage pulses are then converted to channel numbers and stored by the cytometer computer system for analysis by a flow cytometry analysis program (McKinnon, 2018; Mealey & Long, 2018).

Flow cytometry can be used to detect viral proteins such as the SARS-CoV-2 spike protein. Lapuente et al. (2020) used flow cytometry to detect IgG and IgM antibodies against SARS-CoV-2 spike proteins in human serum samples (Lapuente et al., 2020). Antibodies against SARS-CoV-2 present in the serum of patients bound to human embryonic kidney cells expressing the SARS-CoV-2 spike protein (Lapuente et al., 2020). Bound antibodies were then incubated with secondary detection antibodies and analyzed using flow cytometry (Lapuente et al., 2020). Results showed that the flow cytometry assay detected 100% of IgM and IgG positive samples, with similar sensitivity to commercial ELISA and chemiluminescent immunoassays kits (Lapuente et al., 2020). The flow cytometry assay was also highly sensitive in detecting antibodies in PCR-confirmed SARS-CoV-2 positive patients (Lapuente et al., 2020). Thus the flow cytometry–based serological assay developed by Lapuente et al. (2020) can be used in seroepidemiological studies and evaluation of immunity of patients after vaccination.

General flow cytometry protocol

A detailed general protocol on flow cytometry (Fig. 4.6) can be found below. This protocol was adapted from Biorad, and the Yale flow cytometry protocol (Biorad, 2016; Yale, 2019).

Samples are injected in sheath fluid, which focuses samples to the point of laser intercept. Fluorescence signals and forward and side scattered light are recorded for each cell as it passes through the laser beam.

Materials

Cells to be stained, PBS/BSA (containing 1% BSA), Fc blocking solution, antibodies (primary and secondary antibody), Fc block, and 5 mL falcon tubes.

Flow cytometry procedure

1. Harvest, wash cells, and adjust cell number to a concentration between 1×10^5 to 1×10^7 cells/mL in ice-cold PBS/BSA.
2. Place 100 μL of cell suspension in a tube.
3. Add primary antibody to the cell suspension and incubate for 30 minutes.
4. Wash cells with PBS/BSA three times.
5. Centrifuge at 300–400 g for 5 minutes at room temperature and discard the supernatant.
6. Add labeled secondary antibody and incubate in the dark for approximately 1 hour.
7. Wash cells with PBS/BSA three times.
8. Centrifuge at 300–400 g for 5 minutes at room temperature and discard the supernatant.
9. Include control tubes in the experiment, one tube should contain only the secondary antibody and the other tube should not contain an antibody.
10. Resuspend cells in 200 μL to 1 mL of cold (4°C) PBS or with 200 μL of 0.5 mL paraformaldehyde in PBS if needed.
11. Keep samples on ice and run samples on the flow cytometer.

In flow cytometry, unstained cells should be included in the experiment to check for autofluorescence. Fc block should be added to cells that express high levels of Fc receptors to avoid nonspecific binding. This is a general

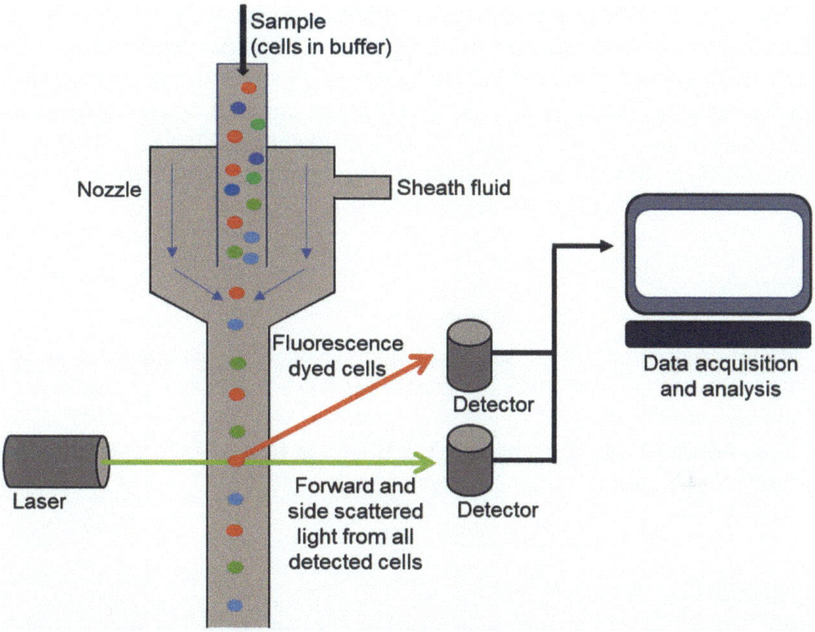

FIGURE 4.6 Schematic representation of flow cytometry.

protocol and hence cell preparations, controls, and antibody dilutions will need to be evaluated before the experiment.

Other methods of in vitro evaluation of immunogenicity

Cytokine analysis

Cytokines, including interferons (e.g., IFN-γ), tumor necrosis factor (e.g., TNF-α), and interleukins (e.g., IL-2), can be analyzed following an experimental vaccination schedule, EITHER DIRECTLY IN TISSUE/BLOOD, or by detecting and measuring their production IN VITRO FROM ISOLATED CELLS in the presence of the antigen. The aim of carrying out cytokine analysis is to determine how immunity is mediated and regulated in the presence of an antigen. Cytokines are secreted by specific cells of the immune system, as discussed in Chapter 1. Lymphoproliferative cellular responses can also be assessed to determine immune competence. Cytokines can be detected intracellularly, where the expression levels are determined by qPCR or ICS. Secreted cytokines can be measured using an enzyme-linked immunosorbent spot (ELISpot) assay, or a cytokine immunoassay kit. We will focus on ICS combined with polychromatic flow cytometry. This approach involves staining of stimulated PBMCs and then analysis with flow cytometry to detect expressed cytokines. It is a cutting-edge technique compared to

ELISpot and ELISA in the assay of cytokines; however, it needs to be customized compared to the fit-for-all ELISA or ELISpot kits (Smith et al., 2015). The method can be combined with ELISpot in the determination of T-cell responses (Bliss et al., 2017). Other techniques, some of which are more comprehensive in assessing T cell–mediated vaccine immunogenicity in terms of phenotypic and multiparametric readout, have been extensively reviewed elsewhere (Flaxman & Ewer, 2018).

Materials

- Freshly isolated whole blood/PBMCs.
- RPMI containing 10% heat-inactivated sterile FBS (screened for reactivity), 1% L-glutamine, and 1% pen/strep. AIM-V serum free medium can also be used.
- FACS buffer (PBS containing 0.1% BSA and 0.01% sodium azide).
- 5 mL FACS tubes.
- Target antigen.

Sample preparation

Freshly drawn whole blood [for use of whole blood, please see Hanekom et al. (2004)] or isolated PBMCs from heparinized whole blood can be used. To obtain PBMCs, use LymphoprepTM to carry out density gradient centrifugation. Resuspend the washed PBMCs in RPMI and use them immediately, or freeze in FBS containing 10% DMSO and store in liquid nitrogen to be assayed later. Perform cell counts and culture cells at 2×10^6 cells/mL in RPMI (below).

Procedure

1. If using frozen PBMCs, thaw rapidly in a water bath at 37°C and add Benzonase 25 U/10^6 PBMCs to allow them to acclimatize. Leave for 2 hours.
   - **a.** To stimulate the PBMCs, add the antigen(s), positive control (1 μg/mL Staphylococcal Enterotoxin B) or negative control (nothing).
   - **b.** Incubate overnight* at 37°C in FACs tubes, adding 1 μg/mL Brefeldin A (BD Biosciences) and 1 μg/mL Monesin (eBioscience) or 3 μg/mL Brefeldin A and Golgistop (1/2000 final concentration) after 2 hours of incubation, and incubating for further 16 hours to inhibit secretion.
   - **c.** Wash cells in FACs buffer and stain for viability with LIVE/DEAD Aqua amine-reactive dye for 20 minutes in the dark and at room temperature.
   - **d.** Wash cells again with FACs buffer and permeabilize for 20 minutes with cytofix/cytoperm. Wash in a 1:10 permeabilization buffer.
   - **e.** Add a cocktail of antibodies** for surface and intracellular staining and incubate for 30 minutes in the dark at room temperature***.

f. Wash again in permeabilization buffer and resuspend in PBS containing 1% paraformaldehyde.
g. Acquire images on a suitable flow cytometer, for example, BDLSR II, on the day of staining, followed by data analysis. In-depth training is required for flow cytometry, including data analysis, especially for ICS (Smith et al., 2015).

*Cells may need to be incubated for longer (up to 5 days) for weaker immunogens or very low-frequency responder cells (a pilot experiment can determine if this is needed). Inhibition of secretion should be done in the last 18 hours. **Co-stimulatory antibodies may need to be added. ***Surface staining may need to be done for longer or at 4°C depending on the antibodies used (see protocol provided with each antibody).

Purification of antibodies (IgGs)

After animals or study subjects have been immunized with a target antigen/vaccine candidate in preclinical or clinical studies, respectively, their immunoglobulins (Ig), specifically total IgG, is purified from sera and used to investigate their ability to inhibit pathogen infection in vitro or in vivo. IgG from the sera of control groups is also collected. Purification involves the use of column chromatography, with various commercial columns available each provided with the manufacturer's instructions. Examples of commercial columns and resins include protein G-Sepharose (Pharmacia), DEAE-trisacryl M, and TALON metal affinity resin, which binds to His-tagged proteins. Fractions from the columns are usually analyzed by SDS-PAGE or dot blot assays. Traditional antibody purification chromatography involves the exploitation of bacterially-derived protein A or G, which bind to Fc domains; a more recently developed approach employs the use of Fc-binding peptides. After the initial IgG separation, antigen-specific IgG can then be purified from the total IgG using antigen affinity chromatography. Specific chromatography columns can be generated by immobilizing antigens with a specific affinity tag including His6 or the four amino acid affinity tag C-tag E-P-E-A, on activated affinity resins. Purified antibodies can subsequently be examined using western blotting as explained in Section "Basic protein studies."

The next three protocols are focused on *Plasmodium* antibodies.

Growth inhibition assay

Growth inhibition assay (GIA), also referred to as invasion inhibition assay, measures the ability of antibodies to inhibit the replication of the merozoite form of the malaria parasite in red blood cells, using parasite lactate dehydrogenase activity as a measure of parasite growth (Illingworth et al., 2019). Slightly different versions of the assay are available, for example estimating inhibition of parasite invasion by microscopy (Mazumdar et al., 2010). The preerythrocytic stage of the parasite's life cycle involves the invasion of hepatocytes by sporozoites and the ability of antibodies to reduce such invasion can be determined using a similar assay (Shi et al., 1999). For a more detailed protocol, please refer to the GIA SOP from the European Malaria Reagent repository.

Materials

- Serum/purified IgG (test and preimmune or normal serum as control). Both test and control IgG concentration should be the same.
- Late trophozoite-staged *P. falciparum*, synchronized and grown in complete media containing 20% human serum and 25 ng/μL gentamycin, and at 0.2%–0.4% parasitemia and 1% Haematocrit (Hct).
- AMA1 purified immunoglobulin antibody pool (BG98) as a positive control may be included.
- Sterile 96-well flat-bottom microculture plates.
- Media: RPMI (also called incomplete media), complete media [RPMI plus some additives for *Plasmodium* culture (human serum at 10%)].
- Uninfected O^+ human erythrocytes.
- ELISA flat-bottomed plates.
- Sterile PBS.

Sample preparation

Buffer exchange the purified IgG three times into incomplete media with Amicon 10K MWCO centrifugal filtration devices. Preadsorb this purified IgG (excluding those purified by the antigen-specific method) by incubation with uninfected O^+ erythrocytes at 100% Hct (100 μL per 1 mL of antibody) for 1 hour to deplete antihuman erythrocyte antibodies. At the end of 1 hour, centrifuge at 1000*g* for 5 minutes and collect the supernatant. The supernatant can be dialyzed with a Pierce MINI dialysis device into RPMI, to remove any contaminating azide. Sterilize the IgG by filtering through a 0.22-μm filter and heat inactivate it at 56°C for 20 minutes. Determine the concentration of the IgG and store it at 4°C until used. Grow parasites in complete media containing 10% human serum before synchronization. After synchronization, estimate the volume of culture needed for the assay and add complete media to the parasites to give the required volume.

Procedure

1. Dilute IgG with preimmune or normal serum in 50 μL volumes, either a single concentration or more than one dilution.
2. 100 μL of incomplete PBS should be placed in the wells at the outer rim of the 96-well plate to prevent drying out of the cultures during incubation.

3. Mix 50 μL/well of the trophozoites with prediluted test or control antibody samples, and culture separately in the different wells 96-well plate (an ELISA plate can be used if materials are scarce). See Fig. 4.7 for an example of a plate set-up. Make a smear of parasite suspension to represent a time 0 smear. Include a control with normal (uninfected) RBCs with preimmune serum/control IgG, at the same Hct.
4. Incubate for 40 hours at 37°C in a similar gas mixture used to culture the parasites. Include a moistened tissue or container of sterile water to provide humidity.
5. At the end of the culture period, detect parasites by making smears and measure growth using the lactate dehydrogenase (LDH) method or flow cytometry. If a lot of assay plates are handled, place those not being analyzed at 4°C until needed.
6. If using the LDH method, fill the required number of wells in a new ELISA plate with 250 μL/well of ice-cold sterile PBS. Keep plate cold.
7. Transfer 50 μL of assay cultures starting from the one with the lowest concentration of IgG. Mix thoroughly before transferring into wells containing the ice-cold PBS. Keep the row and column position consistent. Return the plate to 4°C for flow cytometry (if doing this) and Giemsa staining.
8. Centrifuge the plate at 4°C, 1200 rcp for 10 minutes, and discard 240 μL of supernatant carefully.
9. Store what is left in the wells at −20°C or −80°C for lysis of the RBCs in the LDH assay. The plates should be covered and sealed with Saran wrap.

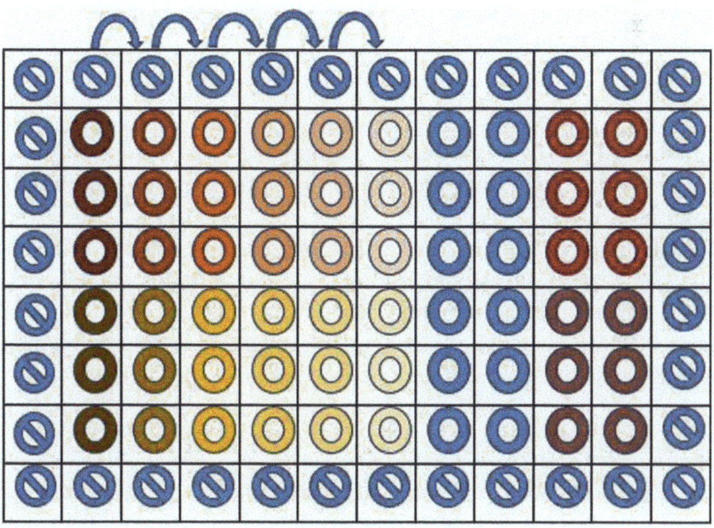

FIGURE 4.7 GIA 96-well plate setup.

Growth inhibition by the test IgGs is carried out in triplicate for five serial dilutions, and the undiluted stock. AMA1 IgGs can be tested alongside in five serial dilutions and stock. Serial dilutions are indicated by curved arrows. Test IgGs are represented in the upper six columns on the left-hand side, consisting of three rows each. AMA1 IgG is in the bottom six columns consisting of three rows each. The upper 2 columns (of three rows) on the right-hand side with the red circles are the controls with uninfected RBCs, while the bottom part is the control with *Plasmodium* iRBCs.

Data analysis

OD650 values obtained with test IgGs should be compared with those obtained with parasites incubated with control serum and with uninfected red cells. Percent inhibition for the test and positive control can be calculated as follows: 100 − [(OD650 of infected RBCs with tested IgG − OD650 of normal RBCs only)/(OD650 of infected RBCs with control IgG − OD650 of normal RBCs only) × 100].

Antibody-dependent cellular inhibition assay

This is an assay that evaluates inhibition of parasite growth by molecules secreted during monocyte phagocytosis. In 1990, Bouharoun-Tayoun et al. demonstrated that antibodies protective against malaria do not act directly but inhibit blood-stage *Plasmodium* growth and invasion through the help of monocytes (Bouharoun-Tayoun et al., 1990). This inhibition may be concentration-dependent (Mazumdar et al., 2010) and has been suggested as a good correlate of protection for vaccine studies (Tiendrebeogo et al., 2015).

Materials

- Purified test IgG (from the sera of immunized individuals)
- Purified control IgG (from the sera of an immune-naive individual)
- Human monocytes
- Schizont-stage *P. falciparum*, synchronized and cultured in complete media containing 10% serum, and at 0.5% parasitemia and 1% Hct
- 96-well flat-bottom microculture plate
- Media: RPMI (also called incomplete media), complete media [RPMI plus additives for *Plasmodium* culture (human serum at 10%)]

Sample preparation

Purified IgGs should be prepared as described earlier. Purify the monocytes from the blood samples of healthy human donors by using CD14 (monocyte/macrophage) microbeads (Miltenyi Biotech) or Lymphoprep.

Procedure

1. Distribute monocytes in each of the wells to be used at 2×10^6 and incubate for 90 minutes at 37°C in a 5% CO_2 gas mixture.
2. At the end of the incubation remove nonadherent cells by washing three times with RPMI.
3. Coculture the adherent monocytes with *P. falciparum-infected red blood cells (RBCs)* at the ratio of 200 RBCs per monocyte in complete media.
4. Add purified test IgG at the test concentration(s) to the cultures and incubate at 37°C for 96 hours in the same gas mixture used to culture *Plasmodium.*
5. After 48 and 72 hours of incubation, add 50 μL of complete media to each well.
6. At the end of the 96 hours, estimate the parasitemia of each well from a Giemsa-stained thin smear of pRBCs counting against 10,000 erythrocytes per smear. Parasitemia can also be determined by other methods (Shi et al., 1999).
7. Controls should include parasite alone, parasite + control IgG, parasite + monocytes, parasite + test IgG without monocytes, parasites + control IgG + monocytes.

Data analysis

Calculate the specific growth inhibitory index (SGI) as follows: SGI = 1 − [(percent parasitemia with monocytes and test IgG/percent parasitemia with test IgG)/(percent parasitemia with monocytes and control IgG/percent parasitemia with control IgG)] × 100.

Antibody-dependent respiratory burst activity

The antibody-dependent respiratory burst (ADRB) assay assesses the ability of antimerozoite antibodies to activate the release of reactive oxygen species (ROS) from human neutrophils via Fc receptors in response to *P. falciparum* blood-stage parasites. The assay complements GIA and antibody-dependent cellular inhibition. The production of ROS can effectively inhibit the growth of intracellular parasites, including *Plasmodium.* Several studies have demonstrated the usefulness of the ADRB activity as a surrogate of protection. This method has recently been standardized (Llewellyn et al., 2015) such that it can now be used comparatively across various laboratories in the malaria research community around the world.

RNA expression analysis

Other studies that can be used to evaluate the response to vaccination include microarray analysis and qPCR, which analyze the transcriptional profile of immune cells. These approaches can help to monitor immunogenicity and microbial burden in challenge studies, and estimate vaccine efficacy in the field, as well as during vaccine development. They can be used to give an insight into markers that correlate with protection following natural infection or controlled infection experiments. For example, PBMCs can be collected before sporozoite challenge, and protection/delay to parasitemia compared to control can be correlated with the transcriptional profile. RNA can be extracted from stimulated PBMCs (see the subsection on cytokine analysis) and used to carry out microarray and/or qPCR (Dunachie et al., 2015). RNA analysis has also been used to determine which factors contribute to the variable response to the hepatitis B vaccine. IL-27 and C-X-C motif chemokine 12 were found to be lower in nonresponders after the boost immunization (Qiu et al., 2018).

Conclusion

After a vaccine candidate has been designed and selected for investigation, the recombinant protein or genetically attenuated live vaccine can then be produced using a varity of approaches. The ability of the subunit protein or whole organism to provide protection against disease manifestation or infection can be evaluated by determining the nature and level of immune responses that are induced in vivo. Various methods of evaluating these responses may either be generally applicable to any vaccine, or specific to the particular pathogen under consideration.

References

Adan, A., Alizada, G., Kiraz, Y., Baran, Y., & Nalbant, A. (2017). Flow cytometry: Basic principles and applications. *Critical Reviews in Biotechnology*, *37*(2), 163–176. Available from https://doi.org/10.3109/07388551.2015.1128876.

Alhajj, M., & Farhana, A. (2020). *Enzyme linked immunosorbent assay. StatPearls*. StatPearls Publishing. Available from http://www.ncbi.nlm.nih.gov/pubmed/32310382.

Annoura, T., Ploemen, I. H. J., van Schaijk, B. C. L., Sajid, M., Vos, M. W., van Gemert, G.-J., Chevalley-Maurel, S., Franke-Fayard, B. M. D., Hermsen, C. C., Gego, A., Franetich, J.-F., Mazier, D., Hoffman, S. L., Janse, C. J., Sauerwein, R. W., & Khan, S. M. (2012). Assessing the adequacy of attenuation of genetically modified malaria parasite vaccine candidates. *Vaccine*, *30*(16), 2662–2670. Available from https://doi.org/10.1016/j.vaccine.2012.02.010.

Aydin, S. (2015). A short history, principles, and types of ELISA, and our laboratory experience with peptide/protein analyses using ELISA. *Peptides*, *72*, 4–15. Available from https://doi.org/10.1016/j.peptides.2015.04.012.

Berg, J., Tymoczko, J., & Stryer, L. (2002). *Immunology provides important techniques with which to investigate proteins. Biochemistry* (5th ed.). W H Freeman.

Bertero, A., Brown, S., Vallier, L., Jalali, M., Saldanha, F. Y. L., & Jalali, M. (2017). *Chapter 2 - Methods of cloning* (pp. 19–39). Academic Press. Available from https://doi.org/10.1016/B978-0-12-803077-6.00002-3.

Biorad. (2016). *Indirect immunofluorescence staining of cells & blood protocol | Bio-Rad*. Available from https://www.bio-rad-antibodies.com/indirect-immunofluorescence-staining-of-cells-blood-for-flow-cytometry.html.

Bisht, H., Roberts, A., Vogel, L., Bukreyev, A., Collins, P., Murphy, B., Subbarao, K., & Moss, B. (2004). Severe acute respiratory syndrome coronavirus spike protein expressed by attenuated vaccinia virus protectively immunizes mice. *Proceedings of the National Academy of Sciences of the United States of America, 101*(17), 6641–6646. Available from https://doi.org/10.1073/pnas.0401939101.

Bliss, C. M., Drammeh, A., Bowyer, G., Sanou, G. S., Jagne, Y. J., Ouedraogo, O., Edwards, N. J., Tarama, C., Ouedraogo, N., Ouedraogo, M., Njie-Jobe, J., Diarra, A., Afolabi, M. O., Tiono, A. B., Yaro, J. B., Adetifa, U. J., Hodgson, S. H., Anagnostou, N. A., Roberts, R., ... Ewer, K. J. (2017). Viral vector malaria vaccines induce high-level T cell and antibody responses in West African children and infants. *Molecular Therapy, 25*(2), 547–559. Available from https://doi.org/10.1016/j.ymthe.2016.11.003.

Bouharoun-Tayoun, H., Attanath, P., Sabchareon, A., Chongsuphajaisiddhi, T., & Druilhe, P. (1990). Antibodies that protect humans against *Plasmodium falciparum* blood stages do not on their own inhibit parasite growth and invasion in vitro, but act in cooperation with monocytes. *Journal of Experimental Medicine, 172*(6), 1633–1641. Available from https://doi.org/10.1084/jem.172.6.1633.

Brinks, V., Jiskoot, W., & Schellekens, H. (2011). Immunogenicity of therapeutic proteins: The use of animal models. *Pharmaceutical Research, 28*, 2379–2385.

Büyükköroğlu, G., Dora, D. D., Özdemir, F., & Hizel, C. (2018). *Techniques for protein analysis. Omics technologies and bio-engineering: Towards improving quality of life* (Vol. 1, pp. 317–351). Elsevier. Available from https://doi.org/10.1016/B978-0-12-804659-3.00015-4.

Cabral, M. P., García, P., Beceiro, A., Rumbo, C., Pérez, A., Moscoso, M., & Bou, G. (2017). Design of live attenuated bacterial vaccines based on D-glutamate auxotrophy. *Nature Communications, 8*, 15480. Available from https://doi.org/10.1038/ncomms15480.

Cafruny, W. A., Heruth, D. P., Jaqua, M. J., & Plagemann, P. G. W. (1986). Immunoglobulins that bind to uncoated ELISA plate surfaces: Appearance in mice during infection with lactate-dehydrogenase-elevating virus and in human anti-nuclear antibody-positive sera. *Journal of Medical Virology, 19*(2), 175–186. Available from https://doi.org/10.1002/jmv.1890190211.

Carrat, F., & Flahault, A. (2007). Influenza vaccine: The challenge of antigenic drift. *Vaccine, 25*(39–40), 6852–6862. Available from https://doi.org/10.1016/j.vaccine.2007.07.027.

Clem, A. S. (2011). Fundamentals of vaccine immunology. *Journal of Global Infectious Diseases, 3*(1), 73–78. Available from https://doi.org/10.4103/0974-777X.77299, Wolters Kluwer – Medknow Publications.

Cossarizza, A., & Chang, H. (2018). Guidelines for the use of flow cytometry and cell sorting in immunological studies (second edition). *Physiology & Behavior, 176*(1), 139–148. Available from https://doi.org/10.1117/12.2549369.Hyperspectral.

Cui, Z., Zhenkui, L., Huiting, C., Yuanyuan, J., Zhenke, Y., Xu, W., Han, G., Cong, L., Shujia, Z., Xin-zhuan, S., Jing, Y., David, S. L., Andrew, W., & Richard, C. (2021). Systematic CRISPR-Cas9-mediated modifications of *Plasmodium yoelii* ApiAP2 genes reveal functional insights into parasite development. *MBio, 8*(6), e01986. Available from https://doi.org/10.1128/mBio.0198617, 17.

Cunningham, A., Garçon, N., Friedland, L., Strugnelle, R., Laupèze, B., Doherty, M., & Sterng, P. (2016). Vaccine development: From concept to early clinical testing. *Vaccine, 34*(52), 6655–6664. Available from https://doi.org/10.1016/j.vaccine.2016.10.016.

den Hartog, G., Schepp, R. M., Kuijer, M., GeurtsvanKessel, C., van Beek, J., Rots, N., Koopmans, M. P. G., van der Klis, F. R. M., & van Binnendijk, R. S. (2020). SARS-CoV-2–specific antibody detection for seroepidemiology: A multiplex analysis approach accounting for accurate seroprevalence. *The Journal of Infectious Diseases*, *222*(9), 1452–1461. Available from https://doi.org/10.1093/infdis/jiaa479.

Desjardins, P., Hansen, J. B., & Allen, M. (2010). Microvolume protein concentration determination using the NanoDrop 2000c Spectrophotometer. *Journal of Visualized Experiments*, *33*, 1610. Available from https://doi.org/10.3791/1610.

Dobaño, C., Sanz, H., Sorgho, H., Dosoo, D., Mpina, M., Ubillos, I., Aguilar, R., Ford, T., Díez-Padrisa, N., Williams, N. A., Ayestaran, A., Traore, O., Nhabomba, A. J., Jairoce, C., Waitumbi, J., Agnandji, S. T., Kariuki, S., Abdulla, S., Aponte, J. J., ... Daubenberger, C. (2019). Concentration and avidity of antibodies to different circumsporozoite epitopes correlate with RTS,S/AS01E malaria vaccine efficacy. *Nature Communications*, *10*(1), 1–13. Available from https://doi.org/10.1038/s41467-019-10195-z.

Dobaño, C., Ubillos, I., Jairoce, C., Gyan, B., Vidal, M., Jiménez, A., Santano, R., Dosoo, D., Nhabomba, A. J., Ayestaran, A., Aguilar, R., Williams, N. A., Díez-padrisa, N., Lanar, D., Chauhan, V., Chitnis, C., Dutta, S., Gaur, D., Angov, E., ... Coppel, R. L. (2019). RTS, S / AS01E immunization increases antibody responses to vaccine-unrelated Plasmodium falciparum antigens associated with protection against clinical malaria in African children: A case-control study. *BMC Medicine*, *17*, 157.

Douradinha, B., van Dijk, M., van Gemert, G. J., Khan, S. M., Janse, C. J., Waters, A. P., Sauerwein, R. W., Luty, A. J., Silva-Santos, B., Mota, M. M., & Epiphanio, S. (2011). Immunization with genetically attenuated P52-deficient *Plasmodium berghei* sporozoites induces a long-lasting effector memory CD8 + T cell response in the liver. *Journal of Immune Based Therapies and Vaccines*, *9*(1), 6. Available from https://doi.org/10.1186/1476-8518-9-6.

Dowlatshahi, S., Shabani, E., & Abdekhodaie, M. J. (2021). Serological assays and host antibody detection in coronavirus-related disease diagnosis. *Archives of Virology*, *166*(3), 715–731. Available from https://doi.org/10.1007/s00705-020-04874-2.

Dunachie, S., Berthoud, T., Hill, A. V. S., & Fletcher, H. A. (2015). Transcriptional changes induced by candidate malaria vaccines and correlation with protection against malaria in a human challenge model. *Systems Vaccinology: How Big Data Can Accelerate Vaccine Development*, *33* (40), 5321–5331. Available from https://doi.org/10.1016/j.vaccine.2015.07.087.

Epp, C., Kauth, C. W., Bujard, H., & Lutz, R. (2003). Expression and purification of *Plasmodium falciparum* MSP-142: A malaria vaccine candidate. *European Contest for Young Scientists on Protein Purification: From Gene to Functional Protein*, *786*(1), 61–72. Available from https://doi.org/10.1016/S1570-0232(02)00722-5.

Finney, O. C., Keitany, G. J., Smithers, H., Kaushansky, A., Kappe, S., & Wang, R. (2014). Immunization with genetically attenuated *P. falciparum* parasites induces long-lived antibodies that efficiently block hepatocyte invasion by sporozoites. *Vaccine*, *32*(19), 2135–2138. Available from https://doi.org/10.1016/j.vaccine.2014.02.055.

Flaxman, A., & Ewer, K. J. (2018). Methods for measuring T-cell memory to vaccination: From mouse to man. *Vaccines*, *6*(3). Available from https://doi.org/10.3390/vaccines6030043.

Hanekom, W. A., Hughes, J., Mavinkurve, M., Mendillo, M., Watkins, M., Gamieldien, H., Gelderbloem, S. J., Sidibana, M., Mansoor, N., Davids, V., Murray, R. A., Hawkridge, A., Haslett, P. A. J., Ress, S., Hussey, G. D., & Kaplan, G. (2004). Novel application of a whole blood intracellular cytokine detection assay to quantitate specific T-cell frequency in field studies. *Journal of Immunological Methods*, *291*(1), 185–195. Available from https://doi.org/10.1016/j.jim.2004.06.010.

Hartman, H., Wang, Y., Schroeder, H. W., Jr, & Cui, X. (2018). Absorbance summation: A novel approach for analyzing high-throughput ELISA data in the absence of a standard. *PLoS One*, *13*(6)e0198528. Available from https://doi.org/10.1371/journal.pone.0198528.

He, Q., Chong, K. H., Chng, H. H., Leung, B., Ling, A. E., Wei, T., Chan, S. W., Ooi, E. E., & Kwang, J. (2004). Development of a western blot assay for detection of antibodies against coronavirus causing severe acute respiratory syndrome. *Clinical and Diagnostic Laboratory Immunology*, *11*(2), 417–422. Available from https://doi.org/10.1128/CDLI.11.2.417-422.2004.

Illingworth, J. J., Alanine, D. G., Brown, R., Marshall, J. M., Bartlett, H. E., Silk, S. E., Labbé, G. M., Quinkert, D., Cho, J. S., Wendler, J. P., Pattinson, D. J., Barfod, L., Douglas, A. D., Shea, M. W., Wright, K. E., de Cassan, S. C., Higgins, M. K., & Draper, S. J. (2019). Functional comparison of blood-stage *Plasmodium falciparum* malaria vaccine candidate antigens. *Frontiers in Immunology*, *10*, 1254. Available from https://doi.org/10.3389/fimmu.2019.01254.

Kaushansky, A., Rezakhani, N., Mann, H., & Kappe, S. H. I. (2012). Development of a quantitative flow cytometry-based assay to assess infection by *Plasmodium falciparum* sporozoites. *Molecular and Biochemical Parasitology*, *183*(1), 100–103. Available from https://doi.org/10.1016/j.molbiopara.2012.01.006.

Kerker, M. (1983). Elastic and inelastic light scattering in flow cytometry. *Cytometry*, *4*(1), 1–10. Available from https://doi.org/10.1002/cyto.990040102.

Kohl, T. O., & Ascoli, C. A. (2017). Direct competitive enzyme-linked immunosorbent assay (ELISA). *Cold Spring Harbor Protocols*, *2017*(7), 564–568. Available from https://doi.org/10.1101/pdb.prot093740.

Kublin, J. G., Mikolajczak, S. A., Sack, B. K., Fishbaugher, M. E., Seilie, A., Shelton, L., VonGoedert, T., Firat, M., Magee, S., Fritzen, E., Betz, W., Kain, H. S., Dankwa, D. A., Steel, R. W. J., Vaughan, A. M., Noah Sather, D., Murphy, S. C., & Kappe, S. H. I. (2017). Complete attenuation of genetically engineered *Plasmodium falciparum* sporozoites in human subjects. *Science Translational Medicine*, *9*(371), eaad9099. Available from https://doi.org/10.1126/scitranslmed.aad9099.

Kumar, S., Zheng, H., Sangweme, D. T., Mahajan, B., Kozakai, Y., Pham, P. T., Morin, M. J., Locke, E., & Kumar, N. (2013). A chemiluminescent-western blot assay for quantitative detection of *Plasmodium falciparum* circumsporozoite protein. *Journal of Immunological Methods*, *390*(1–2), 99–105. Available from https://doi.org/10.1016/j.jim.2013.02.001.

Labaied, M., Harupa, A., Dumpit, R. F., Coppens, I., Mikolajczak, S. A., & Kappe, S. H. I. (2007). *Plasmodium yoelii* sporozoites with simultaneous deletion of P52 and P36 are completely attenuated and confer sterile immunity against infection. *Infection and Immunity*, *75*(8), 3758–3768. Available from https://doi.org/10.1128/IAI.00225-07.

Lapuente, D., Maier, C., Irrgang, P., Hübner, J., Peter, A. S., Hoffmann, M., Ensser, A., Ziegler, K., Winkler, T. H., Birkholz, T., Kremer, A. E., Steininger, P., Korn, K., Neipel, F., Überla, K., & Tenbusch, M. (2020). Rapid response flow cytometric assay for the detection of antibody responses to SARS-CoV-2. *European Journal of Clinical Microbiology and Infectious Diseases*, 1–9. Available from https://doi.org/10.1007/s10096-020-04072-7.

Letko, M., Marzi, A., & Munster, V. (2020). Functional assessment of cell entry and receptor usage for SARS-CoV-2 and other lineage B betacoronaviruses. *Nature Microbiology*, *5*, 562–569.

Liu, W., Liu, L., Kou, G., Zheng, Y., Ding, Y., Ni, W., Wang, Q., Tan, L., Wu, W., Tang, S., Xiong, Z., & Zheng, S. (2020). Evaluation of nucleocapsid and spike protein-based enzyme-linked immunosorbent assays for detecting antibodies against SARS-CoV-2. *Journal of Clinical Microbiology*, *58*(6). Available from https://doi.org/10.1128/JCM.00461-20.

Llewellyn, D., Miura, K., Fay, M. P., Williams, A. R., Murungi, L. M., Shi, J., Hodgson, S. H., Douglas, A. D., Osier, F. H., Fairhurst, R. M., Diakite, M., Pleass, R. J., Long, C. A., & Draper, S. J. (2015). Standardization of the antibody-dependent respiratory burst assay with human neutrophils and *Plasmodium falciparum* malaria. *Scientific Reports*, *5*(1), 14081. Available from https://doi.org/10.1038/srep14081.

Loken, M. R., & Houck, D. W. (1981). Light scattered at two wavelengths can discriminate viable lymphoid cell populations on a fluorescence-activated cell sorter. *Journal of Histochemistry & Cytochemistry*, *29*(5), 609–615. Available from https://doi.org/10.1177/29.5.7252128.

Mahmood, T., & Yang, P. C. (2012). Western blot: Technique, theory, and troubleshooting. *North American Journal of Medical Sciences*, *4*(9), 429–434. Available from https://doi.org/10.4103/1947-2714.100998.

Mazumdar, S., Mukherjee, P., Yazdani, S. S., Jain, S. K., Mohmmed, A., & Chauhan, V. S. (2010). *Plasmodium falciparum* merozoite surface protein 1 (MSP-1)-MSP-3 chimeric protein: Immunogenicity determined with human-compatible adjuvants and induction of protective immune response. *Infection and Immunity*, *78*(2), 872–883. Available from https://doi.org/10.1128/IAI.00427-09.

McKinnon, K. M. (2018). Flow cytometry: An overview. *Current Protocols in Immunology*, *2018*. Available from https://doi.org/10.1002/cpim.40, 5.1.1-5.1.11.

Mealey, R. H., & Long, M. T. (2018). *Mechanisms of disease and immunity. Equine internal medicine* (4th ed., pp. 3–78). Elsevier. Available from https://doi.org/10.1016/B978-0-323-44329-6.00001-2.

Mishra, M., Tiwari, S., & Gomes, A. V. (2017). *Protein purification and analysis: Next generation western blotting techniques. Expert review of proteomics* (Vol. 14, pp. 1037–1053). Taylor and Francis Ltd Issue 11. Available from https://doi.org/10.1080/14789450.2017.1388167.

Miura, K., Orcutt, A. C., Muratova, O. V., Miller, L. H., Saul, A., & Long, C. A. (2008). Development and characterization of a standardized ELISA including a reference serum on each plate to detect antibodies induced by experimental malaria vaccines. *Vaccine*, *26*(2), 193–200. Available from https://doi.org/10.1016/j.vaccine.2007.10.064.

Othman, A. S., Lin, J., Franke-Fayard, B. M., Kroeze, H., van Pul, F. J. A., Chevalley-Maurel, S., Ramesar, J., Marin-Mogollon, C., Jore, M. M., Morin, M. J., Long, C. A., Sauerwein, R., Birkett, A., Miura, K., Janse, C. J., & Khan, S. M. (2018). Expression of full-length *Plasmodium falciparum* P48/45 in *P. berghei* blood stages: A method to express and evaluate vaccine antigens. *Molecular and Biochemical Parasitology*, *224*, 44–49. Available from https://doi.org/10.1016/j.molbiopara.2018.07.009.

Othman, A. S., Marin-Mogollon, C., Salman, A. M., Franke-Fayard, B., Janse, C. J., & Khan, S. M. (2017). The use of transgenic parasites in malaria vaccine research. *Expert Review of Vaccines*, *16*(7), 685–697. Available from https://doi.org/10.1080/14760584.2017.1333426.

Ploemen, I. H., Chakravarty, S., van Gemert, G.-J. J., Annoura, T., Khan, S. M., Janse, C. J., Hermsen, C. C., Hoffman, S. L., & Sauerwein, R. W. (2013). *Plasmodium* liver load following parenteral sporozoite administration in rodents. *Vaccine*, *31*(34), 3410–3416. Available from https://doi.org/10.1016/j.vaccine.2012.09.080.

Prinz, H., Sattler, J. M., Roth, A., Ripp, J., Adams, J. H., & Frischknecht, F. (2018). Immunization efficacy of cryopreserved genetically attenuated *Plasmodium berghei* sporozoites. *Parasitology Research, 117*(8), 2487–2497. Available from https://doi.org/10.1007/s00436-018-5937-0.

Qiu, S., He, P., Fang, X., Tong, H., Lv, J., Liu, J., Zhang, L., Zhai, X., Wang, L., Hu, Z., & Yu, Y. (2018). Significant transcriptome and cytokine changes in hepatitis B vaccine non-responders revealed by genome-wide comparative analysis. *Human Vaccines & Immunotherapeutics, 14*(7), 1763–1772. Available from https://doi.org/10.1080/21645515.2018.1450122.

Quintero-Ronderos, P., Arango, M.-T., Castiblanco, J., Correa, N. E., & Montoya-Ortíz, G. (2013). *Analysis of proteins and antibodies. Autoimmunity: From bench to bedside.* El Rosario University Press. Available from https://www.ncbi.nlm.nih.gov/books/NBK459443/.

Reardon, A. J. F., Elliott, J. A. W., & McGann, L. E. (2014). Fluorescence as an alternative to light-scatter gating strategies to identify frozen-thawed cells with flow cytometry. *Cryobiology, 69*(1), 91–99. Available from https://doi.org/10.1016/j.cryobiol.2014.05.009.

Reggeti, F., & Bienzle, D. (2011). Flow cytometry in veterinary oncology. *Veterinary Pathology, 48*, 223–235. Available from https://doi.org/10.1177/0300985810379435.

Roco, J. A., Mesin, L., Binder, S. C., Nefzger, C., Gonzalez-Figueroa, P., Canete, P. F., Ellyard, J., Shen, Q., Robert, P. A., Cappello, J., Vohra, H., Zhang, Y., Nowosad, C. R., Schiepers, A., Corcoran, L. M., Toellner, K. M., Polo, J. M., Meyer-Hermann, M., Victora, G. D., & Vinuesa, C. G. (2019). Class-switch recombination occurs infrequently in germinal centers. *Immunity, 51*(2), 337–350. Available from https://doi.org/10.1016/j.immuni.2019.07.001, e7.

Rodríguez-Galán, A., Salman, A. M., Bowyer, G., Collins, K. A., Longley, R. J., Brod, F., Ulaszewska, M., Ewer, K. J., Janse, C. J., Khan, S. M., Hafalla, J. C., Hill, A. V. S., & Spencer, A. J. (2017). An in vitro assay to measure antibody-mediated inhibition of *P. berghei* sporozoite invasion against *P. falciparum* antigens. *Scientific Reports, 7*(1), 17011. Available from https://doi.org/10.1038/s41598-017-17274-5.

Rosano, G. L., & Ceccarelli, E. A. (2014). Recombinant protein expression in *Escherichia coli*: Advances and challenges. *Frontiers in Microbiology, 5*, 172. Available from https://doi.org/10.3389/fmicb.2014.00172.

Sakamoto, S., Putalun, W., Vimolmangkang, S., Phoolcharoen, W., Shoyama, Y., Tanaka, H., & Morimoto, S. (2018). Enzyme-linked immunosorbent assay for the quantitative/qualitative analysis of plant secondary metabolites. *Journal of Natural Medicines, 72*(1), 32–42. Available from https://doi.org/10.1007/s11418-017-1144-z.

Salman, A. M., Mogollon, C. M., Lin, J.-W., van Pul, F. J. A., Janse, C. J., & Khan, S. M. (2015). Generation of transgenic rodent malaria parasites expressing human malaria parasite proteins. *Methods in Molecular Biology (Clifton, N.J.), 1325*, 257–286. Available from https://doi.org/10.1007/978-1-4939-2815-6_21.

Salzman, G. C., Wilder, M. E., & Jett, J. H. (1979). Light scattering with stream-in-air flow systems. *Journal of Histochemistry & Cytochemistry, 27*(1), 264–267. Available from https://doi.org/10.1177/27.1.374583.

Shi, Y. P., Hasnain, S. E., Sacci, J. B., Holloway, B. P., Fujioka, H., Kumar, N., Wohlhueter, R., Hoffman, S. L., Collins, W. E., & Lal, A. A. (1999). Immunogenicity and in vitro protective efficacy of a recombinant multistage *Plasmodium falciparum* candidate vaccine. *Proceedings of the National Academy of Sciences of the United States of America, 96*(4), 1615–1620. Available from https://doi.org/10.1073/pnas.96.4.1615.

Siegrist, C.-A. (2008). *The immunology of vaccination. Vaccines* (5th ed., pp. 16–34). Saunders. Available from https://www.who.int/immunization/documents/Elsevier_Vaccine_immunology.pdf.

Smith, S. G., Smits, K., Joosten, S. A., van Meijgaarden, K. E., Satti, I., Fletcher, H. A., Caccamo, N., Dieli, F., Mascart, F., McShane, H., Dockrell, H. M., Ottenhoff, T. H. M., & TBVI TB Biomarker Working Group. (2015). Intracellular cytokine staining and flow cytometry: Considerations for application in clinical trials of novel tuberculosis vaccines. *PLoS One*, *10*(9)e0138042. Available from https://doi.org/10.1371/journal.pone.0138042.

Spring, M., Murphy, J., Nielsen, R., Dowler, M., Bennett, J. W., Zarling, S., Williams, J., de la Vega, P., Ware, L., Komisar, J., Polhemus, M., Richie, T. L., Epstein, J., Tamminga, C., Chuang, I., Richie, N., O'Neil, M., Heppner, D. G., Healer, J., … Kappe, S. H. I. (2013). First-in-human evaluation of genetically attenuated *Plasmodium falciparum* sporozoites administered by bite of Anopheles mosquitoes to adult volunteers. *Vaccine*, *31*(43), 4975–4983. Available from https://doi.org/10.1016/j.vaccine.2013.08.007.

Suthanthiraraj, P., & Graves, S. (2013). Fluidics. *Current protocols in cytometry*, *01*(Suppl. 65), Unit-Unit. Available from https://doi.org/10.1002/0471142956.cy0102s65.

Tan, A., Atack, J., Jennings, M., & Seib, K. (2016). The capricious nature of bacterial pathogens: Phasevarions and vaccine development. *Frontiers in Immunology*, *7*(586), 1–9. Available from https://doi.org/10.3389/fimmu.2016.00586.

Tiendrebeogo, R. W., Adu, B., Singh, S. K., Dziegiel, M. H., Nébié, I., Sirima, S. B., Christiansen, M., Dodoo, D., & Theisen, M. (2015). Antibody-dependent cellular inhibition is associated with reduced risk against febrile malaria in a longitudinal cohort study involving Ghanaian children. *Open Forum Infectious Diseases*, *2*(2). Available from https://doi.org/10.1093/ofid/ofv044, ofv044–ofv044.

van Dijk, M. R., Douradinha, B., Franke-Fayard, B., Heussler, V., van Dooren, M. W., van Schaijk, B., van Gemert, G.-J., Sauerwein, R. W., Mota, M. M., Waters, A. P., & Janse, C. J. (2005). Genetically attenuated, P36p-deficient malarial sporozoites induce protective immunity and apoptosis of infected liver cells. *Proceedings of the National Academy of Sciences of the United States of America*, *102*(34), 12194–12199. Available from https://doi.org/10.1073/pnas.0500925102.

van Schaijk, B. C. L., Janse, C. J., van Gemert, G.-J., van Dijk, M. R., Gego, A., Franetich, J.-F., van de Vegte-Bolmer, M., Yalaoui, S., Silvie, O., Hoffman, S. L., Waters, A. P., Mazier, D., Sauerwein, R. W., & Khan, S. M. (2008). Gene disruption of *Plasmodium falciparum* p52 results in attenuation of malaria liver stage development in cultured primary human hepatocytes. *PLoS One*, *3*(10), e3549. Available from https://doi.org/10.1371/journal.pone.0003549.

van Schaijk, B. C. L., Vos, M. W., Janse, C. J., Sauerwein, R. W., & Khan, S. M. (2010). Removal of heterologous sequences from *Plasmodium falciparum* mutants using FLPe-recombinase. *PLoS One*, *5*(11), e15121. Available from https://doi.org/10.1371/journal.pone.0015121.

Wang, N., Shang, J., Jiang, S., & Lanying, D. (2020). Subunit vaccines against emerging pathogenic human coronaviruses. *Frontiers in Microbiology*, *11*(298), 1–19. Available from https://doi.org/10.3389/fmicb.2020.00298.

World Health Organization. (2020). *World malaria report 2020: 20 years of global progress and challenges*. Available from https://www.who.int/docs/default-source/malaria/world-malaria-reports/9789240015791-double-page-view.pdf?sfvrsn = 2c24349d_5.

World Health Organization. (2021). *WHO Coronavirus (COVID-19) dashboard overview data table*. Available from https://covid19.who.int/.

Yale. (2019). *Indirect immunofluorescence staining of cells and blood for flow cytometry*. Available from https://medicine.yale.edu/immuno/flowcore/protocols/analysis/.

Yang, Z., Kong, W., Huang, Y., Roberts, A., Murphy, B., Subbarao, K., & Nabel, G. (2004). A DNA vaccine induces SARS coronavirus neutralization and protective immunity in mice. *Nature*, *428*, 561–564.

Yoshimoto, F. K. (2020). The proteins of severe acute respiratory syndrome Coronavirus-2 (SARS CoV-2 or n-COV19), the cause of COVID-19. *Protein Journal*, *39*(3), 198–216, Springer. Available from https://doi.org/10.1007/s10930-020-09901-4.

Further reading

Yang, C., Li, H., Lee, W., & Lo, S. (2021). Antibodies targeting two epitopes in SARS-CoV-2 neutralize pseudoviruses with the spike proteins from different variants. *Pathogens*, *10*(869), 1–10. Available from https://doi.org/10.3390/pathogens10070869.

Chapter 5

Harnessing the power of CD8$^+$ T-cells: identification and validation of peptides bound to major histocompatibility complex class I by immunopeptidomics

Camila R.R. Barbosa[1,2] and Paulo J.G. Bettencourt[3,4]
[1]Centre for Inflammation Biology and Cancer Immunology (CIBCI), Peter Gorer Department of Immunobiology, King's College London, London, United Kingdom, [2]Francis Crick Institute, London, United Kingdom, [3]Faculty of Medicine, Catholic University of Portugal, Lisbon, Portugal, [4]Center for Interdisciplinary Research in Health, Catholic University of Portugal, Lisbon, Portugal

CD8$^+$ T cells and major histocompatibility complex class I antigen presentation

Cellular immunity has a fundamental role in the immune response induced by many vaccines to combat disease. This is of particular relevance for diseases such as HIV and tuberculosis (TB), where protection cannot be conferred by antibodies, or to enhance protection when antibodies are insufficient, as is the case for malaria. Cellular immunity is mediated by T lymphocytes, also known as T cells. To participate in the adaptive immune response, a naïve T cell must first encounter an antigen, engage with the antigen presenting cell (APC), and then be induced to differentiate into effector cells capable of removing cells presenting these antigens. This arm of the immune system promotes the killing of infected cells, and in many cases will successfully eliminate reservoirs of infection.

A primary immune response occurs by the activation of naïve T cells in response to antigen with subsequent differentiation and proliferation. This response provides effector T cells and also generates immunological

Vaccinology and Methods in Vaccine Research. DOI: https://doi.org/10.1016/B978-0-323-91146-7.00008-1

memory, conferring protection from subsequent challenge by the same pathogen, which is the fundamental basis of vaccine development. While $CD4^+$ T cells are important to mediate many aspects of the adaptive immune response, $CD8^+$ T or cytotoxic T lymphocytes (CTL) are crucial in priming and killing host cells infected with virus, intracellular bacteria, and protozoans such as malaria (Bahjat et al., 2006; Junqueira et al., 2018; Wiesel et al., 2009). Therefore vaccines should be designed to stimulate a strong $CD8^+$ T cell response in order to improve efficacy, particularly in diseases caused by intracellular pathogens where antibodies are not effective, such as HIV, malaria, and TB (Rappuoli & Aderem, 2011).

Unlike antibodies, which recognize conformational epitopes exposed on the surfaces of proteins with tertiary structure, $CD8^+$ T cells recognize 8–14 amino acid peptide fragments presented at the plasma membrane of cells, associated with major histocompatibility complex class I (MHC-I) molecules (Gomez-Perosanz et al., 2020; Sanchez-Trincado et al., 2017). MHC-I molecules are expressed at the surface of all nucleated cells and can present peptide fragments derived from proteins of intracellular pathogens (Hewitt, 2003), and from the natural protein turnover resulting from cellular metabolism. These peptides can be discriminated as nonself and self-peptides, respectively (Klein et al., 2014), although a partial overlap between them has been hypothesized (Calis et al., 2012; Frankild et al., 2008; Mansurkhodzhaev et al., 2021; Mishto et al., 2021). Source antigens must be processed through a specialized pathway before being presented by the MHC molecule (Fig. 5.1).

To generate optimal peptide epitopes, two key processing events exist within the antigen presentation pathway. Proteins initially undergo fragmentation into peptides in the cytosol facilitated by the ubiquitin proteasome system, or in some cases by other cytoplasmic proteases. This processing event is often responsible for generating the final C-terminal residue of peptides that bind to MHC-I. The transport from the cytoplasm to the ER requires an ER-resident and heterodimeric peptide transporter, the TAP. Within the ER lumen, a macromolecular peptide-loading complex (PLC) is formatted to associate the MHC-I immature heavy chain and β-2-microglobulin. This event is coordinated by chaperones calnexin (CNX), calreticulin (CRT), and ERp57, and the peptide editor tapasin (Tpn) (Reeves & James, 2017). As TAP preferentially transports peptides of 8–16 amino acids, peptides are often too long for stable binding to MHC-I and require further processing. ERAP edits the peptide repertoire by trimming N-terminal extensions of antigenic peptides, creating a pool of peptides with high affinity for MHC-I binding (Yewdell et al., 2003). Finally, the MHC-I heterodimer is directed to the cell surface where it will present the processed antigen to the $CD8^+$ T cell (Fig. 5.1).

Identifying antigenic epitopes generated by the antigen presenting pathway is of great interest for a number of practical reasons, including understanding disease etiology, immune monitoring, developing diagnostic assays, and especially designing epitope-based vaccines (Cosma & Eisenlohr, 2018).

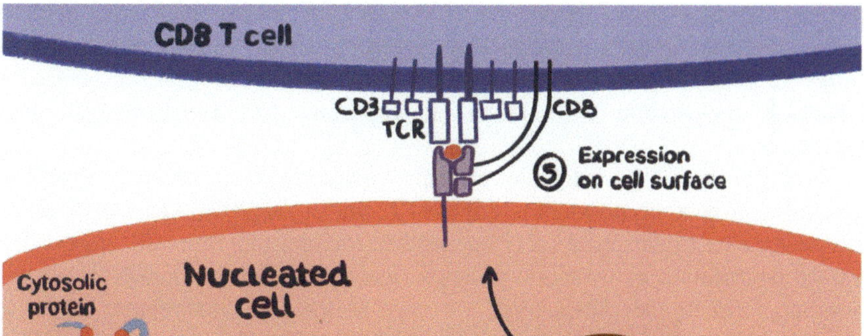

FIGURE 5.1 Antigen processing and presentation to CD8$^+$ T cells. Antigenic peptides are generated in the cytosol (1), cleaved by the proteasome (2), and transported to the lumen of the endoplasmic reticulum (ER) by TAP (3). The heavy chain of MHC-I and -2-mibroglobulin are assembled by chaperones and to form the PCL complex. An extra trimming of the peptides is performed by ERAP. After loading the peptide, the MHC-I heterodimer is released from the ER (4) and exported to the cell surface to present the antigen to CD8$^+$ T cells (5). The TCR/CD3/CD8 complex at the surface of the CD8$^+$ T cell recognizes and binds to the pMHC-I complex at the surface of the nucleated cell, establishing the immunological synapse and starting the cascade of events leading to CTL effector functions.

In this chapter, we will describe advanced tools for the identification of CD8$^+$ T cell-specific peptides, with particular focus on immunopeptidomics, which is of major importance for vaccine design and development.

Proteasome cleavage mechanisms

While the MHC-I antigen processing pathway involves degradation of cytosolic proteins in all nucleated cells, the MHC-II pathway processes and presents peptides derived from the endocytic pathway in APCs, such as dendritic cells, macrophages, and B cells. In the processing and presentation of antigenic peptides bound by the MHC-I molecule, the ubiquitin-proteasome system of eukaryotic cells plays an important role in proteolysis and degradation.

During processing, antigenic proteins covalently bind multiple ubiquitin molecules to become poly-ubiquitinylated (Song et al., 2009). The proteasome is an adenosine triphosphate (ATP)-dependent multisubunit protease, responsible for the degradation of cytosolic proteins (Rock & Goldberg, 1999). The most common form of the proteasome is known as the 26S proteasome, which is composed of a catalytic nucleus (20S) and two regulatory complexes (19S), located on each side of the 20S nucleus (Kloetzel, 2001). The catalytic activity of the proteasome is located in the β5 (X, LMP7), β2 (Z, MECL-1), and β1 (Y, LMP2) subunits of the 20S nucleus. The proteasome cleaves between residues known as P1 and P1′, with the P1 residue forming the C terminus of a peptide subsequently presented by MHC-I (Fig. 5.2) (Diez-Rivero et al., 2010).

Proteasomal degradation in the cytosol was experimentally shown to be the event, which generates the C-termini of most antigenic peptides (Craiu et al., 1997). The distribution of amino acids within peptide epitopes confirms a preference for hydrophobic and basic residues at the C termini (Altuvia & Margalit, 2000).

Interferon-gamma (IFN-γ), an important cytokine in cell-mediated immunity, enhances the efficiency of antigen presentation. In response to IFN-γ receptor activation, immune cells produce the immunoproteasome (Gaczynska et al., 1994). Simultaneously, IFN-γ can induce the expression of the TAP transporter, ERAP1 peptide trimming enzyme, and MHC molecules (Rock & Goldberg, 1999). This sequence of events will eventually induce an adaptive immune response.

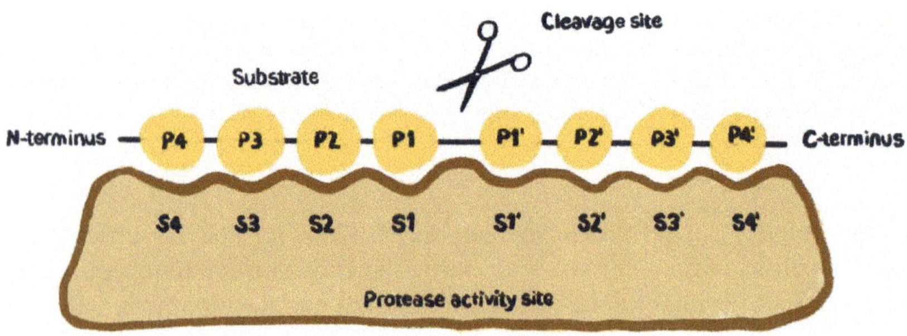

FIGURE 5.2 Peptide cleavage in the proteasome. Amino acid residues are named in function to its proximity of the cleavage site. Amino acid residues from the cleavage site towards the N-terminus are named P1-P2-P3-P4, whereas towards the C-terminus are named P1’-P2’-P3’-P4’. Similarly, the corresponding sites in the protease are named S1-S2-S3-S4 or S1’- S2’- S3’- S3’, respectively.

Pioneering biochemical experimentation suggested that the catalytic subunits of the immunoproteasome have greater ability to generate peptides with C-terminal residues that better match the terminal (anchor) amino acid residue preferences of MHC-I molecules (Toes et al., 2001), although more recent studies proposed that such differences may be related to the number of peptides produced rather than their variety (Calis et al., 2012; Frankild et al., 2008; Mansurkhodzhaev et al., 2021; Mishto et al., 2021). Mice lacking immunoproteasome catalytic subunits were shown to have defects in presenting epitopes via MHC-I (not MHC-II), with substantial differences in the MHC-I repertoire compared to wild type mice, leading to transplant rejection of wild type cells (Kincaid et al., 2012).

Proteasome-mediated peptide hydrolysis generates canonical MHC-I-bound peptides. A second mechanism that generates antigenic peptides has been described over the first two decades of the 21st century: proteasome-catalyzed peptide splicing (PCPS), which gives rise to noncanonical peptides, was described for the first time in vertebrates in 2004 (Hanada et al., 2004; Vigneron et al., 2004). This mechanism entails posttranslational excision of a polypeptide segment followed by ligation of the newly liberated carboxy-terminal and amino-terminal residues in the parental protein (Liepe et al., 2018; Vigneron et al., 2004). The two splice-reactants can be generated inside the proteasome by the cleavage of a single polypeptide (cis-PCPS). This event occurs by an excision of a peptide sequence, followed by ligation of the splice-reactants. When this takes place in the normal order, following the orientation from N terminus to C terminus of the parental protein, is named normal cis-PCPS; however it can also occur in the reverse order, giving rise to reverse cis-PCPS (Mishto & Liepe, 2017). Moreover, PCPS can also arise between two splice-reactants originated from two different proteins, named trans-PCPS (Fig. 5.3) (Berkers et al., 2015; Dalet et al., 2010; Faridi et al., 2018; Mishto et al., 2012).

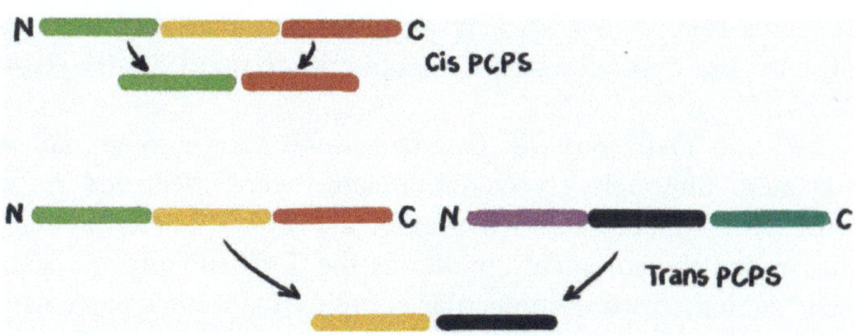

FIGURE 5.3 Mechanisms for generation of spliced peptides. Peptide splicing of cis-PCPS generation is represented by the junction of two non-adjacent peptide sequences (green, N-terminus and red, C-terminus) originated from the same polypeptide (top). Trans-PCPS is represented by the ligation of peptide sequences originated from two distinct polypeptide molecules (yellow and black fragments, bottom).

The importance of spliced epitopes for the immune response against infectious diseases, cancer, and metabolic diseases is a recent topic in the field of immunopeptidomics and antigen discovery, and has attracted enormous interest. It has been shown that spliced peptides can trigger specific CTL responses ex vivo/in vivo against *Listeria monocytogenes* infection (Plattel et al., 2017), HIV infection (Paes et al., 2019), tumor-associated antigens (Ebstein et al., 2016; Faridi et al., 2020; Michaux et al., 2014), and Type 1 diabetes (Gonzalez-Duque et al., 2018). Proteasome-generated cis-spliced peptides have attracted the attention of the scientific community, due to their biological importance and relevance in the development of vaccination strategies based on $CD8^+$ T cells. Spliced peptides may have a potential role in $CD8^+$ T cell tolerance (Mansurkhodzhaev et al., 2021) and viral-induced autoimmunity (Mishto et al., 2021), and clearly the immune response triggered by these peptides needs to be further explored (Faridi et al., 2021).

Intracellular trafficking of peptides

The intracellular trafficking of peptides from the cytosol into the ER is a fundamental part of the MHC-I antigen presentation pathway, and is responsible for the selection of peptides matching in length and sequence specificity for MHC-I molecules. TAP is a key molecule in this process (Abele & Tampé, 1999). Peptides with a length of 8–16 amino acids are preferentially transported by TAP, although longer peptides can also be transported by a heterodimer formed by TAP, but with lower efficiency (Koopmann et al., 1996).

The transport of peptides by TAP is a multistep process. Following the binding of peptides of the preferred length, there is a slow isomerization of the TAP complex. The structural reorganization of the molecule triggers peptide translocation across the membrane, which is ATP-dependent (Knittler et al., 1999). Peptides with low affinity for MHC-I are transported into the ER, but are not rapidly loaded into the MHC molecule. As a result, low affinity peptides are available for export out of the ER into the cytosol via the translocon channel Sec61 (Koopmann et al., 1996).

The TAP1 and TAP2 proteins included in the TAP complex are normally nonglycosylated, although glycosylation sites were identified in both the cytosol and ER of insect cells (Meyer et al., 1994). Later, it was demonstrated that under physiological conditions the TAP subunits are phosphorylated. These subunits form a molecular complex with other proteins: tapasin, MHC-I heavy chain, and a phosphatase, responsible for the de-phosphorylation of the macromolecule. This process is crucial for proper TAP function, because if TAP is phosphorylated it is not able to transport peptides into the ER (Li et al., 2000).

Both proteasomal cleavage and TAP expression are regulated by cytokines. The levels of MHC-I expression can be modulated by cytokines such as IFN-γ (Zhou, 2009), IFN-alpha (IFN-α) (Coomans de Brachène et al., 2018) and tumor necrosis factor-alpha (TNF-α) (Lavi et al., 1988). Similarly to MHC-I, TAP1 and TAP2 are upregulated by these cytokines. For example, TAP showed a significant upregulation after IFN-γ treatment in a programmed cell death-ligand 1 (PD-L1) knock-out mouse model (Stifter et al., 2020), and in IFN-γ-treated human renal carcinoma cells (Seliger et al., 1997). While some cytokines lead to the overexpression of TAP, interleukin (IL)-10 causes downregulation of TAP1 and TAP2, leading to a reduced expression of MHC-I on the cell surface (Kurte et al., 2004). The balance between cytokines involved in pathologies has important implications at the level of MHC-I presentation, with consequences for CD8$^+$ T cell activity and the success of the immune response against pathogens and malignancies.

Peptide binding affinity to major histocompatibility complex class I

The successful activation of CD8$^+$ T cells depends in part on the binding affinity of the MHC-I complex for the peptide on the surface of an APC. This affinity depends on two factors: first, the peptide binding to MHC-I involves a hydrogen bonding network between the N- and C-terminal regions of the peptide and the H chain residues of the MHC, pockets A and F, respectively, which are highly conserved interactions in different MHC-I alleles (Wieczorek et al., 2017). These interactions contribute more binding energy to MHC-I-peptide complexes than the peptide side chain interactions, and they are regarded as the main limiting factor to the size of the peptides presented by MHC-I molecules, which predominantly 8–10 amino acids (Rist et al., 2013). Some longer peptides, between 11 and 16 residues, can also bind to MHC-I, but in this case with lower affinity (Speir et al., 2001).

Second, the MHC antigen binding groove has two (or less commonly three) pockets that show a preference for one to five (more often one or two) of the 20 possible amino acid side chains (Daniel et al., 1998). One of these pockets always accommodates the COOH-terminus of the peptide. The residues lodged in the other pockets vary, depending on the shape, but are almost always the second, third, or fifth amino terminal residue. The residues that make up the crevices have highly variable shapes (Fig. 5.4) (Bjorkman, 1997; Yewdell & Bennink, 1999).

Various bioinformatic tools and resources that allow prediction of peptide binding affinity to MHC-I molecules are provided on several publicly accessible sites, including Immune Epitope Database and Analysis Resource (Kim

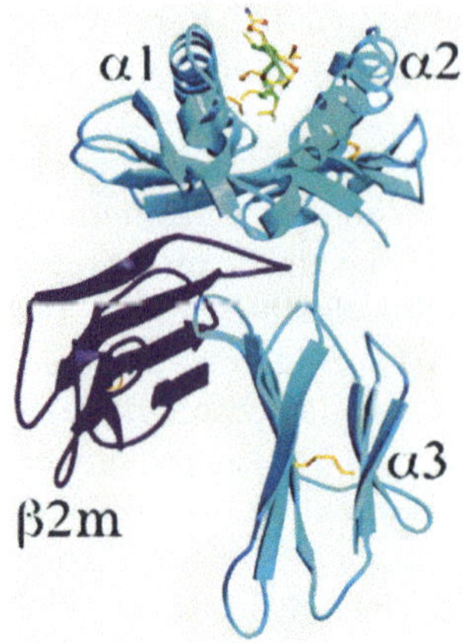

FIGURE 5.4 MHC-I structure. The MHC-I molecule is divided into three globular domains named alpha-1, alpha-2 and alpha-3 (light blue). The second portion of the molecule is the small globular Beta-2 Microglobulin (dark blue). The groove in between alpha-1 and alpha-2 helices accommodates the peptide (yellow-green on top). *Reprinted from Bjorkman, P. J. (1997). MHC restriction in three dimensions: A view of T cell receptor/ligand interactions.* Cell, *89(2), 167–170. https://doi.org/10.1016/S0092-8674(00)80195-6, Copyright (1997) by Cell Press, with permission from Elsevier.*

et al., 2012), BIMAS (Parker et al., 1994), SYFPEITHI (Rammensee et al., 1999), NetMHC (Jurtz et al., 2017), ProPred (Singh & Raghava, 2001), and Multipred (Zhang et al., 2005). In general, MHC-I ligand prediction tools scan the amino acid sequence of a protein to determine the ability of each subsequence to bind to a specific MHC-I allele.

In silico methods for the prediction of proteasome cleavage sites were developed using peptide ligands restricted to MHC-I and their C-terminal flanking regions (Altuvia & Margalit, 2000). The main tools applied for this approach are NetChop 3.1 (Keşmir et al., 2002), Paproc (Kuttler et al., 2000), and improved Proteasome Cleavage Prediction Server (iPCPS) (Gomez-Perosanz et al., 2020). A combination of experimentally validated, in silico algorithms that predict both binding affinity and proteasome cleavage sites may be the most appropriate methodology to identify peptides that are naturally processed and have high affinity for a specific MHC-I allele, and are thus capable of inducing a strong cellular immune response.

However, there remains a considerable discrepancy between peptides that are predicted to be presented by MHC-I, and those that are validated by, for example, mass spectrometry: in general, algorithms predict many peptides that are not naturally presented, and the most immunogenic peptides may not have the highest predicted binding affinity. There is a mismatch between predicted and actual (experimentally measured) peptide-MHC affinity, although a greater issue appears to be accurate prediction of proteasomal cleavage resulting in cell surface presentation on MHC-I molecules.

Tools for discovering immunogenic peptides

The identification of epitopes (MHC-bound peptides with the capacity to induce antigen-specific T cell activation) is very complex and depends on several factors. For example, the MHC molecules are encoded in humans by Human Leucocyte Antigen (HLA) genes that are highly polymorphic, giving rise to different HLA alleles, each of which has unique peptide-binding specificities. These peptide-binding specificities can be either very restricted (i.e., binding a small number of potential peptides) or promiscuous, and some are well characterized whilst others are ill-defined. Additionally, variations in genetics and protein expression, combined with the specific requirements of proteolytic processing during antigen processing and presentation, will all contribute to the diversity of peptide-MHC complexes (pMHC) (Bassani-Sternberg et al., 2015).

The interaction between the TCR and pMHC, leading to a constructive immunological synapse, is the basis of the adaptive immune response (Fig. 5.1). The strength of the interaction between the TCR molecule and the pMHC, known as affinity (Stone & Kranz, 2013; Tan et al., 2015), is governed by the TCR/pMHC transient half-life interactions (Bridgeman et al., 2012), with TCRs that bind self-peptides tending to have lower affinity than those recognizing pathogen-derived peptides (Bridgeman et al., 2012). TCRs are promiscuous in nature, potentially recognizing multiple MHC-presented peptides (Wooldridge et al., 2012).

TCR "deep sequencing" studies have been performed to analyze the T cell repertoires of individuals. There is a large number of specific TCR alpha- and/or beta-chains at a given time in a single individual, estimated to be in the order of 10^{10} (Lythe et al., 2016). To accommodate the diversity and complexity of these repertoires, large databases have been produced, and data-mining efforts have been made to identify interesting TCR clonotypes, such as the iReceptor Scientific Gateway (Corrie et al., 2018).

Affinity-based methods can generate informative data on the recognition of antigens by T cells, such as pMHC-multimer surface staining of T cells; useful reagents include barcode-labeled pMHC multimers (Bentzen et al., 2016) and the commonly used tetramers (Klenerman et al., 2002). This method of T cell recognition analysis, based on the affinity of the TCR/pMHC complex, can identify high-affinity peptides that may be irrelevant and miss low-affinity epitopes that could be of biological importance (Dolton et al., 2018). Moreover, high-affinity peptides could be decoy antigens produced by a pathogen to mislead the immune response. This is the case for the peptide TB10.4$_{4-11}$, which is highly expressed on macrophages infected with *Mycobacterium tuberculosis*

(M.tb), but $CD8^+$ T cells specific for this peptide are not functional in killing infected cells (Carpenter et al., 2016; Yang et al., 2018).

Given the diversity of HLA alleles and the size of the human proteome, the number of short peptide sequences that could be processed and presented by MHC molecules is very large, for self-peptides alone. The presentation of nonself pMHC (e.g., from pathogens) able to induce antigen-specific T cell activation adds to this complexity. The identification of pathogen-derived T cell epitopes, and evaluation of T cells stimulated by these epitopes, including memory $CD8^+$ T cells, is fundamental for successful vaccine design and development. Direct epitope identification through analysis of pMHC complexes by mass spectrometry, known as immunopeptidomics, brings a new dimension to the complex field of antigen presentation and particularly to the discovery of new vaccine targets able to induce $CD8^+$ T cells.

Immunopeptidomics

The group of peptides presented by MHC molecules at the cell surface have been recently designated as the immunopeptidome (Caron et al., 2015), although some studies refer at these peptides as the immunoproteome (Shetty et al., 2011; Testa, Shetty, Hafner, et al., 2012; Testa, Shetty, Sinnathamby, et al., 2012). The immunopeptidome consist of all peptides, self and nonself, processed by the antigen processing machinery and presented by MHC molecules on the surface of host cells, regardless of their biological function. Mass spectrometry–based immunopeptidomics was first developed by Hunt et al. (1992), and since then several groups throughout the world have been developing technology to improve the identification of peptides presented by MHC-I and MHC-II, with or without posttranslational modifications. The source material can be cells, tissues, or organisms, and this method is not only useful in identifying immunogenic epitopes for the design of better vaccines, but also to understand the molecular mechanism of the immunological synapse.

Recent advances in immunopeptidomics based on new mass spectrometry instrumentation such as the Orbitrap and quadrupole time-of-flight (TOF) tandem mass spectrometer (MS/MS) instruments, combined with improvements in data analysis, including the user-friendly software Peaks (Tran et al., 2019), Mascot (Perkins et al., 1999), and MaxQuant (Cox & Mann, 2008), have led to an unprecedented improvement in sensitivity. Currently, it is possible to identify peptide sequences bound to MHC molecules at the femto molar scale (Bettencourt et al., 2020). The immunopeptidomics method is shown here and can be performed with some modifications (Fig. 5.5).

MHC-I-derived peptides are typically nine amino acid long; however, shorter and longer sequences have been described, ranging from 8 to 14 amino acids long (Ritz et al., 2016; Sanchez-Trincado et al., 2017). Even though different cell types express different numbers of MHC molecules at their surface

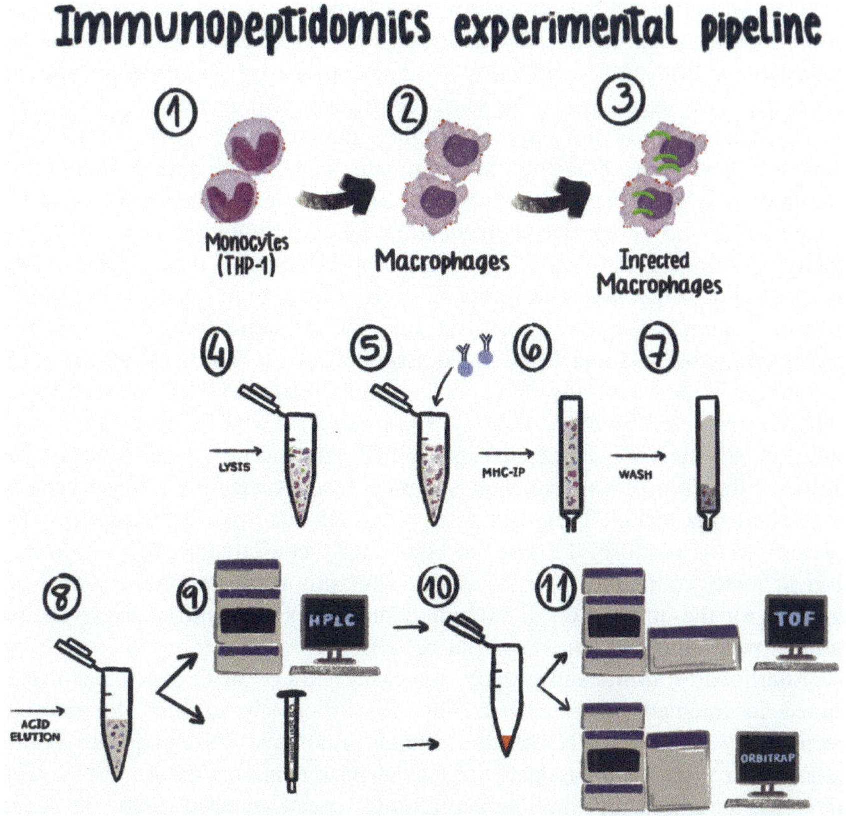

FIGURE 5.5 Immunopeptidomics experimental pipeline. Samples of different origin can be analysed by immunopeptidomics. Here we show a typical infection experiment where a human monocyte cell line (1), is differentiated into macrophages (2), followed by experimental infection with a pathogen of interested (3). (4) Infected cells are lysed. (5) Anti-MHC-I antibody is cross-linked to latex beads and added to the lysate, where the immunoprecipitation between pMHC-I and antibody occurs (MHC-IP). (6) The MHC-IP is transferred to a column, and a sequence of washing steps removes contaminant proteins (7), followed by a mild acid elution (MAE) that will break down the MHC-I alpha and beta-2-microglobulin subunits, releasing the peptides bound in the MHC-I pocket (8). (9) The peptides can be separated from the MHC-I subunits by HPLC or C18 Spin columns. (10) A solution containing pure peptides presented by MHC-I molecules is obtained. (11) The purified peptides can be analysed by mass spectrometry either using Quadrupole Time of Flight Instruments (TOF) or Orbitrap Instruments. Red dots represent peptides presented by MHC-I molecules. Green rods represent microorganisms to illustrate an experimental infection with a pathogen of interest.

at a given time point, based on the current methodology of liquid chromatography (LC) tandem mass spectrometry and bioinformatics, the minimal requirements for the identification of peptide sequences requires a sample size of approximately 5×10^8 cells expressing approximately 2×10^5 MHC molecules per cell (Bettencourt et al., 2020; Caron et al., 2015).

The requirement for a relatively large sample size could be a drawback in the identification of peptides presented by infected cells, as is the case for malaria. As the rate of infection of hepatocytes by *Plasmodium* spp. is extremely low, arguably 5% or less, identification of pathogen peptides by mass spectrometry would require a higher number (in the order of 10^{10}) of infected hepatocytes at a 5% infection rate, making the production of the required sample size extremely laborious and complex (Bettencourt, 2020).

With the aim of epitope identification for improving the efficacy of the bacille Calmette-Guerin (BCG) vaccine, cell cultures of macrophages were infected with *Mycobacterium bovis* strain BCG at a high infection rate (30% in vitro). Samples were sufficient in size for the identification of peptides presented by MHC-I and MHC-II (Bettencourt et al., 2020). Applying mass spectrometry and bioinformatics, the authors identified 94 mycobacterial peptides presented by MHC-II and 43 presented by MHC-I, from 76 and 41 antigens, respectively. Three antigens (glfT2, fas, and iniB) were selected for further validation of experimental vaccines against TB in a murine aerosol M.tb challenge model. These antigens, when included in a heterologous viral vector vaccine platform designed to boost BCG, conferred significant protection in mice, compared to BCG alone. Thus this study represents proof-of-concept for the application of immunopeptidomics to identify novel vaccine antigen candidates (Bettencourt et al., 2020).

Additionally, in the same study, the expression of BCG genes was determined in infected macrophages. This is particularly difficult to address, because mycobacterial RNA is very diluted in the host RNA, requiring application of a protocol developed for the purpose (Tailleux et al., 2008). The antigens identified by immunopeptidomics were mapped onto the gene expression data of all BCG genes. Interestingly, the expression of the peptide antigens identified by immunopeptidomics correlated with the global gene expression pattern in vivo, both for MHC-I and MHC-II presented peptides. The positive correlation between peptide presentation by MHC molecules and the gene expression of their source antigens is of major importance for the shortlisting of antigen candidates in the design of new vaccines, including neo-antigen vaccines in cancer (Bettencourt et al., 2020; Koşaloğlu-Yalçın et al., 2020).

Vaccine design based on immunopeptidomics

Immunopeptidomics technology has allowed the identification of epitopes presented by conventional MHC-I molecules in infectious diseases including dengue (Testa, Shetty, Sinnathamby, et al., 2012), influenza (Testa, Shetty, Hafner, et al., 2012), hepatitis B (Comber et al., 2014), hepatitis C (Wölk et al., 2012), HIV (Ternette et al., 2016), TB (Flyer et al., 2002; McMurtrey et al., 2017), and live attenuated vaccine BCG (Bettencourt et al., 2020).

Interestingly, Leishmania MHC-II-bound peptides have been identified, and vaccines based on these antigens developed (Mou et al., 2015) (Table 5.1).

To benefit from the cytotoxic potential of CTLs, immunopeptidomics has been applied to identify tumor-specific neoantigens, which can then be incorporated into cancer vaccines. Peptides presented by MHC molecules have been identified in ovarian cancer (Ramakrishna et al., 2003), melanoma (Bassani-Sternberg et al., 2016; Ritz et al., 2016), B cell lymphoma (Khodadoust et al., 2017; Lanoix et al., 2018), B cell acute lymphoblastic leukemia (Laumont et al., 2018), lung cancer (Laumont et al., 2018; Shetty et al., 2011), hepatocellular carcinoma (Altuvia & Margalit, 2000; Löffler et al., 2019), colorectal cancer (Chen et al., 2018; Newey et al., 2019), and breast cancer (Ternette et al., 2018) (Table 5.1).

Of note, a recent study in melanoma patients revealed that cancer cells can be infected with several bacterial species and these cells present bacterial peptides on MHC-I and MHC-II. The authors used 16S rRNA gene sequencing in combination with immunopeptidomics to identify 248 MHC-I and 35 MHC-II peptides from 41 bacterial species, presented on melanoma tumors (Kalaora et al., 2021). The presence of microbial peptides can pose challenges for the correct identification of epitopes from cancer tissue by immunopeptidomics, but could offer new treatment strategies to cancer patients (Table 5.1).

Following the exciting and relevant discoveries made in recent years in the field of immunopeptidomics, there is an anticipation of future work identifying peptides presented by MHC-I and MHC-II from tissues, primary cells, and disease models of many other infectious diseases, cancers, metabolic and nontransmissible diseases.

Methods for the validation of immunogenic peptides

Peptide identification by immunopeptidomics should be experimentally validated to verify biological relevance, especially for peptides with low predicted binding affinity for MHC. This is particularly relevant for infectious diseases where a strong immune response is required to counter pathogen attack. Diverse immunological approaches have been employed independently from immunopeptidomics to confirm the validity and relevance of these peptides. Methods commonly used include, but are not limited to, spectral match validation (SMV), IFN-γ ELISpot, intracellular cytokine staining (ICS), tetramer staining, the generation of peptide-specific CTL lines, and CTL binding affinity (Table 5.1).

Briefly, these methods can be divided into: (1) technical validation, with the purpose of confirming the peptide sequences identified by immunopeptidomics were assigned correctly, and (2) biological validation, which is employed to study and confirm the biological and immunological relevance of the peptide sequences identified by immunopeptidomics.

TABLE 5.1 Selected studies of antigen discovery through immunopeptidomics.

Condition	Sample	HLA	Peptides identified[a]	Experimental validation methods	References
Infectious diseases					
Dengue	JY cells and leukapheresis generated dendritic cells infected with DV2 and DV3, respectively	HLA-A,B,C	HLA-A2—3	Spectral match validation, CTL generated against Dengue virus epitopes, ELISpot	Testa, Shetty, Sinnathamby, et al. (2012)
Influenza	Influenza-infected JY, HepG2 cell lines and monocyte-derived human dendritic cells	HLA-A,B,C	HLA-A2—6HLA-B44—10HLA-B7—6	Peptide-specific CTL cell lines, ELISpot and flow cytometry	Testa, Shetty, Hafner, et al. (2012)
Hepatitis B	Hepatitis B infected cell line HepDE19	HLA-A,B,C	MHC-I—8	Peptide-specific CTL cell lines, ELISpot, cytokine secretion analysis, and flow cytometry	Comber et al. (2014)
Hepatitis C	UNS3-4A/A2-27.35 and UHCV/A2-27 human osteosarcoma-derived lines engineered to express HCV	HLA-A,B,C	HLA-A2—2	CTL cell lines raised against two HCV epitopes	Wölk et al. (2012)
HIV	Cell line C8166 and primary $CD8^+$ T-cell-depleted PBMCs	HLA-A,B,C	MHC-I—75	Spectral match validation, ELISpot	Ternette et al. (2016)
Tuberculosis	U937 cell line transfected with gene encoding HLA-A*0201 and infected with M.tb	HLA-A,B,C	MHC-I—3	Generation of peptide-specific CTL lines and CTL binding affinity	Flyer et al. (2002)

Tuberculosis	T cell clone D160 1—23, infected with M.tb	HLA-E	MHC-I—28	ELISpot, generation of T cell line and tetramer staining	McMurtrey et al. (2017)
BCG	THP-1 cells infected with BCG	HLA-A,B,C, DR	MHC-I—43MHC-II—94	Gene expression, spectral match validation, ELIspot, vaccine efficacy	Bettencourt et al. (2020)
Leishmaniasis	Murine bone marrow dendritic cells infected with *Leishmania major*	Mouse MHC-II (I-Ab)	MHC-II—1	Tetramer staining and vaccine efficacy	Mou et al. (2015)
Cancer					
Ovarian cancer	Ovarian adenocarcinoma line SKOV3 transfected with HLA-A2.1 and OVCAR3	HLA-A,B,C	HLA-A2—16	Spectral match validation, peptide-specific CTL cell lines, and CTL binding affinity	Ramakrishna et al. (2003)
Melanoma	Diverse commercial cell lines and soluble HLA-I complexes from sera of melanoma patients	HLA-A,B,C	Identified 22 MHC-I peptides from known tumor-associated antigens	N/A	Ritz et al. (2016)
Melanoma	Tumor tissue samples from melanoma patients	HLA-A,B,C	Numerous MHC-I and MHC-II neoantigens;11 mutated peptide ligands	ELISpot in patient samples; Reactivity of peptide-specific T-cells derived from PBMC and ex vivo expanded tumor-infiltrating T cells;Intracellular cytokine staining on T cell lines coculture with neoantigens	Bassani-Sternberg et al. (2016)

(*Continued*)

TABLE 5.1 (Continued)

Condition	Sample	HLA	Peptides identified[a]	Experimental validation methods	References
B cell lymphoma	Tumor samples of 17 patients and two cell lines of mantle-cell lymphoma	HLA-A,B,C, DR	Numerous neoantigens, mostly from the lymphoma immunoglobulin heavy- or light-chain variable regions	Tetramers of neoantigen peptide-HLA-Dr*0401 revealed peptide-specific $CD4^+$ T cells in one out of three patients;Isolated and expanded neoantigen-specific $CD4^+$ T cells able to mediate killing of autologous lymphoma cells	Khodadoust et al. (2017)
B cell lymphoma	Epstein–Barr virus transformed cell line, and human B acute lymphoblastic leukemia cells expanded in immunodeficient mice	HLA-A,B,C	Numerous peptides identified using different methods	N/A	Lanoix et al. (2018)
B cell acute lymphoblastic leukemia	Cells from four patients with B-lineage acute lymphoblastic leukemia	HLA-A,B,C	MHC-I—12	Spectral match validation	Laumont et al. (2018)
Lung cancer	Cells from three patients with lung cancer	HLA-A,B,C	MHC-I—15	Spectral match validation	Laumont et al. (2018)
Lung cancer	Human lung adenocarcinoma cell lines	HLA-A,B,C	MHC-I—68	Spectral match validation, peptide stability experiments, generation of peptide-specific CTL lines, ELISpot	Shetty et al. (2011)

Hepatocellular carcinoma	Samples from 16 patients with hepatocellular carcinomas	HLA-A,B,C	Large numbers of MHC-I peptides were identified; no naturally presented mutated peptides were detected	N/A	Löffler et al. (2019)
Colorectal cancer	Patient-derived organoids from colorectal cancers	HLA-A,B,C, Dr, DP, DQ	Three mutated neoantigens	N/A	Newey et al. (2019)
Colorectal cancer	Colorectal carcinoma HCT 116 cell line	HLA-A,B,C	10 mutated peptides	N/A	Chen et al. (2018)
Breast Cancer	HLA-A2-positive triple-negative breast cancer samples	HLA-A,B,C	Increased abundance of 130 MHC-I peptides in tumor samples	N/A	Ternette et al. (2018)
Microbiome					
Melanoma microbiome	Melanoma	HLA-A,B,C, DR	MHC-I—248MHC-II—35	Bacterial peptides were fed into B cells. These were cocultured with tumor-infiltrating lymphocytes. Lymphocyte responses were measured by flow cytometry	Kalaora et al. (2021)

Studies focusing on MHC-I peptide identification in human samples were included. *CTL*, Cytotoxic T lymphocytes; *MHC*, major histocompatibility complex; *M.tb*, *Mycobacterium tuberculosis*; *N/A*, not available.
[a]*Pathogenic peptides only, host peptides excluded.*

The most common method for technical validation is the SMV, where *synthetic peptides* are compared to a selection of peptides identified by immunopeptidomics (*biological peptides*). Briefly, the method consists of running the synthetic peptides through the same experimental conditions as the biological peptides, to analyze the following parameters: (1) the charge state of each peptide, (2) mass over charge [*m*/*z*] of each peptide, (3) the intensity and distribution of each peak from each peptide sequence, and (4) the peptide specific retention time on LC. Synthetic peptides matching these four parameters can confirm that the biological peptide sequences were assigned correctly.

The most common biological validation methods are ELISpot, ICS, and tetramer staining. The IFN-γ ELISpot (or ELISPOT) assay is a relatively quick and simple method to evaluate the ability of peripheral blood mononuclear cells (PBMC), or T cell subpopulations, to secrete IFN-γ upon contact with a peptide of interest. Briefly, an antibody recognizing IFN-γ is coated on a supporting membrane and a suspension of T cells is put in contact with the membrane. T cells are "pulsed" with peptides or proteins of interest, and upon recognition of specific peptides presented by MHC, IFN-γ is secreted and captured on the membrane. It is detected by, for example, a biotinylated anti-IFN-γ antibody and streptavidin-alkaline phosphatase. The resulting colored spots, each of which represents an activated IFN-γ-producing T cell, can be quantified.

The ICS method requires a multiparameter flow cytometer, and is described in more detail in Chapter 4: Vaccine Development: From the Laboratory to the Field. In this method, subpopulations of immune cells are sorted and intracellular cytokines are stained with specific antibodies. The most common cytokines studied upon activation of $CD8^+$ T cells are IFN-γ, TNF-γ, and IL-2. Tetramer staining also requires a flow cytometer: as discussed in the Section Tools for discovering immunogenic peptides, tetramers are protein complexes composed of four stabilized soluble MHC molecules bound to a peptide of interest. They are used to detect and quantify T cells specific for that peptide.

The generation of peptide-specific T cell lines allows the recognition and killing of cells presenting the target peptide to be studied; several strategies have been employed to generate such lines (Sharma et al., 2019; Theaker et al., 2016). All of these methods provide different types of information that can be used alone or in combination to give additional insight into the biological function of a peptide sequence identified by immunopeptidomics.

Conclusions

The development of peptide-based vaccines to exploit the power of cytotoxic $CD8^+$ T cells is in its infancy, and immunopeptidomics will have a crucial role in the identification of the best immunogenic candidates. Due to the reduced size of T cell-specific peptides compared to whole antigens, peptide-based vaccines could be designed to include epitopes from several antigens, against a multitude of diseases. Broadening the range of pathogen epitopes in a vaccine could be a strategy to prevent the emergence of immune escape variants, a phenomenon described in diverse pathogens, for example SARS-CoV-2 (Di Caro et al., 2021). For this reason, vaccines should be designed to target highly conserved T cell epitopes that are processed and presented by MHC molecules.

The successful design of peptide-based vaccines requires the consideration of several characteristics, including high affinity binding of peptides to the MHC molecules, as well as the presence of cleavage sites for the proteasome (Doytchinova & Flower, 2006). Recently, new multidisciplinary approaches combining the analysis of human immunology, structural biology, and microbial genomics, termed "reverse vaccinology 2.0," have been applied to the discovery and design of vaccine antigens (Rappuoli et al., 2016). Ullah et al. for example, utilized a combination of different immunoinformatic tools to design subunit vaccines against Ebola virus. Potential MHC-I, MHC-II, and B cell epitopes from three viral antigens underwent robust analysis of antigenicity, allergenicity, toxicity, conservancy, and molecular docking analysis; EV-1, EV-2, and EV-3 were then constructed as vaccine candidates (Ullah et al., 2020). Additional in vitro and in vivo validation is likely to be required for the candidates generated by reverse vaccinology studies.

In cancer, the challenge of peptide identification by immunopeptidomics has additional complexity at several levels. The identification of peptide sequences bound to MHC molecules by mass spectrometry requires a database containing the proteome of interest to facilitate the correct identification of a given peptide. Each cancer patient sample is a genetic entity in its own right, due to mutations developed throughout the disease process. Therefore RNA sequencing must be performed in the tumor sample, to generate a database from which peptides will be identified. Moreover, the recently published phenomenon of spliced peptides generated by the proteasome, which can account for up to one-third of all peptides presented to CD8$^+$ T cells (Liepe et al., 2016), brings additionally complexity to the peptide identification process from cancer patients.

Immunopeptidomics is undoubtedly one of the most promising approaches in improving future vaccine design. Using this method to better understand the biology of peptide presentation by APCs and to unleash the power of CD8$^+$ T cells could be instrumental for the treatment of a plethora of infectious and nontransmissible diseases.

Acknowledgments

The authors would like to acknowledge the artwork kindly produced by Miss Madalena Martins de Bettencourt, Cascais, Portugal, and Dr. Michele Mishto, King's College London, and the Francis Crick Institute, London, United Kingdom, for critical review of the manuscript. Camila Barbosa was supported by the Cancer Research UK grant [C67500, A29686] held by M. Mishto (Kings College London).

References

Abele, R., & Tampé, R. (1999). Function of the transport complex TAP in cellular immune recognition. *Biochimica et Biophysica Acta (BBA) – Biomembranes*, 405–419. Available from https://doi.org/10.1016/s0005-2736(99)00171-6.

Altuvia, Y., & Margalit, H. (2000). Sequence signals for generation of antigenic peptides by the proteasome: Implications for proteasomal cleavage mechanism. *Journal of Molecular Biology*, *295*(4), 879–890. Available from https://doi.org/10.1006/jmbi.1999.3392.

Bahjat, K. S., Liu, W., Lemmens, E. E., Schoenberger, S. P., Portnoy, D. A., Dubensky, T. W., & Brockstedt, D. G. (2006). Cytosolic entry controls CD8 + -T-cell potency during bacterial infection. *Infection and Immunity*, 6387–6397. Available from https://doi.org/10.1128/IAI.01088-06.

Bassani-Sternberg, M., Bräunlein, E., Klar, R., Engleitner, T., Sinitcyn, P., Audehm, S., Straub, M., Weber, J., Slotta-Huspenina, J., Specht, K., Martignoni, M. E., Werner, A., Hein, R., Busch, D. H., Peschel, C., Rad, R., Cox, J., Mann, M., & Krackhardt, A. M. (2016). Direct identification of clinically relevant neoepitopes presented on native human melanoma tissue by mass spectrometry. *Nature Communications*, *7*. Available from https://doi.org/10.1038/ncomms13404.

Bassani-Sternberg, M., Pletscher-Frankild, S., Jensen, L. J., & Mann, M. (2015). Mass spectrometry of human leukocyte antigen class I peptidomes reveals strong effects of protein abundance and turnover on antigen presentation. *Molecular and Cellular Proteomics*, *14*(3), 658–673. Available from https://doi.org/10.1074/mcp.M114.042812.

Bentzen, A. K., Marquard, A. M., Lyngaa, R., Saini, S. K., Ramskov, S., Donia, M., Such, L., Furness, A. J. S., McGranahan, N., Rosenthal, R., Straten, P. T., Szallasi, Z., Svane, I. M., Swanton, C., Quezada, S. A., Jakobsen, S. N., Eklund, A. C., & Hadrup, S. R. (2016). Large-scale detection of antigen-specific T cells using peptide-MHC-I multimers labeled with DNA barcodes. *Nature Biotechnology*, *34*(10), 1037–1045. Available from https://doi.org/10.1038/nbt.3662.

Berkers, C. R., De Jong, A., Schuurman, K. G., Linnemann, C., Meiring, H. D., Janssen, L., Neefjes, J. J., Schumacher, T. N. M., Rodenko, B., & Ovaa, H. (2015). Definition of proteasomal peptide splicing rules for high-efficiency spliced peptide presentation by MHC class I molecules. *Journal of Immunology*, *195*(9), 4085–4095. Available from https://doi.org/10.4049/jimmunol.1402455.

Bettencourt, P. (2020). Current challenges in the identification of pre-erythrocytic malaria vaccine candidate antigens. *Frontiers in Immunology*. Available from https://doi.org/10.3389/fimmu.2020.00190.

Bettencourt, P., Müller, J., Nicastri, A., Cantillon, D., Madhavan, M., Charles, P. D., Fotso, C. B., Wittenberg, R., Bull, N., Pinpathomrat, N., Waddell, S. J., Stylianou, E., Hill, A. V. S., Ternette, N., & McShane, H. (2020). Identification of antigens presented by MHC for vaccines against tuberculosis. *npj Vaccines*, *5*(1). Available from https://doi.org/10.1038/s41541-019-0148-y.

Bjorkman, P. J. (1997). MHC restriction in three dimensions: A view of T cell receptor/ligand interactions. *Cell*, *89*(2), 167–170. Available from https://doi.org/10.1016/S0092-8674(00)80195-6.

Bridgeman, J. S., Sewell, A. K., Miles, J. J., Price, D. A., & Cole, D. K. (2012). Structural and biophysical determinants of αβ T-cell antigen recognition. *Immunology*, *135*(1), 9–18. Available from https://doi.org/10.1111/j.1365-2567.2011.03515.x.

Calis, J. J. A., de Boer, R. J., & Keşmir, C. (2012). Degenerate T-cell recognition of peptides on MHC molecules creates large holes in the T-cell repertoire. *PLoS Computational Biology*, *8*(3). Available from https://doi.org/10.1371/journal.pcbi.1002412.

Caron, E., Kowalewski, D. J., Koh, C. C., Sturm, T., Schuster, H., & Aebersold, R. (2015). Analysis of major histocompatibility complex (MHC) immunopeptidomes using mass spectrometry. *Molecular and Cellular Proteomics, 14*(12), 3105–3117. Available from https://doi.org/10.1074/mcp.O115.052431.

Carpenter, S. M., Nunes-Alves, C., Booty, M. G., Way, S. S., & Behar, S. M. (2016). A higher activation threshold of memory CD8 + T cells has a fitness cost that is modified by TCR affinity during tuberculosis. *PLoS Pathogens, 12*(1). Available from https://doi.org/10.1371/journal.ppat.1005380.

Chen, R., Fauteux, F., Foote, S., Stupak, J., Tremblay, T. L., Gurnani, K., Fulton, K. M., Weeratna, R. D., Twine, S. M., & Li, J. (2018). Chemical derivatization strategy for extending the identification of MHC Class I immunopeptides. *Analytical Chemistry, 90*(19), 11409–11416. Available from https://doi.org/10.1021/acs.analchem.8b02420.

Comber, J. D., Karabudak, A., Shetty, V., Testa, J. S., Huang, X., & Philip, R. (2014). MHC Class I presented T cell epitopes as potential antigens for therapeutic vaccine against HBV chronic infection. *Hepatitis Research and Treatment*, 1–11. Available from https://doi.org/10.1155/2014/860562.

Coomans de Brachène, A., Dos Santos, R. S., Marroqui, L., Colli, M. L., Marselli, L., Mirmira, R. G., Marchetti, P., & Eizirik, D. L. (2018). IFN-α induces a preferential long-lasting expression of MHC class I in human pancreatic beta cells. *Diabetologia, 61*(3), 636–640. Available from https://doi.org/10.1007/s00125-017-4536-4.

Corrie, B. D., Marthandan, N., Zimonja, B., Jaglale, J., Zhou, Y., Barr, E., Knoetze, N., Breden, F. M. W., Christley, S., Scott, J. K., Cowell, L. G., & Breden, F. (2018). iReceptor: A platform for querying and analyzing antibody/B-cell and T-cell receptor repertoire data across federated repositories. *Immunological Reviews, 284*(1), 24–41. Available from https://doi.org/10.1111/imr.12666.

Cosma, G., & Eisenlohr, L. (2018). CD8 + T-cell responses in vaccination: Reconsidering targets and function in the context of chronic antigen stimulation. *F1000 Research, 7*, 1–8.

Cox, J., & Mann, M. (2008). MaxQuant enables high peptide identification rates, individualized p.p.b.-range mass accuracies and proteome-wide protein quantification. *Nature Biotechnology, 26*(12), 1367–1372. Available from https://doi.org/10.1038/nbt.1511.

Craiu, A., Akopian, T., Goldberg, A., & Rock, K. L. (1997). Two distinct proteolytic processes in the generation of a major histocompatibility complex class I-presented peptide. *Proceedings of the National Academy of Sciences of the United States of America*, 10850–10855. Available from https://doi.org/10.1073/pnas.94.20.10850.

Dalet, A., Vigneron, N., Stroobant, V., Hanada, K. I., & Van Den Eynde, B. J. (2010). Splicing of distant peptide fragments occurs in the proteasome by transpeptidation and produces the spliced antigenic peptide derived from fibroblast growth factor-5. *Journal of Immunology, 184*(6), 3016–3024. Available from https://doi.org/10.4049/jimmunol.0901277.

Daniel, C., Horvath, S., & Allen, P. M. (1998). A basis for alloreactivity: MHC helical residues broaden peptide recognition by the TCR. *Immunity, 8*(5), 543–552. Available from https://doi.org/10.1016/S1074-7613(00)80559-2.

Di Caro, A., Cunha, F., Petrosillo, N., Beeching, N. J., Ergonul, O., Petersen, E., & Koopmans, M. P. G. (2021). SARS-CoV-2 escape mutants and protective immunity from natural infections or immunizations. *Clinical Microbiology and Infection*. Available from https://doi.org/10.1016/j.cmi.2021.03.011.

Diez-Rivero, C. M., Lafuente, E. M., & Reche, P. A. (2010). Computational analysis and modeling of cleavage by the immunoproteasome and the constitutive proteasome. *BMC Bioinformatics, 11*. Available from https://doi.org/10.1186/1471-2105-11-479.

Dolton, G., Zervoudi, E., Rius, C., Wall, A., Thomas, H. L., Fuller, A., Yeo, L., Legut, M., Wheeler, S., Attaf, M., Chudakov, D. M., Choy, E., Peakman, M., & Sewell, A. K. (2018). Optimized peptide-MHC multimer protocols for detection and isolation of autoimmune T-cells. *Frontiers in Immunology*, *9*. Available from https://doi.org/10.3389/fimmu.2018.01378.

Doytchinova, I. A., & Flower, D. R. (2006). Class I T-cell epitope prediction: Improvements using a combination of proteasome cleavage, TAP affinity, and MHC binding. *Molecular Immunology*, *43*(13), 2037–2044. Available from https://doi.org/10.1016/j.molimm.2005.12.013.

Ebstein, F., Textoris-Taube, K., Keller, C., Golnik, R., Vigneron, N., Van Den Eynde, B. J., Schuler-Thurner, B., Schadendorf, D., Lorenz, F. K. M., Uckert, W., Urban, S., Lehmann, A., Albrecht-Koepke, N., Janek, K., Henklein, P., Niewienda, A., Kloetzel, P. M., & Mishto, M. (2016). Proteasomes generate spliced epitopes by two different mechanisms and as efficiently as non-spliced epitopes. *Scientific Reports*, *6*. Available from https://doi.org/10.1038/srep24032.

Faridi, P., Dorvash, M., & Purcell, A. W. (2021). Spliced HLA bound peptides; a Black-Swan event in immunology. *Clinical and Experimental Immunology*.

Faridi, P., Li, C., Ramarathinam, S. H., Vivian, J. P., Illing, P. T., Mifsud, N. A., Ayala, R., Song, J., Gearing, L. J., Hertzog, P. J., Ternette, N., Rossjohn, J., Croft, N. P., & Purcell, A. W. (2018). A subset of HLA-I peptides are not genomically templated: Evidence for cis- and trans-spliced peptide ligands. *Science Immunology*, *3*(28). Available from https://doi.org/10.1126/sciimmunol.aar3947.

Faridi, P., Woods, K., Ostrouska, S., Deceneux, C., Aranha, R., Duscharla, D., Wong, S. Q., Chen, W., Ramarathinam, S. H., Lim Kam Sian, T. C. C., Croft, N. P., Li, C., Ayala, R., Cebon, J. S., Purcell, A. W., Schittenhelm, R. B., & Behren, A. (2020). Spliced peptides and cytokine-driven changes in the immunopeptidome of melanoma. *Cancer Immunology Research*, *8*(10), 1322–1334. Available from https://doi.org/10.1158/2326-6066.CIR-19-0894.

Flyer, D. C., Ramakrishna, V., Miller, C., Myers, H., McDaniel, M., Root, K., Flournoy, C., Engelhard, V. H., Canaday, D. H., Marto, J. A., Ross, M. M., Hunt, D. F., Shabanowitz, J., & White, F. M. (2002). Identification by mass spectrometry of CD8 + -T-cell *Mycobacterium tuberculosis* epitopes within the Rv0341 gene product. *Infection and Immunity*, *70*(6), 2926–2932. Available from https://doi.org/10.1128/IAI.70.6.2926-2932.2002.

Frankild, S., de Boer, R. J., Lund, O., Nielsen, M., & Kesmir, C. (2008). Amino acid similarity accounts for T cell cross-reactivity and for \holes\ in the T cell repertoire. *PLoS One*, *3*(3). Available from https://doi.org/10.1371/journal.pone.0001831.

Gaczynska, M., Rock, K. L., Spies, T., & Goldberg, A. L. (1994). Peptidase activities of proteasomes are differentially regulated by the major histocompatibility complex-encoded genes for LMP2 and LMP7. *Proceedings of the National Academy of Sciences of the United States of America*, *91*(20), 9213–9217. Available from https://doi.org/10.1073/pnas.91.20.9213.

Gomez-Perosanz, M., Ras-Carmona, A., Lafuente, E. M., & Reche, P. A. (2020). Identification of CD8 + T cell epitopes through proteasome cleavage site predictions. *BMC Bioinformatics*, *21*. Available from https://doi.org/10.1186/s12859-020-03782-1.

Gonzalez-Duque, S., Azoury, M. E., Colli, M. L., Afonso, G., Turatsinze, J.-V., Nigi, L., Lalanne, A. I., Sebastiani, G., Carré, A., Pinto, S., Culina, S., Corcos, N., Bugliani, M., Marchetti, P., Armanet, M., Diedisheim, M., Kyewski, B., Steinmetz, L. M., Buus, S., & Mallone, R. (2018). Conventional and neo-antigenic peptides presented by β cells are targeted by circulating naïve CD8 + T cells in type 1 diabetic and healthy donors. *Cell Metabolism*, 946–960.e6. Available from https://doi.org/10.1016/j.cmet.2018.07.007.

Hanada, K. I., Yewdell, J. W., & Yang, J. C. (2004). Immune recognition of a human renal cancer antigen through post-translational protein splicing. *Nature*, *427*(6971), 252–256. Available from https://doi.org/10.1038/nature02240.

Hewitt, E. W. (2003). The MHC class I antigen presentation pathway: Strategies for viral immune evasion. *Immunology*, *110*(2), 163–169. Available from https://doi.org/10.1046/j.1365-2567.2003.01738.x.

Hunt, D. F., Henderson, R. A., Shabanowitz, J., Sakaguchi, K., Michel, H., Sevilir, N., Cox, A. L., Appella, E., & Engelhard, V. H. (1992). Characterization of peptides bound to the class I MHC molecule HLA-A2.1 by mass spectrometry. *Science (New York, N.Y.)*, *255*(5049), 1261–1263. Available from https://doi.org/10.1126/science.1546328.

Junqueira, C., Barbosa, C. R. R., Costa, P. A. C., Teixeira-Carvalho, A., Castro, G., Sen Santara, S., Barbosa, R. P., Dotiwala, F., Pereira, D. B., Antonelli, L. R., Lieberman, J., & Gazzinelli, R. T. (2018). Cytotoxic CD8 + T cells recognize and kill *Plasmodium vivax*–infected reticulocytes. *Nature Medicine*, *24*(9), 1330–1336. Available from https://doi.org/10.1038/s41591-018-0117-4.

Jurtz, V., Paul, S., Andreatta, M., Marcatili, P., Peters, B., & Nielsen, M. (2017). NetMHCpan-4.0: Improved peptide-MHC class I interaction predictions integrating eluted ligand and peptide binding affinity data. *Journal of Immunology*, *199*(9), 3360–3368. Available from https://doi.org/10.4049/jimmunol.1700893.

Kalaora, S., Nagler, A., Nejman, D., Alon, M., Barbolin, C., Barnea, E., Steven, L. C., Ketelaars, K., Cheng, K., Vervier, N., Shental, Y., Bussi, R., Rotkopf, R., Levy, G., Benedek, S., Trabish, T., Dadosh, S., Levin-Zaidman, L. T., Geller, K., ... Roitman. (2021).

Keşmir, C., K., N. A., Hansjörg, S., Vincent, D., & Søren, B. (2002). Prediction of proteasome cleavage motifs by neural networks. *Protein Engineering, Design and Selection*, 287–296. Available from https://doi.org/10.1093/protein/15.4.287.

Khodadoust, M. S., Olsson, N., Wagar, L. E., Haabeth, O. A. W., Chen, B., Swaminathan, K., Rawson, K., Liu, C. L., Steiner, D., Lund, P., Rao, S., Zhang, L., Marceau, C., Stehr, H., Newman, A. M., Czerwinski, D. K., Carlton, V. E. H., Moorhead, M., Faham, M., & Alizadeh, A. A. (2017). Antigen presentation profiling reveals recognition of lymphoma immunoglobulin neoantigens. *Nature*, *543*(7647), 723–727. Available from https://doi.org/10.1038/nature21433.

Kim, Y., Ponomarenko, J., Zhu, Z., Tamang, D., Wang, P., Greenbaum, J., Lundegaard, C., Sette, A., Lund, O., Bourne, P. E., Nielsen, M., & Peters, B. (2012). Immune epitope database analysis resource. *Nucleic Acids Research*, *40*(1), W525–W530. Available from https://doi.org/10.1093/nar/gks438.

Kincaid, E. Z., Che, J. W., York, I., Escobar, H., Reyes-Vargas, E., Delgado, J. C., Welsh, R. M., Karow, M. L., Murphy, A. J., Valenzuela, D. M., Yancopoulos, G. D., & Rock, K. L. (2012). Mice completely lacking immunoproteasomes show major changes in antigen presentation. *Nature Immunology*, *13*(2), 129–135. Available from https://doi.org/10.1038/ni.2203.

Klein, L., Kyewski, B., Allen, P. M., & Hogquist, K. A. (2014). Positive and negative selection of the T cell repertoire: What thymocytes see (and don't see). *Nature Reviews. Immunology*, *14*(6), 377–391. Available from https://doi.org/10.1038/nri3667.

Klenerman, P., Cerundolo, V., & Dunbar, P. R. (2002). Tracking T cells with tetramers: New tales from new tools. *Nature Reviews. Immunology*, *2*(4), 263–272. Available from https://doi.org/10.1038/nri777.

Kloetzel, P.-M. (2001). Antigen processing by the proteasome. *Nature Reviews. Molecular Cell Biology*, 179–188. Available from https://doi.org/10.1038/35056572.

Knittler, M. R., Alberts, P., Deverson, E. V., & Howard, J. C. (1999). Nucleotide binding by TAP mediates association with peptide and release of assembled MHC class I molecules. *Current Biology*, *9*(18), 999–1008. Available from https://doi.org/10.1016/S0960-9822(99)80448-5.

Koopmann, J.-O., Post, M., Neefjes, J. J., Hämmerling, G. J., & Momburg, F. (1996). Translocation of long peptides by transporters associated with antigen processing (TAP). *European Journal of Immunology*, 1720–1728. Available from https://doi.org/10.1002/eji.1830260809.

Koşaloğlu-Yalçın, Z., Lee, J., Nielsen, M., Greenbaum, J., Schoenberger, S. P., Miller, A., Kim, Y.J., Sette, A., & Peters, B. (2020). Combined assessment of MHC binding and antigen expression improves T cell epitope predictions. *BioRxiv*.

Kurte, M., López, M., Aguirre, A., Escobar, A., Aguillón, J. C., Charo, J., Larsen, C. G., Kiessling, R., & Salazar-Onfray, F. (2004). A synthetic peptide homologous to functional domain of human IL-10 down-regulates expression of MHC Class I and transporter associated with antigen processing 1/2 in human melanoma cells. *The Journal of Immunology*, 1731–1737. Available from https://doi.org/10.4049/jimmunol.173.3.1731.

Kuttler, C., Nussbaum, A. K., Dick, T. P., Rammensee, H. G., Schild, H., & Hadeler, K. P. (2000). An algorithm for the prediction of proteasomal cleavages. *Journal of Molecular Biology*, *298*(3), 417–429. Available from https://doi.org/10.1006/jmbi.2000.3683.

Lanoix, J., Durette, C., Courcelles, M., Cossette, É., Comtois-Marotte, S., Hardy, M. P., Côté, C., Perreault, C., & Thibault, P. (2018). Comparison of the MHC I immunopeptidome repertoire of B-cell lymphoblasts using two isolation methods. *Proteomics*, *18*(12). Available from https://doi.org/10.1002/pmic.201700251.

Laumont, C. M., Vincent, K., Hesnard, L., Audemard, É., Bonneil, É., Laverdure, J. P., Gendron, P., Courcelles, M., Hardy, M. P., Côté, C., Durette, C., St-Pierre, C., Benhammadi, M., Lanoix, J., Vobecky, S., Haddad, E., Lemieux, S., Thibault, P., & Perreault, C. (2018). Noncoding regions are the main source of targetable tumor-specific antigens. *Science Translational Medicine*, *10*(470). Available from https://doi.org/10.1126/scitranslmed.aau5516.

Lavi, E., Suzumura, A., Murasko, D. M., Murray, E. M., Silberger, D. H., & Weiss, S. R. (1988). Tumor necrosis factor induces expression of MHC class I antigens on mouse astrocytes. *Journal of Neuroimmunology*, 245–253. Available from https://doi.org/10.1016/0165-5728(88)90102-6.

Li, Y., Luisa, S.-C., Antonella, V., Tobias, P., Jiing-Dwan, L., Ana, A., Zeling, C., A., P. P., & Young, Y. (2000). Regulation of transporter associated with antigen processing by phosphorylation. *Journal of Biological Chemistry*, 24130–24135. Available from https://doi.org/10.1074/jbc.M003617200.

Liepe, J., Marino, F., Sidney, J., Jeko, A., Bunting, D. E., Sette, A., Kloetzel, P. M., Stumpf, M. P. H., Heck, A. J. R., & Mishto, M. (2016). A large fraction of HLA class I ligands are proteasome-generated spliced peptides. *Science (New York, N.Y.)*, *354*(6310), 354–358. Available from https://doi.org/10.1126/science.aaf4384.

Liepe, J., Ovaa, H., & Mishto, M. (2018). Why do proteases mess up with antigen presentation by re-shuffling antigen sequences? *Current Opinion in Immunology*, *52*, 81–86. Available from https://doi.org/10.1016/j.coi.2018.04.016.

Löffler, M. W., Mohr, C., Bichmann, L., Freudenmann, L. K., Walzer, M., Schroeder, C. M., Trautwein, N., Hilke, F. J., Zinser, R. S., Mühlenbruch, L., Kowalewski, D. J., Schuster, H., Sturm, M., Matthes, J., Riess, O., Czemmel, S., Nahnsen, S., Königsrainer, I., Thiel, K., . . . Hepavac Consortium. (2019). Multi-omics discovery of exome-derived neoantigens in hepatocellular carcinoma. *Genome Medicine*, *11*(1), 28. Available from https://doi.org/10.1186/s13073-019-0636-8.

Lythe, G., Callard, R. E., Hoare, R. L., & Molina-París, C. (2016). How many TCR clonotypes does a body maintain? *Journal of Theoretical Biology*, *389*, 214–224. Available from https://doi.org/10.1016/j.jtbi.2015.10.016.

Mansurkhodzhaev, A., Barbosa, C. R. R., Mishto, M., & Liepe, J. (2021). Proteasome-generated cis-spliced peptides and their potential role in CD8+ T cell tolerance. *Frontiers in Immunology*, *12*. Available from https://doi.org/10.3389/fimmu.2021.614276.

McMurtrey, C., Harriff, M. J., Swarbrick, G. M., Duncan, A., Cansler, M., Null, M., Bardet, W., Jackson, K. W., Lewinsohn, D. A., Hildebrand, W., & Lewinsohn, D. M. (2017). T cell recognition of *Mycobacterium tuberculosis* peptides presented by HLA-E derived from infected human cells. *PLoS One*, *12*(11). Available from https://doi.org/10.1371/journal.pone.0188288.

Meyer, T. H., van Endert, P. M., Uebel, S., Ehring, B., & Tampé, R. (1994). Functional expression and purification of the ABC transporter complex associated with antigen processing (TAP) in insect cells. *FEBS Letters*, 443–447. Available from https://doi.org/10.1016/0014-5793(94)00908-2.

Michaux, A., Larrieu, P., Stroobant, V., Fonteneau, J.-F., Jotereau, F., Van den Eynde, B. J., Moreau-Aubry, A., & Vigneron, N. (2014). A spliced antigenic peptide comprising a single spliced amino acid is produced in the proteasome by reverse splicing of a longer peptide fragment followed by trimming. *The Journal of Immunology*, 1962–1971. Available from https://doi.org/10.4049/jimmunol.1302032.

Mishto, M., Goede, A., Taube, K. T., Keller, C., Janek, K., Henklein, P., Niewienda, A., Kloss, A., Gohlke, S., Dahlmann, B., Enenkel, C., & Kloetzel, P. M. (2012). Driving forces of proteasome-catalyzed peptide splicing in yeast and humans. *Molecular and Cellular Proteomics*, *11*(10), 1008–1023. Available from https://doi.org/10.1074/mcp.M112.020164.

Mishto, M., & Liepe, J. (2017). Post-translational peptide splicing and T cell responses. *Trends in Immunology*, 904–915. Available from https://doi.org/10.1016/j.it.2017.07.011.

Mishto, M., Mansurkhodzhaev, A., Rodriguez-Calvo, T., & Liepe, J. (2021). Potential mimicry of viral and pancreatic β cell antigens through non-spliced and cis-spliced Zwitter epitope candidates in type 1 diabetes. *Frontiers in Immunology*, *12*. Available from https://doi.org/10.3389/fimmu.2021.656451.

Mou, Z., Li, J., Boussoffara, T., Kishi, H., Hamana, H., Ezzati, P., Hu, C., Yi, W., Liu, D., Khadem, F., Okwor, I., Jia, P., Shitaoka, K., Wang, S., Ndao, M., Petersen, C., Chen, J., Rafati, S., Louzir, H., & Uzonna, J. E. (2015). Identification of broadly conserved cross-species protective Leishmania antigen and its responding CD4+ T cells. *Science Translational Medicine*, *7*(310). Available from https://doi.org/10.1126/scitranslmed.aac5477.

Newey, A., Griffiths, B., Michaux, J., Pak, H. S., Stevenson, B. J., Woolston, A., Semiannikova, M., Spain, G., Barber, L. J., Matthews, N., Rao, S., Watkins, D., Chau, I., Coukos, G., Racle, J., Gfeller, D., Starling, N., Cunningham, D., Bassani-Sternberg, M., & Gerlinger, M. (2019). Immunopeptidomics of colorectal cancer organoids reveals a sparse HLA class I neoantigen landscape and no increase in neoantigens with interferon or MEK-inhibitor treatment. *Journal for ImmunoTherapy of Cancer*, *7*(1). Available from https://doi.org/10.1186/s40425-019-0769-8.

Paes, W., Leonov, G., Partridge, T., Chikata, T., Murakoshi, H., Frangou, A., Brackenridge, S., Nicastri, A., Smith, A. G., Learn, G. H., Li, Y., Parker, R., Oka, S., Pellegrino, P., Williams, I., Haynes, B. F., McMichael, A. J., Shaw, G. M., Hahn, B. H., ... Borrow, P. (2019). Contribution of proteasome-catalyzed peptide cis-splicing to viral targeting by CD8+ T cells in HIV-1 infection. *Proceedings of the National Academy of Sciences of the United States of America*, 24748–24759. Available from https://doi.org/10.1073/pnas.1911622116.

Parker, K. C., Bednarek, M. A., & Coligan, J. E. (1994). Scheme for ranking potential HLA-A2 binding peptides based on independent binding of individual peptide side-chains. *Journal of Immunology*, *152*(1), 163–175.

Perkins, D. N., Pappin, D. J. C., Creasy, D. M., & Cottrell, J. S. (1999). Probability-based protein identification by searching sequence databases using mass spectrometry data. *Electrophoresis*, *20*(18), 3551–3567, https://doi.org/10.1002/(SICI)1522-2683(19991201)20:18 < 3551::AID-ELPS3551 > 3.0.CO;2-2.

Plattel, A. C. M., Liepe, J., Textoris-Taube, K., Keller, C., Henklein, P., Schalkwijk, H. H., Cardoso, R., Kloetzel, P. M., Mishto, M., & Sijts, A. J. A. M. (2017). Multi-level strategy for identifying proteasome-catalyzed spliced epitopes targeted by CD8 + T cells during bacterial infection. *Cell Reports*, 1242–1253. Available from https://doi.org/10.1016/j.celrep.2017.07.026.

Ramakrishna, V., Ross, M. M., Petersson, M., Gatlin, C. C., Lyons, C. E., Miller, C. L., Myers, H. E., McDaniel, M., Karns, L. R., Kiessling, R., Parmiani, G., & Flyer, D. C. (2003). Naturally occurring peptides associated with HLA-A2 in ovarian cancer cell lines identified by mass spectrometry are targets of HLA-A2-restricted cytotoxic T cells. *International Immunology*, *15*(6), 751–763. Available from https://doi.org/10.1093/intimm/dxg074.

Rammensee, H. G., Bachmann, J., Emmerich, N. P. N., Bachor, O. A., & Stevanović, S. (1999). SYFPEITHI: Database for MHC ligands and peptide motifs. *Immunogenetics*, *50*(3–4), 213–219. Available from https://doi.org/10.1007/s002510050595.

Rappuoli, R., & Aderem, A. (2011). A 2020 vision for vaccines against HIV, tuberculosis and malaria. *Nature*, *473*(7348), 463–469. Available from https://doi.org/10.1038/nature10124.

Rappuoli, R., Bottomley, M. J., D'Oro, U., Finco, O., & De Gregorio, E. (2016). Reverse vaccinology 2.0: Human immunology instructs vaccine antigen design. *Journal of Experimental Medicine*, *213*(4), 469–481. Available from https://doi.org/10.1084/JEM.20151960.

Reeves, E., & James, E. (2017). Antigen processing and immune regulation in the response to tumours. *Immunology*, 16–24. Available from https://doi.org/10.1111/imm.12675.

Rist, M. J., Theodossis, A., Croft, N. P., Neller, M. A., Welland, A., Chen, Z., Sullivan, L. C., Burrows, J. M., Miles, J. J., Brennan, R. M., Gras, S., Khanna, R., Brooks, A. G., McCluskey, J., Purcell, A. W., Rossjohn, J., & Burrows, S. R. (2013). HLA peptide length preferences control CD8 + T cell responses. *Journal of Immunology*, *191*(2), 561–571. Available from https://doi.org/10.4049/jimmunol.1300292.

Ritz, D., Gloger, A., Weide, B., Garbe, C., Neri, D., & Fugmann, T. (2016). High-sensitivity HLA class I peptidome analysis enables a precise definition of peptide motifs and the identification of peptides from cell lines and patients' sera. *Proteomics*, *16*(10), 1570–1580. Available from https://doi.org/10.1002/pmic.201500445.

Rock, K. L., & Goldberg, A. L. (1999). Degradation of cell proteins and the generation of MHC Class I-presented peptides. *Annual Review of Immunology*, 739–779. Available from https://doi.org/10.1146/annurev.immunol.17.1.739.

Sanchez-Trincado, J. L., Gomez-Perosanz, M., & Reche, P. A. (2017). Fundamentals and methods for T- and B-cell epitope prediction. *Journal of Immunology Research*, 1–14. Available from https://doi.org/10.1155/2017/2680160.

Seliger, B., Hammers., Höhne., Zeidler., Knuth, C. D., Gerharz, C., & Huber. (1997). IFN-gamma-mediated coordinated transcriptional regulation of the human TAP-1 and LMP-2 genes in human renal cell carcinoma. *Clinical Cancer Research*, *3*, 573–578.

Sharma, G., Rive, C. M., & Holt, R. A. (2019). Rapid selection and identification of functional CD8 + T cell epitopes from large peptide-coding libraries. *Nature Communications*, *10*(1). Available from https://doi.org/10.1038/s41467-019-12444-7.

Shetty, V., Sinnathamby, G., Nickens, Z., Shah, P., Hafner, J., Mariello, L., Kamal, S., Vlahovic', G., Lyerly, H. K., Morse, M. A., & Philip, R. (2011). MHC class I-presented lung cancer-associated tumor antigens identified by immunoproteomics analysis are targets for cancer-specific T cell response. *Journal of Proteomics*, *74*(5), 728–743. Available from https://doi.org/10.1016/j.jprot.2011.02.020.

Singh, H., & Raghava, G. P. S. (2001). ProPred: Prediction of HLA-DR binding sites. *Bioinformatics (Oxford, England)*, 1236–1237. Available from https://doi.org/10.1093/bioinformatics/17.12.1236.

Song, Z., Liu, T., Jiao, C., Liu, W., Zhu, M., Guo, Y., & Wang, X. (2009). Specificity of the proteasome cleavage to the antigen protein. *Science in China Series B: Chemistry*, *52*, 39–47. Available from https://doi.org/10.1007/s11426-008-0107-3.

Speir, J. A., Stevens, J., Joly, E., Butcher, G. W., & Wilson, I. A. (2001). Two different, highly exposed, bulged structures for an unusually long peptide bound to rat MHC class I RT1-Aa. *Immunity*, *14*(1), 81–92. Available from https://doi.org/10.1016/S1074-7613(01)00091-7.

Stifter, K., Krieger, J., Ruths, L., Gout, J., Mulaw, M., Lechel, A., Kleger, A., Seufferlein, T., Wagner, M., & Schirmbeck, R. (2020). IFN-γ treatment protocol for MHC-I^{lo}/PD-L1$^+$ pancreatic tumor cells selectively restores their TAP-mediated presentation competence and CD8 T-cell priming potential. *Journal for ImmunoTherapy of Cancer*, *8*(2). Available from https://doi.org/10.1136/jitc-2020-000692.

Stone, J. D., & Kranz, D. M. (2013). Role of T cell receptor affinity in the efficacy and specificity of adoptive T cell therapies. *Frontiers in Immunology*, *4*. Available from https://doi.org/10.3389/fimmu.2013.00244.

Tailleux, L., Waddel, S. J., Pelizzola, M., Mortellaro, A., Withers, M., Tanne, A., Castagnoli, P. R., Gicquel, B., Stoker, N. G., Butcher, P. D., Foti, M., & Neyrolles, O. (2008). Probing host pathogen cross-talk by transcriptional profiling of both *Mycobacterium tuberculosis* and infected human dendritic cells and macrophages. *PLoS One*, *3*(1). Available from https://doi.org/10.1371/journal.pone.0001403.

Tan, M. P., Gerry, A. B., Brewer, J. E., Melchiori, L., Bridgeman, J. S., Bennett, A. D., Pumphrey, N. J., Jakobsen, B. K., Price, D. A., Ladell, K., & Sewell, A. K. (2015). T cell receptor binding affinity governs the functional profile of cancer-specific CD8 + T cells. *Clinical and Experimental Immunology*, *180*(2), 255–270. Available from https://doi.org/10.1111/cei.12570.

Ternette, N., Olde Nordkamp, M. J. M., Müller, J., Anderson, A. P., Nicastri, A., Hill, A. V. S., Kessler, B. M., & Li, D. (2018). Immunopeptidomic profiling of HLA-A2-positive triple negative breast cancer identifies potential immunotherapy target antigens. *Proteomics*, *18*(12). Available from https://doi.org/10.1002/pmic.201700465.

Ternette, N., Yang, H., Partridge, T., Llano, A., Cedeño, S., Fischer, R., Charles, P. D., Dudek, N. L., Mothe, B., Crespo, M., Fischer, W. M., Korber, B. T. M., Nielsen, M., Borrow, P., Purcell, A. W., Brander, C., Dorrell, L., Kessler, B. M., & Hanke, T. (2016). Defining the HLA class I-associated viral antigen repertoire from HIV-1-infected human cells. *European Journal of Immunology*, *46*(1), 60–69. Available from https://doi.org/10.1002/eji.201545890.

Testa, J. S., Shetty, V., Hafner, J., Nickens, Z., Kamal, S., Sinnathamby, G., & Philip, R. (2012). MHC class I-presented T cell epitopes identified by immunoproteomics analysis are targets for a cross reactive influenza-specific T cell response. *PLoS One*, *7*(11). Available from https://doi.org/10.1371/journal.pone.0048484.

Testa, J. S., Shetty, V., Sinnathamby, G., Nickens, Z., Hafner, J., Kamal, S., Zhang, X., Jett, M., & Philip, R. (2012). Conserved MHC class I-presented dengue virus epitopes identified by immunoproteomics analysis are targets for cross-serotype reactive T-cell response. *Journal of Infectious Diseases*, *205*(4), 647–655. Available from https://doi.org/10.1093/infdis/jir814.

Theaker, S. M., Rius, C., Greenshields-Watson, A., Lloyd, A., Trimby, A., Fuller, A., Miles, J. J., Cole, D. K., Peakman, M., Sewell, A. K., & Dolton, G. (2016). T-cell libraries allow simple parallel generation of multiple peptide-specific human T-cell clones. *Journal of Immunological Methods*, *430*, 43–50. Available from https://doi.org/10.1016/j.jim.2016.01.014.

Toes, R. E. M., Nussbaum, A. K., Degermann, S., Schirle, M., Emmerich, N. P. N., Kraft, M., Laplace, C., Zwinderman, A., Dick, T. P., Müller, J., Schönfisch, B., Schmid, C., Fehling, H. J., Stevanovic, S., Rammensee, H. G., & Schild, H. (2001). Discrete cleavage motifs of constitutive and immunoproteasomes revealed by quantitative analysis of cleavage products. *Journal of Experimental Medicine*, *194*(1), 1–12. Available from https://doi.org/10.1084/jem.194.1.1.

Tran, N. H., Qiao, R., Xin, L., Chen, X., Liu, C., Zhang, X., Shan, B., Ghodsi, A., & Li, M. (2019). Deep learning enables de novo peptide sequencing from data-independent-acquisition mass spectrometry. *Nature Methods*, *16*(1), 63–66. Available from https://doi.org/10.1038/s41592-018-0260-3.

Ullah, M. A., Sarkar, B., & Islam, S. S. (2020). Exploiting the reverse vaccinology approach to design novel subunit vaccines against Ebola virus. *Immunobiology*, *225*(3). Available from https://doi.org/10.1016/j.imbio.2020.151949.

Vigneron, N., Stroobant, V., Chapiro, J., Ooms, A., Degiovanni, G., Morel, S., Van Der Bruggen, P., Boon, T., & Van Den Eynde, B. J. (2004). An antigenic peptide produced by peptide splicing in the proteasome. *Science (New York, N.Y.)*, *304*(5670), 587–590. Available from https://doi.org/10.1126/science.1095522.

Wieczorek, M., Abualrous, E. T., Sticht, J., Álvaro-Benito, M., Stolzenberg, S., Noé, F., & Freund, C. (2017). Major histocompatibility complex (MHC) class I and MHC class II proteins: Conformational plasticity in antigen presentation. *Frontiers in Immunology*, *8*. Available from https://doi.org/10.3389/fimmu.2017.00292.

Wiesel, M., Walton, S., Richter, K., & Oxenius, A. (2009). Virus-specific CD8 T cells: Activation, differentiation and memory formation. *APMIS: Acta Pathologica, Microbiologica, et Immunologica Scandinavica*, *117*(5–6), 356–381. Available from https://doi.org/10.1111/j.1600-0463.2009.02459.x.

Wölk, B., Trautwein, C., Büchele, B., Kersting, N., Blum, H. E., Rammensee, H.-G., Cerny, A., Stevanovic, S., Moradpour, D., Brass, V., & Bertoletti, A. (2012). Identification of naturally processed hepatitis C virus-derived major histocompatibility complex class I ligands. *PLoS One*e29286. Available from https://doi.org/10.1371/journal.pone.0029286.

Wooldridge, L., Ekeruche-Makinde, J., Van Den Berg, H. A., Skowera, A., Miles, J. J., Tan, M. P., Dolton, G., Clement, M., Llewellyn-Lacey, S., Price, D. A., Peakman, M., & Sewell, A. K. (2012). A single autoimmune T cell receptor recognizes more than a million different peptides. *Journal of Biological Chemistry*, *287*(2), 1168–1177. Available from https://doi.org/10.1074/jbc.M111.289488.

Yang, J., Mott, D., Sutiwisesak, R., Lu, Y. J., Raso, F., Stowell, B., Babunovic, G., Lee, J., Carpenter, S., Way, S. S., Fortune, S., & Behar, S. (2018). *Mycobacterium tuberculosis*-specific CD4 + and CD8 + T cells differ in their capacity to recognize infected macrophages. *BioRxiv*. https://doi.org/10.1101/262493.

Yewdell, J. W., & Bennink, J. R. (1999). Immunodominance in major histocompatibility complex class I–restricted T lymphocyte responses. *Annual Review of Immunology*, 51–88. Available from https://doi.org/10.1146/annurev.immunol.17.1.51.

Yewdell, J. W., Reits, E., & Neefjes, J. (2003). Making sense of mass destruction: Quantitating MHC class I antigen presentation. *Nature Reviews Immunology*, 952–961. Available from https://doi.org/10.1038/nri1250.

Zhang, G. L., Khan, A. M., Srinivasan, K. N., August, J. T., & Brusic, V. (2005). MULTIPRED: A computational system for prediction of promiscuous HLA binding peptides. *Nucleic Acids Research, 33*(2), W172–W179. Available from https://doi.org/10.1093/nar/gki452.

Zhou, F. (2009). Molecular mechanisms of IFN-γ to up-regulate MHC class I antigen processing and presentation. *International Reviews of Immunology, 28*(3–4), 239–260. Available from https://doi.org/10.1080/08830180902978120.

Chapter 6

Virus-like particle vaccines

Ekta Mukhopadhyay, Adrian V.S. Hill and Ahmed M. Salman
Jenner Institute, University of Oxford, Oxford, United Kingdom

Introduction

Virus-like particles (VLPs) are noninfectious nanoparticles that are multimeric in nature and do not contain any genetic material (Qian et al., 2020), expressing the desired antigen on the surface. Hence, VLPs morphologically resemble viruses (Syomin & Ilyin, 2019). As many as 110 viral proteins from over 30 viruses that can self-assemble into VLPs have been explored as vaccines. A few examples of these are hepatitis B virus (HBV; Salman, Montoya-Díaz, et al., 2017), Papilloma virus, and bacteriophage Qβ (Alves et al., 2017; Bárcena & Blanco, 2013). The resulting VLPs have been modified, for example, expression of repeated array of antigenic peptides on the VLP surface to generate novel vaccines, and encapsulation of drugs and nucleic acids inside the VLP for targeted delivery (Qian et al., 2020). As VLPs lack genetic material (Roldão et al., 2017), they have no replication capacity and are considered safe. Currently, there are several VLP-based licensed vaccines for use in humans including Recombivax HB and Engerix-B for HBV, Gardasil, Cervarix, and Gardasil-9 for human papillomavirus (HPV), and Hecolin for hepatitis E virus (Qian et al., 2020). VLPs have been successfully combined for example in Gardasil 9 (Yusupov et al., 2019), a nine-valent HPV vaccine, providing protection against multiple strains of HPV in a single formulation (Garg et al., 2020). The applications of VLPs are not limited to use as vaccines and drug delivery vehicles, as they have been explored for use in diagnostic imaging (Manchester & Singh, 2006), tools for antiviral drug screening (Zhang et al., 2020) and catalytically active antimicrobial systems (Patterson et al., 2014).

Immune response to virus-like particles

The biggest advantage of VLP vaccines is the expression of an immunodominant epitope on the surface, eliciting high antibody titers (Caldeira et al., 2020), and effective CD4 proliferative and cytotoxic T lymphocyte responses

Vaccinology and Methods in Vaccine Research. DOI: https://doi.org/10.1016/B978-0-323-91146-7.00007-X

(Noad & Roy, 2003). As VLPs commonly have a diameter of 20–200 nm, which is optimal for recognition by dendritic cells and other antigen-presenting cells (APCs), they are transported to lymph nodes (LNs), and able to induce a strong immune response (Qian et al., 2020). The VLPs that reach the LNs are endocytosed by APCs, enter the T cell zone, and are processed and efficiently presented by major histocompatibility complex class I, eliciting cell-mediated immune responses (Mohsen et al., 2018). Unmodified VLPs entering the B cell zone of the LNs are able to induce B cell receptor cross-linking, thus triggering endocytosis, B cell proliferation, and activation. Following activation, the B cells interact with Th cells leading to immunoglobulin class switch and generation of B cell memory (Zepeda-Cervantes et al., 2020).

Production of virus-like particles

For the production of VLPs with desired yield and purity, a number of parameters are crucial and need optimization. These include choice of expression vector, upstream process parameters like media and growth conditions, and time of induction (for an inducible protein subunit). High yields of VLP with minimal contaminants, for example, host cell protein (HCP) and host cell DNA require efficient reduction or removal steps to enhance purity of the final product. Fig. 6.1 provides an overview of the VLP production process. Further information relating to the manufacturing process is provided in Chapter 10, GMP manufacture and quality control.

Expression systems

Various expression systems, such as bacteria, yeast, insect, mammalian, avian, and plant, have been used for production of VLPs commercially (Ding et al., 2018). This section describes some of the expression systems in widespread use.

Insect cells

An insect cell and baculovirus expression vector system (BEVS) have the advantages of high cell densities and protein expression, with eukaryotic post-translational modifications (Fernandes et al., 2013). This expression system is mainly free from adventitious agents as insect cells do not permit growth of many mammalian viruses, with the exception of arboviruses (Kim & Kim, 2017). It allows expression of individual proteins by coinfection with multiple baculoviruses, each expressing a single protein, or via a single virus designed to express multiple proteins (Kost et al., 2005). Insect cells can grow in serum-free media, eliminating the risk of introduction of adventitious viruses requiring additional purification steps (Vicente et al., 2011). The first VLP produced using this system was a Gag

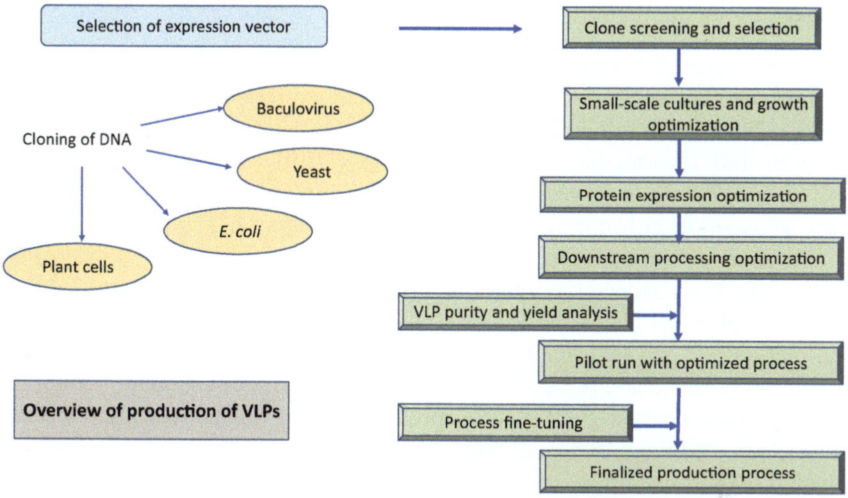

FIGURE 6.1 Overview of VLP production process and steps needing optimization: After selection of antigen, a DNA expression vector is constructed, and transfected into the cell system of choice (e.g., *Pichia pastoris* yeast). Multiple clones are selected, propagated in small cultures, and induced for expression analysis. The clone(s) with the ideal growth density and protein expression capacity is expanded further and processed through steps of downstream processing (VLP purification). Each stage requires optimization in terms of media and growth conditions, ideal time of induction, and removal of impurities to improve product purity. After process and analytical procedures are optimized, a pilot scale run is performed to review the whole process and identify any steps that require fine-tuning, before the process is finalized and GMP manufacture can be initiated.

VLP for Feline immunodeficiency virus, used for analytical applications (Gopal & Schneemann, 2018). GSK's CervarixTM is a licensed VLP vaccine for HPV, to protect against cervical cancer; it was produced using insect cells (Liu et al., 2013), demonstrating that the expression system is scalable and highly productive (Matsuda et al., 2020).

Spodoptera frugiperda (*Sf9* or *Sf21, Sf2*) and *Trichoplusia ni* (*T. ni* or High Five) are insect cells most commonly used for production of VLPs (Fuenmayor et al., 2017; Lee et al., 2000). *Sf2* holds an advantage over the other types of cells, as they do not lyse on infection with baculovirus and provide comparable high protein expression levels (Lee et al., 2000). There are a few issues associated with BEVS: one of these is the similar size and density of VLPs to budding baculoviruses released into the medium, making separation in the downstream processing challenging. This is being addressed by creation of a genetically modified VP80-null baculovirus lacking a key structural protein (Yee et al., 2018). Although insect cells can add eukaryotic glycosylation, they are less complex and less human-like; if glycosylation is important for potency, insect cells can potentially be engineered to express mammalian glycotransferases (Zitzmann et al., 2017).

Yeast

Yeast is a unicellular eukaryote and has been widely used in production of proteins and VLPs (Kim & Kim, 2017). As an expression system, yeast provides advantages including eukaryotic post-translational modifications, fast growth, and low media and operation costs (Bill, 2015; Saraswat et al., 2016). Commonly used yeast for VLP production are *Saccharomyces cerevisiae*, *Hansenula polymorpha*, and *P. pastoris* (Kumar & Kumar, 2019), with *P. pastoris* providing more human-like glycosylation and *H. polymorpha* providing less hyper-glycosylation, which helps correct protein folding (Manfrão-Netto et al., 2019; Wang et al., 2013). *H. polymorpha* is also thermotolerant and can withstand temperatures up to 50°C (Manfrão-Netto et al., 2019). Over the years, these strains have been modified by genetic manipulation to obtain desired glycosylation profiles, which might be critical for the activity of certain proteins or antigens contained in VLPs (Vieira Gomes et al., 2018). Examples of licensed VLP vaccines produced using yeast include Engerix-B, a Hepatitis B vaccine from GSK, and Gardasil for HPV, from Merck and Co. (Shiyu Dai & Fei, 2018). In recent years, there has been interest in using cell-free protein synthesis (CFPS) for proteins that are difficult to express, with promising early results for VLP production using a *P. pastoris* CFPS (Spice et al., 2020).

We have growing in-house experience in Oxford of the use of a wide range of VLPs to target malaria and other diseases by vaccination. Important examples include the improved biosimilar of the RTS,S *Plasmodium falciparum* vaccine developed at Oxford, called R21 (Collins et al., 2017; Datoo et al., 2021; Venkatraman et al., 2019) that has shown immunogenicity and efficacy exceeding that of RTS,S (RTS,S Clinical Trials Partnership et al., 2011, 2012) and the Rv21 vaccine for *Plasmodium vivax* (Salman, Montoya-Diaz, et al., 2017) that is currently awaiting GMP manufacture before clinical trials. Both VLP-based vaccines are made of hepatitis B fusion proteins with most of the malaria circumsporozoite protein expressed in *P. pastoris* and have proven their ability to induce very potent neutralizing antibody responses. Recent field trial data for R21 suggest that, when used with a saponin adjuvant, it induces the highest level of antigen-specific antibody induced by any subunit vaccine for any indication (Collins et al., 2021; Datoo et al., 2021).

Escherichia coli

Escherichia coli has been successfully used for the production of biopharmaceuticals; of over 170 candidate VLP vaccines produced, approximately 28% have been made using bacterial systems, mainly *E. coli*. These include two licensed VLP vaccines, HeberNasvac for hepatitis B and Hecolin for hepatitis E (Huang et al., 2017). Although lacking capacity for eukaryotic post-translational modifications, and issues with expressing proteins over 60 kDa (Rosano & Ceccarelli, 2014), *E. coli* is often the preferred choice for protein and VLP production due to the high biomass achievable in commercial manufacture, especially for proteins that can self-assemble into VLPs in vitro (Liew et al., 2010).

Plant cells

Plant cells have been used for production of proteins and VLPs as they provide an economical, speedy, and scalable platform (Chen & Lai, 2013). With the availability of online immunoinformatic tools for antigen epitope prediction, validation, and allergenicity evaluation, Ghorbani et al. developed a platform to design a SARS-CoV-2 vaccine in plant cells (Ghorbani et al., 2020). A number of VLPs are in early phase clinical trials and are produced using various plants including potato, tobacco, spinach, and lettuce (Chen & Lai, 2013), for example a Norwalk virus vaccine (Lai & Chen, 2012).

Manufacturing: upstream process

The initiation of the upstream process begins with starting a small-scale seed culture of the selected clone (with optimal expression yield), in a shake flask (125 mL−1 L) (Mignaqui et al., 2020) or cryobags (Bogli et al., 2012). This is expanded to a large culture, which can be (1) batch mode, where no further addition of media or additives takes place, (2) fed-batch mode, where media components are added during culture to give high cell densities, or (3) continuous cultivation, where fresh medium is continuously added (Fuenmayor et al., 2017). For expansion of cells to large cultures using these modes, systems such as wave bioreactors with capacity of 10−100 L, and disposable stirred tank bioreactors up to 1000 L have been used (Durous et al., 2019). Parameters including pH, aeration, temperature, dissolved oxygen, cell density, time of induction, and harvest (Lai et al., 2019) are critical, and often optimized using small-scale laboratory bioreactors (Tripathi & Shrivastava, 2019).

High cell densities are achievable, for example, *E. coli* can grow up to 14 g/L wet weight in Terrific medium broth in a baffled shake flask, and up to 172 g/L wet weight in a 30 L fermenter using modified minimal medium in a fed batch mode (Kangwa et al., 2015). Published data suggest yeast can grow over 100 g/L in a fed batch mode, while insect cells can reach a cell density of over 10^6 cells/mL in batch and fed batch modes (Shojaosadati et al., 2008). For production of VLPs in plant cells, the gene of interest is carried by plant-specific vectors, for example, pEAQ, using gram-negative bacterium *Agrobacterium tumefaciens*. Infection of plant cells is initiated in a culture or intact leaves using agroinfiltration, where transformed *A. tumefaciens* are delivered using a syringe or vacuum. A protein yield of about 400 mg/kg of leaf weight was obtained with Norwalk virus capsid protein

VLPs when plant cells were harvested 7 days postinfection (Lai & Chen, 2012). In the case of yeast, for example, *P. pastoris*, protein expression is often driven by the strong AOX1 (Alcohol oxidase 1) promoter that is induced by transferring cells from growth inducing media to media containing methanol. Hepatitis B surface antigen VLPs have been produced using this system in a fed batch mode with a yield of 2.3 g/L after 90–100 hours in media containing methanol (Gurramkonda et al., 2009).

Manufacturing: purification of virus-like particles (downstream process)

The choice of expression system determines the first steps of purification. When expressed proteins are capable of self-assembly, VLPs are harvested from the cell cytoplasm or culture medium. At the optimum harvest time, clarification is often the first step to obtain cells or media containing the VLPs, and remove contaminants like low-molecular-weight HCPs and DNA (Pattenden et al., 2005); this is achieved by dead end and depth filters of 0.2–3 μm pore size (Moleirinho et al., 2020). For isolation of VLPs from cells, the cell pellet is resuspended in buffer containing detergent, for example, Triton X-100, and subjected to high pressure homogenization (Mortola & Roy, 2004). This leads to cell breakage and release of self-assembled VLPs (Milburn & Dunnill, 1994). In some cases, the proteins translocate into the endoplasmic reticulum forming multilayer lamellar structures, as seen in hepatitis B surface antigen produced in *P. pastoris*, and require an additional step using detergents, to release VLPs comprising both lipid and protein (Lünsdorf et al., 2011). For purification of VLPs released into the culture media, postclarification, the supernatant is subjected to density gradient or rate zonal ultracentrifugation to collect VLPs by sedimentation (Arevalo et al., 2016; Van Zyl & Hitzeroth, 2016). As this approach is not scalable, chromatography is considered to be a better option when large quantities are required.

VLPs can be purified using ion exchange chromatography by selective adsorption of VLPs on the basis of charge (Goodridge et al., 2004) and elution using high salt concentration (Silva et al., 2020). Affinity chromatography employing a classic bind-wash-elute approach is an alternative purification approach for VLPs containing a tag, for example, immunoaffinity chromatography (antigen–antibody interaction), or a poly-Histidine tag binds to resins containing immobilized nickel ions (Zhao et al., 2019). The bound VLPs are eluted using low pH (e.g., 0.1 M glycine-HCl, pH 2.5) for antibody interactions, high concentration of imidazole for His-tagged proteins (>200 mM) or strong chelators (e.g., EDTA) (ThermoFisher Scientific, 2011). Size exclusion chromatography is sometimes used as a final polishing step for obtaining homogeneous population of VLPs, and buffer exchange into a stable formulation buffer (Lagoutte et al., 2016; Moleirinho et al., 2020).

Major contaminants and removal strategies

Intermediate purification steps are often included for reduction or removal of contaminants including HCP, host cell DNA, and endotoxin, to obtain a final VLP product of desired purity. The type and levels of contaminants depend on the particular expression system employed. In *E. coli*, HCP is less of a concern than nucleic acid, and a digestion step using benzonase is often incorporated at early stages of purification to reduce DNA levels (Hillebrandt et al., 2020). While this might be more applicable at smaller scales, Immobilized Metal Affinity Chromatography (IMAC) has been found scalable and efficient in removing host cell nucleic acids and endotoxin (Ongkudon et al., 2012). Ultrafiltration and ion-exchange chromatography are also commonly used techniques for endotoxin removal (Magalhaes et al., 2007). A recently developed CaptoTM Core700 resin allows exclusion of large molecular weight contaminants ($>$700,000 kDa), while smaller impurities including partially assembled VLPs, and smaller HCPs, are retained in the octylamine resin beads (Lagoutte et al., 2016). A anion exchange chromatography using a strong exchanger is the most widely used method for removal or reduction of DNA, as this contaminant carries a negative charge. Some resins with proven successful reduction in host cell DNA levels are Q Sepharose Fast Flow, axial flow monolithic strong anion exchanger (CIM QA), and axial flow membrane strong anion exchanger (Sartobind Q pico) (Hernández et al., 2016).

Analytical tools for virus-like particles

Analytical qualification and quantification of VLPs is performed to determine the yield and purity at end of the process. These tools are also used during process optimization steps as in Fig. 6.1, to quantitate the reduction of a contaminant. Commonly used tools for analysis of VLPs are described below.

Total protein and purity analysis

Coomassie or silver staining of proteins separated by sodium dodecyl sulfate-polyacrylamide gel electrophoresis (SDS-PAGE) provides a visual representation of the total proteins present in VLPs, along with contaminating host cell or media proteins, separated by size. Western blotting (see Chapter 4) is normally used for identification of specific protein sequences in VLPs (Thompson et al., 2013). Bicinchoninic acid assay is commonly used for quantifying total protein, reacting with peptides to produce a colorimetric readout (Tang et al., 2011). Techniques such as SDS-PAGE and peptide fingerprinting are used to assess the purity and confirm the composition of the VLP (McCraw et al., 2018). For quantification of host cell DNA in the in-process controls, and final VLP formulation, real-time PCR is a highly sensitive method with a limit of detection in the subpicogram range, while cell-specific kits, for example, enzyme-linked immunoassays are useful tools for quantifying residual HCPs (Weihong Wang, 2016).

Size and morphology confirmation

Transmission electron microscopy or scanning electron microscopy is used for confirming the presence of VLPs, their morphology and size (González-Domínguez et al., 2020; Pushko et al., 2005). This is usually performed by adsorption of a sample on copper grids, and negative staining with uranyl acetate (Krammer et al., 2010). Cryo-electron microscopy is used for detailed study and structural guided evaluation of VLPs (McCraw et al., 2018). Dynamic light scattering is a quick and highly sensitive, widely employed method for determining the size and homogeneity of VLP population ($\sim$0.1 nm$-$1 μm); it is based on light scattering caused by Brownian motion of particles (Challener, 2015). Other techniques that provide high-throughput quantification of VLPs include super-resolution fluorescence microscopy, nanoparticle tracking analysis, and flow virometry (González-Domínguez et al., 2020). High-performance liquid chromatography (HPLC)-based methods including reverse phase HPLC and high-performance size exclusion chromatography (HP-SEC) are frequently used for estimating molecular mass and size uniformity of VLPs, with HP-SEC able to separate monomers, aggregates, and fragments depending on the pore size of the matrix (Ramesh Kumar & Rathore, 2020; Yang et al., 2020). Analytical ultracentrifuge is used for determining broad range size distribution of particles and can analyze particles ranging from 100 Da to 10^9 Da (Berkowitz, 2015).

Composition analysis

Western blotting with antibodies specific for a particular protein sequence can confirm the presence of an antigen band of the correct size. Mass spectrometry is a robust and highly sensitive method that can confirm protein identity and provide additional information on amino acid composition and post-translational modifications (Ramesh Kumar & Rathore, 2020). Differential scanning calorimetry and differential scanning fluorescence use thermodynamic parameters to analyze critical structural changes, to predict the thermal stability of VLPs (Yang et al., 2020).

Conclusion

This chapter reviews the application of VLPs as vaccines, and the process involved in production and analysis. There is a need for process development and optimization at each stage of VLP production, to obtain a final product of the desired yield and purity, with system- and process-related contaminant quantities within an acceptable range, defined by regulatory guidelines for vaccines used in clinical trials. For upstream steps, the choice of the expression system, media, and scale-up parameters is crucial as it defines the downstream processes needed for the effective removal of contaminants, without affecting the overall yield and product quality.

References

Alves, E., Salman, A. M., Leoratti, F., Lopez-Camacho, C., Viveros-Sandoval, M. E., Lall, A., El-Turabi, A., Bachmann, M. F., Hill, A. V. S., Janse, C. J., Khan, S. M., Reyes-Sandoval, A., & Wilkins, P. P. (2017). Evaluation of *Plasmodium vivax* cell-traversal protein for ookinetes and sporozoites as a preerythrocytic *P. vivax* vaccine. *Clinical and Vaccine Immunology*, *24*(4), e00501–e00516.

Arevalo, M. T., Wong, T. M., & Ross, T. M. (2016). Expression and purification of virus-like particles for vaccination. *Journal of Visualized Experiments,* (112)). Available from https://doi.org/10.3791/54041.

Bárcena, J., & Blanco, E. (2013). Design of novel vaccines based on virus-like particles or chimeric virions. In M. G. Mateu Dordrecht (Ed.), *Structure and physics of viruses: An integrated textbook* (pp. 631–665). Netherlands: Springer.

Berkowitz, S. A., & Philo, J. S. (2015). Chapter 9 - Characterizing biopharmaceuticals using analytical ultracentrifugation. In: Biophysical Characterization of Proteins in Developing Biopharmaceuticals, D. J. Houde and S. A. Berkowitz. Amsterdam, Elsevier: 211–260.

Bill, R. M. (2015). Recombinant protein subunit vaccine synthesis in microbes: A role for yeast? *The Journal of Pharmacy and Pharmacology*, *67*(3), 319–328.

Bogli, N. C., Ries, C., Adams, T., Greller, G., Eibi, D., & Eibi, R. (2012). Large-scale, insect-cell–based vaccine development. *BioProcess International.*

Caldeira, J. C., Perrine, M., Pericle, F., & Cavallo, F. (2020). Virus-like particles as an immunogenic platform for cancer vaccines. *Viruses*, *12*, 5.

Challener, C. A. (2015). Dynamic light scattering for non-destructive, rapid analysis of virus-like particles. *BioPharm International Webpage*, *28*(10), 46–49.

Chen, Q., & Lai, H. (2013). Plant-derived virus-like particles as vaccines. *Human Vaccines & Immunotherapeutics*, *9*(1), 26–49.

Collins, K. A., Brod, F., Snaith, R., Ulaszewska, M., Longley, R. J., Salman, A. M., Gilbert, S. C., Spencer, A. J., Franco, D., Ballou, W. R., & Hill, A. V. S. (2021). Ultra-low dose immunization and multi-component vaccination strategies enhance protection against malaria in mice. *Scientific Reports*, *11*(1), 10792.

Collins, K. A., Snaith, R., Cottingham, M. G., Gilbert, S. C., & Hill, A. V. S. (2017). Enhancing protective immunity to malaria with a highly immunogenic virus-like particle vaccine. *Scientific Reports*, *7*, 46621.

Datoo, M. S., Natama, M. H., Some, A., Traore, O., Rouamba, T., Bellamy, D., Yameogo, P., Valia, D., Tegneri, M., Ouedraogo, F., Soma, R., Sawadogo, S., Sorgho, F., Derra, K., Rouamba, E., Orindi, B., Ramos Lopez, F., Flaxman, A., Cappuccini, F., … Tinto, H. (2021). Efficacy of a low-dose candidate malaria vaccine, R21 in adjuvant Matrix-M, with seasonal administration to children in Burkina Faso: A randomised controlled trial. *Lancet*, *397*(10287), 1809–1818.

Ding, X., Liu, D., Booth, G., Gao, W., & Lu, Y. (2018). Virus-like particle engineering: From rational design to versatile applications. *Biotechnology Journal*, *13*(5), e1700324.

Durous, L., Rosa-Calatrava, M., & Petiot, E. (2019). Advances in influenza virus-like particles bioprocesses. *Expert Review of Vaccines, 18*(12), 1285–1300.

Fernandes, F., Teixeira, A. P., Carinhas, N., Carrondo, M. J., & Alves, P. M. (2013). Insect cells as a production platform of complex virus-like particles. *Expert Review of Vaccines, 12*(2), 225–236.

Fuenmayor, J., Gòdia, F., & Cervera, L. (2017). Production of virus-like particles for vaccines. *New Biotechnology, 39*(B), 174–180.

Garg, H., Mehmetoglu-Gurbuz, T., & Joshi, A. (2020). Virus like particles (VLP) as multivalent vaccine candidate against Chikungunya, Japanese Encephalitis, Yellow Fever and Zika Virus. *Scientific Reports, 10*(1), 4017.

Ghorbani, A., Zare, F., Sazegari, S., Afsharifar, A., Eskandari, M. H., & Pormohammad, A. (2020). Development of a novel platform of virus-like particle (VLP)-based vaccine against COVID-19 by exposing epitopes: An immunoinformatics approach. *New Microbes and New Infections, 38*, 100786.

González-Domínguez, I., Puente-Massaguer, E., Cervera, L., & Gòdia, F. (2020). Quality assessment of virus-like particles at single particle level: A comparative study. *Viruses, 12*, 2.

Goodridge, L., Goodridge, C., Wu, J., Griffiths, M., & Pawliszyn, J. (2004). Isoelectric point determination of norovirus virus-like particles by capillary isoelectric focusing with whole column imaging detection. *Analytical Chemistry, 76*(1), 48–52.

Gopal, R., & Schneemann, A. (2018). Production and application of insect virus-based VLPs. *Methods in Molecular Biology, 1776*, 125–141.

Gurramkonda, C., Adnan, A., Gabel, T., Lunsdorf, H., Ross, A., Nemani, S. K., Swaminathan, S., Khanna, N., & Rinas, U. (2009). Simple high-cell density fed-batch technique for high-level recombinant protein production with *Pichia pastoris*: Application to intracellular production of Hepatitis B surface antigen. *Microbial Cell Factories, 8*, 13.

Hernández, L., Rodríguez, A., Peña, V., Zumalacárregui, L., & Amaro, D. (2016). DNA removal on different chromatography supports used for EPO purification by spike studies. *Journal of Liquid Chromatography & Related Technologies, 39*(16), 762–767.

Hillebrandt, N., Vormittag, P., Bluthardt, N., Dietrich, A., & Hubbuch, J. (2020). Integrated process for capture and purification of virus-like particles: Enhancing process performance by cross-flow filtration. *Frontiers in Bioengineering and Biotechnology, 8*, 489.

Huang, X., Wang, X., Zhang, J., Xia, N., & Zhao, Q. (2017). *Escherichia coli*-derived virus-like particles in vaccine development. *NPJ Vaccines, 2*, 3.

Kangwa, M., Yelemane, V., Polat, A. N., Gorrepati, K. D., Grasselli, M., & Fernandez-Lahore, M. (2015). High-level fed-batch fermentative expression of an engineered Staphylococcal protein A based ligand in *E. coli*: Purification and characterization. *AMB Express, 5*(1), 70.

Kim, H. J., & Kim, H. J. (2017). Yeast as an expression system for producing virus-like particles: What factors do we need to consider? *Letters in Applied Microbiology, 64*(2), 111–123.

Kost, T. A., Condreay, J. P., & Jarvis, D. L. (2005). Baculovirus as versatile vectors for protein expression in insect and mammalian cells. *Nature Biotechnology, 23*(5), 567–575.

Krammer, F., Schinko, T., Palmberger, D., Tauer, C., Messner, P., & Grabherr, R. (2010). *Trichoplusia ni* cells (High Five) are highly efficient for the production of influenza A virus-like particles: A comparison of two insect cell lines as production platforms for influenza vaccines. *Molecular Biotechnology, 45*(3), 226–234.

Kumar, R., & Kumar, P. (2019). Yeast-based vaccines: New perspective in vaccine development and application. *FEMS Yeast Research, 19*, 2.

Lagoutte, P., Mignon, C., Donnat, S., Stadthagen, G., Mast, J., Sodoyer, R., Lugari, A., & Werle, B. (2016). Scalable chromatography-based purification of virus-like particle carrier for epitope based influenza A vaccine produced in *Escherichia coli*. *Journal of Virological Methods*, *232*, 8–11.

Lai, C. C., Cheng, Y. C., Chen, P. W., Lin, T. H., Tzeng, T. T., Lu, C. C., Lee, M. S., & Hu, A. Y. (2019). Process development for pandemic influenza VLP vaccine production using a baculovirus expression system. *Journal of Biological Engineering*, *13*, 78.

Lai, H., & Chen, Q. (2012). Bioprocessing of plant-derived virus-like particles of Norwalk virus capsid protein under current good manufacture practice regulations. *Plant Cell Reports*, *31* (3), 573–584.

Lee, D. F., Chen, C. C., Hsu, T. A., & Juang, J. L. (2000). A baculovirus superinfection system: Efficient vehicle for gene transfer into Drosophila S2 cells. *Journal of Virology*, *74*(24), 11873–11880.

Liew, M. W., Rajendran, A., & Middelberg, A. P. (2010). Microbial production of virus-like particle vaccine protein at gram-per-litre levels. *Journal of Biotechnology*, *150* (2), 224–231.

Liu, F., Wu, X., Li, L., Liu, Z., & Wang, Z. (2013). Use of baculovirus expression system for generation of virus-like particles: Successes and challenges. *Protein Expression and Purification*, *90*(2), 104–116.

Lünsdorf, H., Gurramkonda, C., Adnan, A., Khanna, N., & Rinas, U. (2011). Virus-like particle production with yeast: Ultrastructural and immunocytochemical insights into *Pichia pastoris* producing high levels of the Hepatitis B surface antigen. *Microbial Cell Factories*, *10*(1), 48.

Magalhaes, P. O., Lopes, A. M., Mazzola, P. G., Rangel-Yagui, C., Penna, T. C., & Pessoa, A., Jr. (2007). Methods of endotoxin removal from biological preparations: A review. *Journal of Pharmacy & Pharmaceutical Sciences: A Publication of the Canadian Society for Pharmaceutical Sciences, Societe Canadienne des Sciences Pharmaceutiques*, *10*(3), 388–404.

Manchester, M., & Singh, P. (2006). Virus-based nanoparticles (VNPs): Platform technologies for diagnostic imaging. *Advanced Drug Delivery Reviews*, *58*(14), 1505–1522.

Manfrão-Netto, J. H. C., Gomes, A. M. V., & Parachin, N. S. (2019). Advances in using *Hansenula polymorpha* as chassis for recombinant protein production. *Frontiers in Bioengineering and Biotechnology*, *7*, 94.

Matsuda, T., Tanijima, T., Hirose, A., Masumi-Koizumi, K., Katsuda, T., & Yamaji, H. (2020). Production of influenza virus-like particles using recombinant insect cells. *Biochemical Engineering Journal*, *163*, 107757.

McCraw, D. M., Gallagher, J. R., Torian, U., Myers, M. L., Conlon, M. T., Gulati, N. M., & Harris, A. K. (2018). Structural analysis of influenza vaccine virus-like particles reveals a multicomponent organization. *Scientific Reports*, *8*(1), 10342.

Mignaqui, A. C., Ferella, A., Cass, B., Mukankurayija, L., L'Abbe, D., Bisson, L., Sanchez, C., Scian, R., Cardillo, S. B., Durocher, Y., & Wigdorovitz, A. (2020). Foot-and-mouth disease: Optimization, reproducibility, and scalability of high-yield production of virus-like particles for a next-generation vaccine. *Frontiers in Veterinary Science*, *7*, 601.

Milburn, P. T., & Dunnill, P. (1994). The release of virus-like particles from recombinant *Saccharomyces cerevisiae*: Effect of freezing and thawing on homogenization and bead milling. *Biotechnology and Bioengineering*, *44*(6), 736–744.

Mohsen, M. O., Gomes, A. C., Vogel, M., & Bachmann, M. F. (2018). Interaction of viral capsid-derived virus-like particles (VLPs) with the innate immune system. *Vaccines (Basel)*, *6*(3).

Moleirinho, M. G., Silva, R. J. S., Alves, P. M., Carrondo, M. J. T., & Peixoto, C. (2020). Current challenges in biotherapeutic particles manufacturing. *Expert Opinion on Biological Therapy*, *20*(5), 451–465.

Mortola, E., & Roy, P. (2004). Efficient assembly and release of SARS coronavirus-like particles by a heterologous expression system. *FEBS Letters*, *576*(1–2), 174–178.

Noad, R., & Roy, P. (2003). Virus-like particles as immunogens. *Trends in Microbiology*, *11*(9), 438–444.

Ongkudon, C. M., Chew, J. H., Liu, B., & Danquah, M. K. (2012). Chromatographic removal of endotoxins: A bioprocess engineer's perspective. *ISRN Chromatography*, *2012*, 649746.

Pattenden, L. K., Middelberg, A. P., Niebert, M., & Lipin, D. I. (2005). Towards the preparative and large-scale precision manufacture of virus-like particles. *Trends in Biotechnology*, *23* (10), 523–529.

Patterson, D. P., McCoy, K., Fijen, C., & Douglas, T. (2014). Constructing catalytic antimicrobial nanoparticles by encapsulation of hydrogen peroxide producing enzyme inside the P22 VLP. *Journal of Materials Chemistry B*, *2*(36), 5948–5951.

Pushko, P., Tumpey, T. M., Bu, F., Knell, J., Robinson, R., & Smith, G. (2005). Influenza virus-like particles comprised of the HA, NA, and M1 proteins of H9N2 influenza virus induce protective immune responses in BALB/c mice. *Vaccine*, *23*(50), 5751–5759.

Qian, C., Liu, X., Xu, Q., Wang, Z., Chen, J., Li, T., Zheng, Q., Yu, H., Gu, Y., Li, S., & Xia, N. (2020). Recent progress on the versatility of virus-like particles. *Vaccines (Basel)*, *8*(1).

Ramesh Kumar, M. B., & Rathore, A. S. (2020). Virus-like particles as therapeutic moieties of the future. *BioPharm International Webpage*, *33*(12), 21–29.

Roldão, A., Silva, A. C., Mellado, M. C. M., Alves, P. M., & Carrondo, M. J. T. (2017). Viruses and virus-like particles in biotechnology: Fundamentals and applications. *Comprehensive Biotechnology*, 633–656.

Rosano, G. L., & Ceccarelli, E. A. (2014). Recombinant protein expression in *Escherichia coli*: Advances and challenges. *Frontiers in Microbiology*, *5*, 172.

RTS,S Clinical Trials Partnership., Agnandji, S. T., Lell, B., Fernandes, J. F., Abossolo, B. P., Methogo, B. G., Kabwende, A. L., Adegnika, A. A., Mordmuller, B., Issifou, S., Kremsner, P. G., Sacarlal, J., Aide, P., Lanaspa, M., Aponte, J. J., Machevo, S., Acacio, S., Bulo, H., Sigauque, B., ... Vansadia, P. (2012). A phase 3 trial of RTS, S/AS01 malaria vaccine in African infants. *The New England Journal of Medicine*, *367*(24), 2284–2295.

RTS,S Clinical Trials Partnership., Agnandji, S. T., Lell, B., Soulanoudjingar, S. S., Fernandes, J. F., Abossolo, B. P., Conzelmann, C., Methogo, B. G., Doucka, Y., Flamen, A., Mordmuller, B., Issifou, S., Kremsner, P. G., Sacarlal, J., Aide, P., Lanaspa, M., Aponte, J. J., Nhamuave, A., Quelhas, D., ... Vansadia, P. (2011). First results of phase 3 trial of RTS,S/AS01 malaria vaccine in African children. *The New England Journal of Medicine*, *365*(20), 1863–1875.

Salman, A. M., Montoya-Diaz, E., West, H., Lall, A., Atcheson, E., Lopez-Camacho, C., Ramesar, J., Bauza, K., Collins, K. A., Brod, F., Reis, F., Pappas, L., Gonzalez-Ceron, L., Janse, C. J., Hill, A. V. S., Khan, S. M., & Reyes-Sandoval, A. (2017). Rational development of a protective *P. vivax* vaccine evaluated with transgenic rodent parasite challenge models. *Scientific Reports*, *7*, 46482.

Salman, A. M., Montoya-Díaz, E., West, H., Lall, A., Atcheson, E., Lopez-Camacho, C., Ramesar, J., Bauza, K., Collins, K. A., Brod, F., Reis, F., Pappas, L., González-Cerón, L., Janse, C. J., Hill, A. V. S., Khan, S. M., & Reyes-Sandoval, A. (2017). Rational development of a protective *P. vivax* vaccine evaluated with transgenic rodent parasite challenge models. *Scientific Reports*, *7*(1), 46482.

Saraswat, S., Athmaram, T. N., Parida, M., Agarwal, A., Saha, A., & Dash, P. K. (2016). Expression and characterization of yeast derived chikungunya virus like particles (CHIK-VLPs) and its evaluation as a potential vaccine candidate. *PLoS Neglected Tropical Diseases*, *10*(7), e0004782.

Shiyu Dai, H. W., & Fei, D. (2018). Advances and challenges in enveloped virus-like particle (VLP)-based vaccines. *Journal of Immunological Sciences*, *2*(2), 36–41.

Shojaosadati, S. A., Varedi Kolaei, S. M., Babaeipour, V., & Farnoud, A. M. (2008). Recent advances in high cell density cultivation for production of recombinant protein. *Iranian Journal of Biotechnology*, *6*(2), 63–84.

Silva, R. J. S., Moleirinho, M. G., Moreira, A. S., Xenopoulos, A., Alves, P. M., Carrondo, M. J. T., & Peixoto, C. (2020). A flow-through chromatographic strategy for hepatitis C virus-like particles purification. *Processes*, *8*(1), 85.

Spice, A. J., Aw, R., Bracewell, D. G., & Polizzi, K. M. (2020). Synthesis and assembly of hepatitis B virus-like particles in a *Pichia pastoris* cell-free system. *Frontiers in Bioengineering and Biotechnology*, *8*, 72.

Syomin, B. V., & Ilyin, Y. V. (2019). Virus-like particles as an instrument of vaccine production. *Molecular Biology*, *53*(3), 323–334.

Tang, X. C., Lu, H. R., & Ross, T. M. (2011). Baculovirus-produced influenza virus-like particles in mammalian cells protect mice from lethal influenza challenge. *Viral Immunology*, *24*(4), 311–319.

ThermoFisher Scientific. (2011). *His-tagged proteins—Production and purification*. Author.

Thompson, C. M., Petiot, E., Lennaertz, A., Henry, O., & Kamen, A. A. (2013). Analytical technologies for influenza virus-like particle candidate vaccines: Challenges and emerging approaches. *Virology Journal*, *10*(1), 141.

Tripathi, N. K., & Shrivastava, A. (2019). Recent developments in bioprocessing of recombinant proteins: Expression hosts and process development. *Frontiers in Bioengineering and Biotechnology*, *7*, 420.

Van Zyl, A. R., & Hitzeroth., II (2016). Purification of virus-like particles (VLPs) from plants. *Methods in Molecular Biology*, *1404*, 569–579.

Venkatraman, N., Tiono, A., Bowyer, G., Powlson, J., Collins, K., Coulibaly, S., Datoo, M., Silman, D., Ouedraogo, A., Nebie, I., Imoukhuede, E., Brod, F., Folegatti, P., Dickinson, E., Jamieson, S., Bougouma, E., Wright, D., Bellamy, D., Diarra, A., ... Hill, A. (2019). Phase I assessments of first-in-human administration of a novel malaria anti-sporozoite vaccine candidate, R21 in matrix-M adjuvant, in UK and Burkinabe volunteers. *medRxiv*, 19009282.

Vicente, T., Roldão, A., Peixoto, C., Carrondo, M. J., & Alves, P. M. (2011). Large-scale production and purification of VLP-based vaccines. *Journal of Invertebrate Pathology*, *107*(Suppl), S42–S48.

Vieira Gomes, A. M., Souza Carmo, T., Silva Carvalho, L., Mendonça Bahia, F., & Parachin, N. S. (2018). Comparison of yeasts as hosts for recombinant protein production. *Microorganisms*, *6*(2).

Wang, H., Song, H. L., Wang, Q., & Qiu, B. S. (2013). Expression of glycoproteins bearing complex human-like glycans with galactose terminal in *Hansenula polymorpha*. *World Journal of Microbiology and Biotechnology*, *29*(3), 447–458.

Wang, W., & Hauri, P. (2016). Process contaminants in biopharmaceuticals: Host cell DNA & protein. *Eurofins BioPharma Webpage*.

Yang, Y., Su, Z., Ma, G., & Zhang, S. (2020). Characterization and stabilization in process development and product formulation for super large proteinaceous particles. *Engineering in Life Sciences*, *20*(11), 451–465.

Yee, C. M., Zak, A. J., Hill, B. D., & Wen, F. (2018). The coming age of insect cells for manufacturing and development of protein therapeutics. *Industrial & Engineering Chemistry Research*, *57*(31), 10061–10070.

Yusupov, A., Popovsky, D., Mahmood, L., Kim, A. S., Akman, A. E., & Yuan, H. (2019). The nonavalent vaccine: A review of high-risk HPVs and a plea to the CDC. *American Journal of Stem Cells*, *8*(3), 52–64.

Zepeda-Cervantes, J., Ramírez-Jarquín, J. O., & Vaca, L. (2020). Interaction between virus-like particles (VLPs) and pattern recognition receptors (PRRs) from dendritic cells (DCs): Toward better engineering of VLPs. *Frontiers in Immunology*, *11*, 1100.

Zhang, N., Li, C., Jiang, S., & Du, L. (2020). Recent advances in the development of virus-like particle-based flavivirus vaccines. *Vaccines (Basel)*, *8*(3).

Zhao, M., Vandersluis, M., Stout, J., Haupts, U., Sanders, M., & Jacquemart, R. (2019). Affinity chromatography for vaccines manufacturing: Finally ready for prime time? *Vaccine*, *37*(36), 5491–5503.

Zitzmann, J., Sprick, G., Weidner, T., Schreiber, C., Czermak, C. S., & Peter, C. (2017). Process optimization for recombinant protein expression in insect cells. New Insights into Cell Culture Technology. Dimentions. *Chapter*, *2*, 43–97. Available from https://doi.org/10.5772/67849.

Chapter 7

Approaches to cancer vaccination

Ezinne Janefrances Nwonu
Medical Biotechnology Department, National Biotechnology Development Agency, Abuja, Nigeria

Introduction

Over the past decades, extensive research in immuno-oncology has given insights into how immune responses can be manipulated to overcome the immunosuppressive and evasive mechanisms of tumor cells and the tumor microenvironment (TME), and thus serve as an effective strategy for cancer therapy (Malonis et al., 2020). Targeted tumor therapy aims to identify specific antigens on tumor cells that can be recognized by the immune system, thus potentially producing an effective antitumor immune response and immunologic memory that ensures a long-lasting antitumor immunity (Qin et al., 2019). Unlike healthy cells, most tumor cells are characterized by overexpression or mutation of certain endogenous proteins. Any overexpressed or mutated gene product is a potential cancer vaccine target (Malonis et al., 2020).

Vaccination is a feasible therapeutic strategy for the treatment of cancers (Yu & Liu, 2018). The goal of a cancer vaccine is to activate a patient's immune system to recognize and effectively eliminate tumor cells (Vergati et al., 2010; Zhang et al., 2019) without damaging healthy cells (Lee et al., 2018). The potential target for an immune response can be either tumor-specific antigens (TSAs) found exclusively on tumor cells, or tumor-associated antigens (TAAs) overexpressed on tumor cells and also present at some level in healthy tissues. Theoretically, TSAs are the ideal target for cancer immunotherapy because of their tumor specificity, and, in some cases, their role in tumor development and progression. However, one major challenge with targeting TSAs is the fact that most mutations are unique to a specific tumor, thus requiring personalized immunotherapy to be developed for each patient. The main problem with targeting TAAs is that they are poor

Vaccinology and Methods in Vaccine Research. DOI: https://doi.org/10.1016/B978-0-323-91146-7.00003-2

immunogens and can easily evade immune surveillance, due to acquired tolerance for self-antigens (Vergati et al., 2010).

In principle, a cancer vaccine is designed to stimulate a patient's dendritic cells (DCs), which are professional antigen-processing cells (APCs). The DCs engulf and process the tumor antigens introduced by the vaccine, then migrate to lymph nodes where they present antigen-derived peptides on their cell surface complexed with major histocompatibility complex (MHC) I or II molecules, to activate T cells. Activated T cells proliferate and differentiate into effector CD8 + cytotoxic T and CD4 + T helper cells (Zhang et al., 2019). These cells exit the lymph nodes and then recognize specific MHC-peptide complexes on the tumor cell surface, resulting in tumor cell killing (Zhang et al., 2019). In addition, costimulatory receptors are induced and cytokines are released, enhancing antitumor immune responses (Qin et al., 2019). In theory, the immune response induced by the vaccine could either eliminate the tumor or keep it under continuous immune surveillance, delaying relapse and prolonging patient's survival. Unlike other standard treatments for cancer such as surgery, chemotherapy, radiotherapy, and immune checkpoint inhibitors, vaccine-based cancer therapy may be the only treatment that can induce life-long immunity (Vergati et al., 2010).

Another major goal for cancer vaccine strategies is to overcome tumor-related immunosuppression. Preclinical studies have revealed that vaccines alone may not be powerful enough to surmount the complex immunosuppressive mechanisms found within the TME (Yu & Liu, 2018). The TME is made up of various cell types (see Fig. 7.1), and extracellular molecules expressed by tumor cells, stromal cells, and immune cells (Labani-Motlagh et al., 2020). Immunosuppression within the TME plays a key role in tumor progression, metastasis, and drug resistance and is triggered by several factors such as hypoxia, myeloid-derived suppressor cells (MDSCs), production of nitric oxide (NO), increased secretion of immune inhibitory cytokines including interleukin (IL)-10 and transforming growth factor (TGF)-β, T cell exhaustion, histocompatibility antigen (human leukocyte antigen, HLA)-G secretion by cancer cells, inhibition of natural killer (NK) or NKT cells, expression of monocyte chemoattractant proteins like C−C motif chemokine ligand 2 (CCL2), and infiltration of tumors by Tregs (Labani-Motlagh et al., 2020; Tsai et al., 2014).

In addition to cancer cells, various immune cells involved in the antitumor immune response are also found in the TME. They include CD8 T cells, NK cells, Treg cells, NKT cells, DCs, macrophages, and MDSCs.

Alongside vaccines, various immunological products have been developed for the treatment of cancer (Fig. 7.2). Cancer vaccines can be combined with these products and other conventional cancer therapy, and this may amplify the clinical response via synergistic pathways (Fig. 7.3). For instance, vaccines combined with immune checkpoint blockade therapy (e.g., anti-programmed cell death 1 (PD1) antibodies) have been successful both in

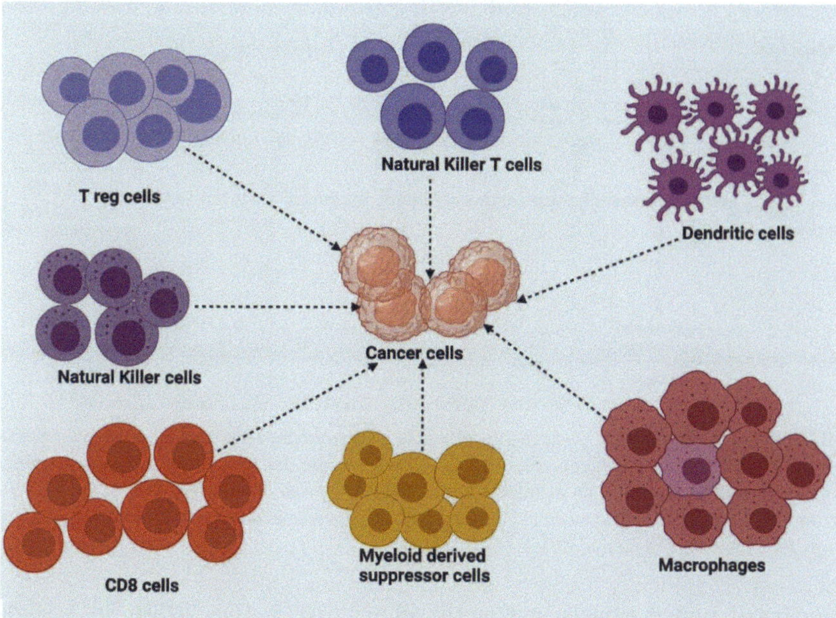

FIGURE 7.1 Various cell types found in the tumor microenvironment.

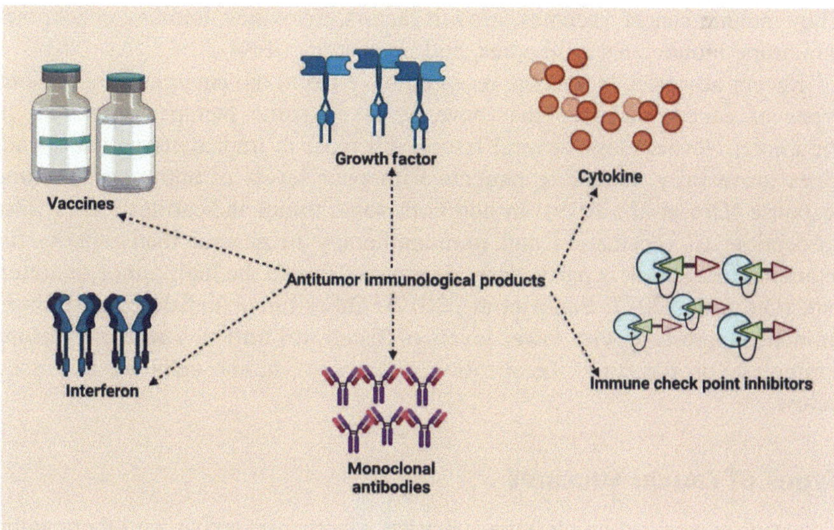

FIGURE 7.2 Antitumor immunological products.

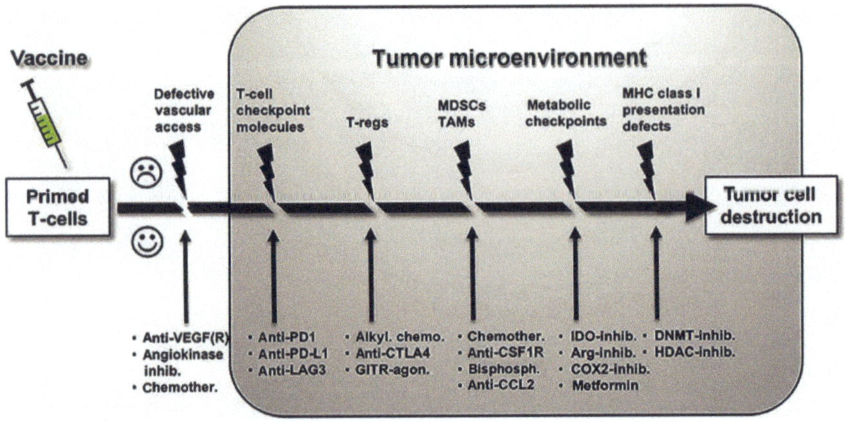

FIGURE 7.3 The multiple obstacles to effective antitumor immune responses following successful priming of tumor antigen-specific T cells by a vaccine. Each obstacle offers opportunities for therapeutic intervention to increase vaccine efficacy. *From Vermaelen, K. (2019). Vaccine strategies to improve anti-cancer cellular immune responses. Frontiers in Immunology, 10, 8. https://doi.org/10.3389/fimmu.2019.00008.*

preclinical tumor models and in the clinic (Yu & Liu, 2018). In a study by Ott et al. (2017), two out of six melanoma patients who received a neoantigen-based vaccine experienced tumor recurrence but after subsequent treatment with anti-PD1 therapy, there was total regression of the tumor with clonal expansion of the repertoire of neoantigen-specific T cells.

There are various products currently used in cancer immunotherapy, either alone or in combination to achieve effective antitumor immunity. They include cancer vaccines, growth factors, cytokines, immune checkpoint inhibitors, monoclonal antibodies, and interferons (IFNs).

Recent advances in vaccine research have led to the emergence of various types of cancer vaccines that have achieved some therapeutic effects in the clinic. Nevertheless, several issues still exist in implementing these vaccines universally, including patients with poor levels of antitumor immune response (Qin et al., 2019). In addition, some tumor indications are far less susceptible to vaccination and immunotherapy in general than others—for example, melanoma is particularly responsive due to the high tumor mutation rate (Ott et al., 2017; Sahin et al., 2017). Other tumor indications are more immunosuppressive with lower levels of T cell infiltration. There are various strategies for targeting therapeutic vaccines to tumor cells as discussed below.

Types of cancer vaccines

Cancer vaccines can be classified as prophylactic/preventive, and therapeutic (Qin et al., 2019; Vergati et al., 2010). Preventive vaccines are given as prophylaxis for cancers caused by oncogenic pathogens including human papilloma virus (HPV) and hepatitis B virus (HBV) (Qin et al., 2019). Currently, HPV and HBV vaccines are commercially available and have greatly helped to reduce the incidence of cervical cancer and liver cancer, respectively. In contrast, therapeutic vaccines can target tumor cells by "inducing autoimmune responses in a targeted manner" (Qin et al., 2019). Different approaches have been utilized in the development of therapeutic cancer vaccine candidates, including whole tumor cell vaccines, DC vaccines, DNA vaccines, mRNA-based vaccines, peptide-based vaccines, and viral vector vaccines (Qin et al., 2019; Yu & Liu, 2018).

Whole tumor cell vaccines

Since tumor cells undergo high rates of mutation affecting multiple antigens, an optimal approach would elicit a wide range of polyclonal tumor-specific responses against numerous different epitopes. Whole tumor cell vaccines offer this specific advantage, as they comprise multiple tumor antigens that can be processed and presented by DCs to stimulate a strong polyclonal T cell response against the tumor (Chiang et al., 2010, 2015; Yaddanapudi et al., 2013). Stimulated CD4 + T helper cells enhance the CD8 + T cell response, leading to a more robust antitumor immunity and long-term immunologic memory (Chiang et al., 2015). Another advantage is that this approach is suitable for use in all cancer patients irrespective of their HLA types (Chiang et al., 2015; Vergati et al., 2010). The diversity of epitopes presented by MHC I and MHC II may also stimulate innate immune responses (i.e., NK cells, macrophages, and eosinophils) (Vergati et al., 2010).

Whole tumor cell vaccines can be classified as allogeneic (nonpatient-specific) cell line vaccines and autologous (patient-specific) vaccines (Qin et al., 2019; Vergati et al., 2010). The benefit of allogeneic vaccines is that the tumor cells are re-engineered to remove oncogenicity but still retain all of the immunogenic antigens, which can then induce antitumor immune responses (Qin et al., 2019; Yu & Liu, 2018). Whole tumor cells can be unmodified or engineered to express MHC, costimulatory molecules, or cytokines (Vergati et al., 2010). However, due to their weak immunogenicity, clinical responses have not been observed using tumor cell vaccines alone. A stronger antitumor immune response can be induced when cell vaccines are used in combination with adjuvants (Qin et al., 2019; Vergati et al., 2010) such as granulocyte-macrophage colony stimulating factor (GM-CSF), Bacille Calmette-Guerin (BCG) (Vergati et al., 2010), and cytokines (Qin et al., 2019). Whole tumor cells can be administered in the form of whole tumor cell lysates (Chiang et al., 2015; Vergati et al., 2010). Issues in using autologous tumor cells include difficulty in harvesting large amounts of patient-specific tumor tissue and producing vaccine preparations free of oncogenic contaminants (Qin et al., 2019; Yaddanapudi et al., 2013).

Conversely, the use of tumor cell lysates reduces the oncogenic potential compared to whole cells (Qin et al., 2019).

Several vaccines derived from whole tumor cells or tumor cell lysates have been tested in preclinical models and clinical trials over the last few decades. These include (1) OncoVAX, consisting of autologous irradiated tumor cells combined with or without BCG as an adjuvant; (2) Reniale, a vaccine for treating renal cell carcinoma (RCC), composed of a lysate of autologous tumor cells preincubated with IFN-γ to improve the antigenicity of these cells, and tocopherol acetate to protect cell membranes during the incubation; (3) GVAX prostate cancer vaccine, which contains two irradiated allogeneic cell lines of human prostate cancer, LNCaP, and PC-3, altered by ex vivo transduction with an adeno-associated viral vector that encodes the human GM-CSF gene (Vergati et al., 2010). In a phase III clinical trial, Melacine, an allogeneic melanoma tumor cell lysate vaccine, was evaluated for minimal residual disease in melanoma patients. The vaccine was reported to have a significant beneficial effect on relapse-free and total survival for patients who express HLA class I antigens A2 and/or C3 (Qin et al., 2019; Sondak & Sosman, 2003). In RCC patients treated with autologous tumor vaccine, it was noticed that survival of patients with progressive RCC was significantly extended (Qin et al., 2019). In general, there is a low response rate with whole cell tumor vaccines, indicating a need to further develop the approach (Qin et al., 2019), for example by loading tumor lysates onto DCs. In line with this, Cornelissen et al. (2016) and Wen et al. (2002) demonstrated that loading DCs with autologous tumor lysates is capable of inducing antitumor immune responses against mesothelioma and myeloma cells, respectively — further detail is provided in the section on dendritic cell vaccines in this chapter.

Peptide-based vaccines

T cells identify their target antigens as small peptide fragments presented in complex with MHC molecules on the cell surface; peptide epitopes can therefore potentially be utilized as antitumor therapeutic vaccines. Since peptide-based therapeutic vaccines are aimed at eliciting specific T cell immune responses against tumor antigens (Tsang et al., 2015; Yang & Zhang, 2020), the peptides used should be both immunogenic and antigenic (Qin et al., 2019). Peptide vaccines contain an MHC-binding antigenic epitope, and are administered with a T cell–inducing adjuvant (Slingluff, 2011; Yang & Zhang, 2020). The quality of the immune response generated is dependent on the peptides and adjuvants used in the vaccine. For instance, vaccines containing multiple peptides (compared to a single peptide) may yield better immune responses; in one example, administration of multi-peptide melanoma vaccines resulted in a nearly 100% response rate (Tsang et al., 2015).

Compared to whole tumor cell vaccines, peptide-based vaccines have the advantages of specificity, reduced risk of autoimmune response, ease and low cost of production in large quantities, and absence of neoplastic seeding risk (Qin et al., 2019; Vergati et al., 2010). The success of this therapeutic strategy thus far has been limited by: (1) the choice of antigenic peptides, (2) poor immunogenicity, (3) escape through antigen mutation or loss, and (4) their use is often HLA-restricted and limited to a subset of patients (usually HLA-A2 +) (Vergati et al., 2010; Yang & Zhang, 2020). The ability to induce a balanced activation of CD4 + and CD8 + T cells is thought to be vital for an effective, long-term antitumor immune response (Vergati et al., 2010). Strategies for increasing the immunogenicity of peptide-based vaccines include the use of (1) peptides with amino acid substitutions at anchor residues and heteroclitic analogs to increase MHC binding affinity, (2) multiple variance long peptides, (3) a mixture of peptides derived from several specific TAAs, (4) class I and II epitope hybrid vaccines, (5) DCs pulsed with peptide, and (6) peptide vaccines in combination with other therapies (Tsang et al., 2015), adjuvants, and cytokines such as GM-CSF (Qin et al., 2019).

Although there are doubts that peptide vaccines alone could significantly affect disease progression, clinical trials have shown that combining these vaccines with appropriate adjuvants and/or immunomodulatory agents may result in improved clinical outcomes (Tsang et al., 2015). An example is the HLA*A0201-restricted peptide vaccine, gp100:209-217(210M). In a phase III clinical trial involving 185 patients with advanced melanoma at 21 centers, one of the treatment groups was given gp100:209-217(210 M) mixed with incomplete Freund's adjuvant, followed by a high-dose IL-2. This group had significantly improved general clinical outcomes and longer progression-free survival (PFS; 2.2 months vs 1.6 months; $P = .008$), and longer median overall survival (17.8 months vs 11.1 months; $P = .06$), compared to the treatment group that received only IL-2 (Schwartzentruber et al., 2011).

A vaccine composed of a mixture of a dozen peptides restricted to HLA-A1, -A2, -A3, -A11, and -A24 can theoretically treat 85% of cancer patients, thereby overcoming the problem of MHC restriction (Tsang et al., 2015). To date, most epitope-based vaccines induce HLA-A2-restricted responses that can kill tumor cells but are characterized by a limited lifespan in the absence of CD4 helper T cells. Protein-based vaccines are capable of generating stronger CD4 responses (MHC class II-restricted), but at the cost of inducing cytotoxic T lymphocytes (CTLs) less effectively (Vergati et al., 2010). Post-2010 most peptide vaccines target neoantigens. Ott et al. demonstrated that neoantigen-specific peptide vaccines for melanoma are effective in combination with a T cell–inducing adjuvant, TLR3 agonist poly-ICLC (Hiltonol) (Ott et al., 2017).

Dendritic cell vaccines

DCs are professional APCs that play a vital role in stimulating cell-mediated immunity, and thus are key targets in cancer vaccine design (Qin et al.,

2019; Yu & Liu, 2018). DC maturation is often impaired in cancer patients, partly caused by cytokines such as vascular endothelial growth factor (VEGF), which induces STAT3 expression and inhibits maturation. Consequently, antigen presentation can be impaired, resulting in weak T cell responses in these patients. To overcome this problem, DC vaccines are prepared using DC precursors, such as elutriated monocytes (isolated from blood using counter-flow centrifugation elutriation) that are matured in vitro, away from the negative influences of the tumor environment. When DCs are matured in vitro, they are induced to express higher levels of costimulatory molecules and/or MHC molecules, which in turn boosts their immunogenicity. To induce tumor-specific responses, DCs can be transduced ex vivo with viral vectors that express tumor antigens (Maeng & Berzofsky, 2019).

DC vaccines are prepared by collecting patient monocytes and culturing them in media containing cytokines such as GM-CSF and IL-4, which stimulate the monocytes to differentiate into DCs. The patient's tumor antigens are loaded into the DCs, thus forming an autologous antigen-pulsed DC vaccine (Maeng & Berzofsky, 2019; Qin et al., 2019). This can be achieved through several methods including transfection, the fusion of DCs and tumor cells, and loading with peptides, tumor RNA, DNA, whole cell lysate, and apoptotic cells (Fry et al., 2009). Lipopolysaccharide, IFN, or a mixture of cytokines is then used to induce maturation of the DCs, after which matured DCs are harvested and re-administered to the patient (Maeng & Berzofsky, 2019). (Maeng & Berzofsky, 2019; Qin et al., 2019) DC vaccines are particularly efficacious in stimulating antitumor immune response through their ability to enhance antigen presentation, prime naïve CD8 + T cells, and stimulate differentiation of T cells into memory T cells (Qin et al., 2019).

Sipuleucel-T is a DC-based vaccine approved by the Food and Drug Administration (FDA) that has achieved favorable results in the treatment of metastatic prostate cancer (Maeng & Berzofsky, 2019; Qin et al., 2019; Yu & Liu, 2018). Sipuleucel-T is prepared by incubating DCs with a recombinant fusion protein combining the tumor antigen prostate acid phosphatase (PAP) and GM-CSF (Razi & Rezaei, 2019; Yu & Liu, 2018). In a clinical trial involving 512 patients with castration-resistant prostate cancer, it was observed that the group of patients given Sipuleucel-T treatment attained a 4.1-month improvement in median survival and 22% decreased death risk compared to the placebo group (Kantoff et al., 2010; Qin et al., 2019). Many clinical and preclinical studies have also supported the potential of DC vaccines for further clinical development. For example, a DC vaccine generated by pulsing DCs with tumor antigens survivin and MUC1, silenced with suppressor of cytokine signaling 1 (SOCS1), and immune stimulating with TLR5 agonist flagellin, was evaluated in a phase 1 open-label uncontrolled, dose-escalation clinical trial for its safety and efficacy in 15 patients with resected stage I–IIIA non-small cell lung cancer at three dose levels. Patients tolerated the vaccine well, without dose-limiting toxicity and serious adverse event other than grade 1 flu-like symptoms. CD3 + CD4 + CD25 + Foxp3 + T regulatory (Treg) cell levels were significantly lower posttreatment. In the long-term follow-up of the maximum dose group, one patient had died, one patient had progressive disease, and the remaining seven patients had no recurrent disease (Ge et al., 2017).

Another pilot clinical trial involving 25 platinum-treated, immunotherapy-naïve, recurrent epithelial ovarian cancer patients evaluated the safety and effectiveness of a personalized DC-based vaccine generated by differentiation of autologous DCs pulsed with oxidized autologous whole-tumor cell lysate (OCDC), administered intranodally. The OCDC vaccine was administered either alone (cohort 1, $n = 5$), in combination with bevacizumab (anti-VEGF-A antibody) (cohort 2, $n = 10$) or bevacizumab plus low-dose intravenous cyclophosphamide (cohort 3, $n = 10$) until disease progression or vaccine exhaustion. The vaccine was well tolerated, induced T cell responses to autologous tumor antigens including neoantigens, and prolonged patient survival significantly, with 100% two-year overall survival rates for responders. The greatest improvements were achieved with the triple combination of the vaccine plus bevacizumab and cyclophosphamide (Mastelic-Gavillet et al., 2019; Tanyi et al., 2018).

Despite promising results, several factors could limit the effectiveness of DC vaccines. This type of vaccine requires that the patient's immune system is functioning sufficiently (for instance, the absence of bone marrow depression induced by chemotherapy) before an adequate number of functional DCs can be isolated from the patient (Qin et al., 2019; Yu & Liu, 2018). It is also costly and laborious to manufacture personalized DC vaccines (Maeng & Berzofsky, 2019; Qin et al., 2019). The route of injection and migration to lymph nodes are also factors affecting the potency of DC vaccines (Yu & Liu, 2018). Inconsistencies in the production and quality control of DC vaccines mean that analysis of small-scale studies is also difficult. There is no consensus for measuring immune correlates and biomarkers to assess the in vivo functionality of DC vaccines. For this approach to advance in the clinic, innovative advances in DC vaccine production techniques and quality control, and methods for optimizing vaccine delivery to enhance cross-priming and overcome the suppressive TME are still needed (Maeng & Berzofsky, 2019).

Nucleic acid vaccines

Nucleic acid vaccines contain DNA or RNA encoding target antigens. This is an attractive vaccine strategy that makes it easy to administer multiple antigens with a single vaccine, and activates both humoral and cellular immune responses, thus reducing the chances of tumor evasion. One of the benefits of nucleic acid vaccines over peptide-based vaccines is that they are not limited by the patient's HLA type. Nucleic acid vaccines have been proven to be nontoxic and tolerable (McNamara et al., 2015). Strategies for developing DNA and mRNA cancer vaccines are further explained in the following subsections.

DNA vaccines

DNA vaccines are genetically engineered vaccines prepared by inserting genes encoding tumor antigens (capable of stimulating both CD4 + and CD8 + T cells responses) into a plasmid or recombinant viral vector (Maeng & Berzofsky, 2019; Qin et al., 2019; Vergati et al., 2010). DNA vaccines can also be modified to incorporate genes that encode immunostimulatory molecules and enhance antitumor immunity (Qin et al., 2019). It has been proven that DNA vaccines can elicit a strong immune response against weak TAAs. The vaccine can be administered to the patient via intradermal, subcutaneous, or intramuscular routes (Vergati et al., 2010). When administered, the plasmid penetrates the "residential" APCs or surrounding cells like myocytes, and the genes encoded in the plasmid are transcribed and translated into the protein of interest (Maeng & Berzofsky, 2019).

DNA vaccines are advantageous over other approaches because they are easy to manufacture at a large scale at low cost. Many phase I and II clinical trials involving patients with prostate cancer, melanoma, colorectal cancer, and head and neck carcinoma have been conducted to assess the effectiveness of DNA vaccines targeting various TAAs [prostate-specific antigen (PSA), PAP, gp100, carcinoembryonic antigen (CEA), hsp65]. The vaccines were either administered alone or with a specific delivery system or adjuvants; however, the sample sizes in some of these studies were too small to detect any significant relationship between generating an immune response and clinical outcomes in vaccinated patients (Vergati et al., 2010). A randomized, double-blind, placebo-controlled phase IIb trial of a DNA vaccine targeting HPV-16 and HPV-18 E6 and E7 proteins revealed that the vaccine may be effective against cervical intraepithelial neoplasia 2/3 (Qin et al., 2019). Only a few DNA vaccines have progressed to phase III clinical trials (Qin et al., 2019).

Although DNA vaccines have promising characteristics, they tend to induce an inferior immune response compared to peptide, cell-based, and RNA vaccines. This could be as a result of ineffective delivery of DNA into cells as a result of variable expression of nucleic acid-sensing pattern recognition receptors (DNA PRRs). The DNA needs to penetrate both cell and nuclear membranes of target cells to be transcribed. Due to the relatively weak immunogenicity of DNA vaccines and doubts over their oncogenic risk following integration into the host genome, there is now a greater focus on RNA vaccines, which may be a safer alternative (McNamara et al., 2015).

mRNA vaccines

mRNA-based vaccines are considered safe because the transfected mRNA is not incorporated into the host genome (Vergati et al., 2010). The mRNA is composed of a cap structure at the 5' end, the coding RNA for a target antigen, and a poly-A tail. The mRNA can encode a single antigen, for example, PSA or CEA, a mixture of antigens derived from allogeneic cancer cell lines or autologous tumor mRNA (Vergati et al., 2010), or a string of epitopes derived from mutated neoantigens (Pardi et al., 2018).

There are two types of mRNA used in vaccines: non-replicating mRNA and self-amplifying mRNA. Self-amplifying mRNA has the advantage of prolonged, elevated expression of the antigen due to its ability to replicate intracellularly, resulting in better vaccine-induced immune responses. Nonreplicating mRNA encodes only the target antigen, while both the target antigen and viral replication machinery are encoded in self-amplifying mRNA. Self-amplifying mRNA has the advantage of prolonged, elevated expression of the antigen due to its ability to replicate intracellularly, resulting in better vaccine-induced immune responses (Heine et al., 2021; Pardi et al., 2018). Both types of mRNA are used in producing vaccines for the prevention of infectious diseases, while nonreplicating mRNA is used for cancer vaccines (Heine et al., 2021).

Briefly, mRNA vaccine production starts with the identification and design of the target antigen using computational methods. The gene encoding the target antigen is cloned into a DNA vector and multiple copies of in vitro transcribed (IVT) mRNA are then produced from the linearized DNA template using bacteriophage-derived RNA polymerase such as T3, T7, or SP6 RNA polymerase (Esprit et al., 2020; Miao et al., 2021). Various strategies can be applied during or after the in vitro transcription process to enhance the immunogenicity of IVT mRNA . These include:

1. The use of chemically modified nucleotides as building blocks decreases innate immune sensing by RNA PRRs and enhances protein translation (Esprit et al., 2020; Pardi et al., 2018).
2. Addition of a synthetic 5′ cap and poly-A tail for stability and increased protein translation (Esprit et al., 2020; Pardi et al., 2018).
3. Addition of regulatory elements in the 5′and 3′- untranslated regions enhances mRNA stability and translation efficacy (Esprit et al., 2020; Pardi et al., 2018).
4. Incorporation of an adjuvant to boost the immunogenicity of the mRNA. This can be achieved by using traditional adjuvants or novel strategies that leverage the intrinsic immunogenicity of mRNA or its ability to encode immuno-stimulatory proteins (Esprit et al., 2020; Pardi et al., 2018).
5. Separation and purification of the synthesized mRNA, which entail adding DNase enzyme to degrade carryover DNA template; purifying the

mRNA using LiCl/NaCl-EtOH precipitation or dT microbeads; removal of contaminants like double-stranded RNA by RNA III treatment or chromatographic methods such as reverse-phase fast protein liquid chromatography or high-performance liquid chromatography. All of these approaches also decrease PRR sensing and enhance protein translation (Esprit et al., 2020; Pardi et al., 2018).

Once the mRNA vaccine is administered and internalized by host cells (DCs), the mRNA transcripts are translated directly into proteins in the cytoplasm. As with DNA vaccines, the resulting antigens are processed into peptide fragments that bind to MHC I molecules and are presented on the APC cell surface, thereby stimulating specific CTL responses (Lee et al., 2018; McNamara et al., 2015; Vergati et al., 2010). The vaccines also induce increased expression of cytokines and chemokines in the mature DCs (Lee et al., 2018). Epitopes processed from the RNA-encoded antigens in the cytoplasm are presented by MHC I instead of MHC II molecules on the surface of DCs (Lee et al., 2018; McNamara et al., 2015), which means that vaccines may fail to induce $CD4^+$ (helper) T cell responses. However, research has shown that when a lysosomal targeting signal is incorporated in the antigen-encoding mRNA sequence, helper T cell responses are stimulated (McNamara et al., 2015).

There are two major approaches for delivering mRNA vaccines to DCs:

1. *Ex vivo delivery*: There are various techniques to deliver mRNA vaccines encoding TAAs into a patient's DCs in vitro, for example, physical transfection methods such as electroporation and microinjection, and chemical transfection agents such as positively charged lipids or polymers that form lipoplexes through electrostatic interaction with negatively charged mRNA. This delivery technique allows for precise control of the DC-mediated immune reaction; however, it is time-consuming and costly (Lee et al., 2018).
2. *In vivo delivery*: This involves direct delivery of mRNA into the patient; several different delivery strategies have been developed to prevent deterioration of the mRNA in the serum and for safe delivery to target cells (Lee et al., 2018). Studies suggest that mRNA vaccines delivered systemically induce more potent immune responses than when delivered locally via intramuscular or intradermal injections. Naked mRNA is unstable (Sayour et al., 2018) and weakly immunogenic when administered via i.m or i.d injections due to poor APC targeting and can be degraded quickly when administered intravenously (Lee et al., 2018). A more efficient response is achieved by combining the vaccine with a delivery system, for example liposomes. Liposomal delivery systems are used for encapsulating mRNA to shield it from destruction by nucleases, enhance cellular uptake, and also ensure its delivery to specific targets in vivo (McNamara et al., 2015; Sayour et al., 2018). In addition to stabilizing

mRNA, liposomes can also activate immune cells, especially when they are also loaded with immunostimulatory molecules, for example cytokines, and this intensifies the immune response to the vaccine (Bringmann et al., 2010; Gu et al., 2020). An example is a cationic liposome, *N*-[1-(2,3-dioleoloxy) propyl]-*N*,*N*,*N*-trimethylammonium chloride 1, otherwise known as DOTAP, which is the most widely used encapsulating agent for mRNA (McNamara et al., 2015).

Clinical trials have assessed mRNA-transfected DCs or in vivo delivered mRNA in patients with prostate cancer, RCC, ovarian cancer, lung cancer, breast cancer, pediatric brain cancer, neuroblastoma, and melanoma, with promising results. For instance, in a phase I clinical trial involving patients with metastatic prostate cancer, PSA-mRNA-transfected DCs were administered repeatedly and this candidate vaccine induced an increase in PSA-specific CTL responses (Vergati et al., 2010).

RNA vaccines, like DNA vaccines, are easy to manufacture in a cell-free system and have some additional benefits: unlike DNA, which needs to cross both cell and nuclear membranes to be transcribed, mRNA only needs to enter into the cytoplasm, where it is translated into the target antigen. RNA has no oncogenic risk because it cannot integrate into the host genome. Lastly, RNA can act as an adjuvant by stimulating costimulatory signals via toll-like receptors TLR3, TLR7, and TLR8 (McNamara et al., 2015).

Novel trends in cancer vaccine development

Cancer stem cell vaccines

Cancer stem cells (CSCs) are a subpopulation of tumor cells having similar features to normal stem cells, including the capability of self-renewal and differentiation into multiple cell types within the tumor (Ayob & Ramasamy, 2018). CSCs are responsible for cancer initiation, maintenance, and progression (Islam et al., 2019). Additionally, they are not susceptible to chemotherapy and radiation therapy and so play significant roles in tumor metastasis and relapse after treatment (Lu et al., 2015; Yu et al., 2012; Yu & Liu, 2018).

Many cancer vaccines target "tumor differentiation" antigens expressed by differentiated cells rather than the CSCs, which lack these antigens and so escape immune attack. Therefore there is a rationale for developing CSC-specific vaccines (Lu et al., 2015). It is thought that vaccines directed against stemness-associated genes may be beneficial in the treatment of various cancer types, and good candidates for a universal cancer vaccine. Despite some stemness targets (i.e., genes or pathways) being specific to a given tumor type, others may overlap and be shared by various cancer indications (Dhodapkar & Dhodapkar, 2011). There are, however, potential risks associated with CSC vaccines such as immune tolerance to pathways shared between CSCs and normal adult stem cells; shared antigens between CSCs and embryonic (fetal) stem cells may therefore be the most desirable targets for generating tumor-specific immune responses (Dhodapkar & Dhodapkar, 2011).

A recent preclinical study using irradiated autologous induced pluripotent stem cells (iPSCs) as a vaccine reported marked prophylactic and therapeutic benefits, suggesting that oncofetal antigens may play important roles in antitumor immunity. The iPSCs were irradiated to inhibit teratoma formation and then coupled with the immunostimulatory adjuvant CpG. Immunizing mice with this candidate vaccine induced the production of antibodies that bound to iPSCs and tumor cells, increased levels of APCs, and also induced tumor-specific CD4 + and CD8 + T cells (Ouyang et al., 2019). In another murine model, ALDH high SCC7 squamous carcinoma and D5 melanoma cells were highly enriched for CSCs, and vaccination with an ALDH high CSC lysate–pulsed DC (CSC-DC) vaccine was able to induce a CSC-specific cytotoxic T cell response. Further studies in this model demonstrated that dual blockade of PD-L1 and CTLA-4 in combination with a CSC-DC vaccine induced a more significant response than CSC-DC vaccine alone, with the elimination of ALDH high CSCs in vivo (Zheng et al., 2018).

Neoantigen-based vaccines

Tumor cells are prone to somatic mutation due to their genetic instability, often leading to the expression of novel or mutated tumor-specific proteins known as neoantigens (Kiyotani et al., 2018; Peng et al., 2019; Shindo et al., 2019). Although neoantigens are mostly unique [RA1] [Z2] to each patient, shared neoantigens also exist in many patients, for example the bcr-abl gene fusion occurring in CML patients (Biernacki & Bleakley, 2020), and common mutations seen in melanoma, thyroid carcinoma, and pancreatic cancer (Zhao et al., 2020), including oncogenic mutations in RAS and p53. Neoantigens are highly tumor-specific, usually immunogenic, and can escape central (thymic) immune tolerance (Aldous & Dong, 2018; Kiyotani et al., 2018; Shindo et al., 2019). This is in contrast to more conventional TAA-directed cancer vaccine targets that can potentially induce central and peripheral tolerance, since TAAs can also be expressed by normal cells, leading to low vaccination efficacy (Shindo et al., 2019). Neoantigens can stimulate both CD4 + and CD8 + T cells immune responses (Peng et al., 2019); hence they are regarded as ideal tumor-specific targets for personalized cancer vaccines (Aldous & Dong, 2018). In addition, murine models have indicated that efficacy is increased synergistically when a neoantigen-based cancer vaccine is used in combination with immune checkpoint inhibitors (Shindo et al., 2019).

Advances in next-generation sequencing technologies, bioinformatics, and immuno-pharmacogenomics have all contributed greatly to the precise

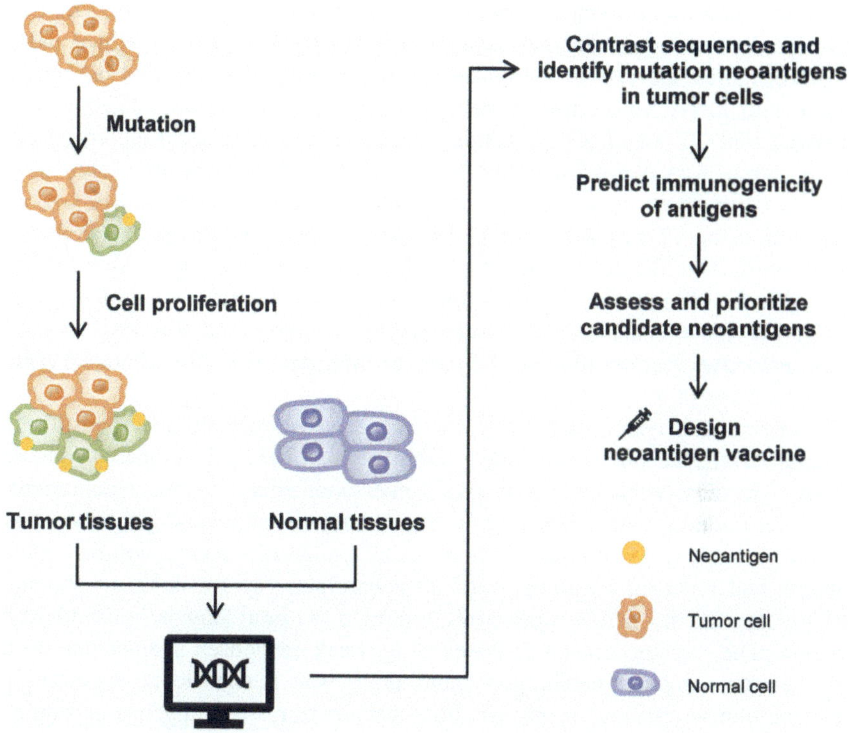

FIGURE 7.4 Mutations in tumor tissue produce neoantigens. Clonal neoantigens can be expressed by a large number of proliferating tumor cells. Various software packages were used to compare the sequence differences between tumor cells and normal cells, and to predict and prioritize the immunogenicity of antigens for screening the optimal tumor neoantigens. Brief illustration of neoantigen discovery and vaccine design. *From Peng, M., Mo, Y., Wang, Y., Wu, P., Zhang, Y., Xiong, F., Guo, C., Wu, X., Li, Y., Li, X., Li, G., Xiong, W., & Zeng, Z. (2019). Neoantigen vaccine: An emerging tumor immunotherapy. Molecular Cancer, 18(1), 128. https://doi.org/10.1186/s12943-019-1055-6.*

identification of cancer neoantigens, which is the basis for the success of this therapeutic strategy (Fang et al., 2020; Kiyotani et al., 2018; Li et al., 2017; Peng et al., 2019). Fig. 7.4 summarizes the process of neoantigen vaccine design.

So far, a few clinical studies have demonstrated the antitumor potential of neoantigen-based personalized vaccines in cancers such as melanoma and glioblastoma (Fang et al., 2020). Clinical trials using neoantigen-targeting cancer vaccines alone or in combination with immune checkpoint inhibitors revealed that these vaccines can generate potent T cell responses against those neoantigens, as well as antitumor responses in patients (Aldous & Dong, 2018). Fang et al. (2020) conducted a clinical study to evaluate the efficacy of neoantigen vaccine iNeo-Vac-P01, comprised of about 5–20 peptides, as a monotherapy in patients with advanced solid tumors. The personalized neoantigen vaccines evaluated in their study were designed using bioinformatics analysis of whole-exome sequencing of tumor and peripheral blood cell DNA, to identify patient-specific tumor mutations. Their results showed that 20 out of the 22 patients enrolled had no or mild adverse reactions, while two exhibited grade 3 or 4 acute allergic reactions only after their sixth booster dose. The disease control rate, median PFS, and overall survival at the 12th month were 71.4%, 4.6 months, and 55.1%, respectively. Approximately 80% of individual peptides or peptide pools were able to trigger specific (T cell—mediated) immune responses against neoantigens. This study showed the neoantigen-based peptide vaccine, iNeo-Vac-P01, to be a safe, effective vaccine that can be used to treat patients with advanced solid tumors (Fang et al., 2020).

Personalized neoantigen RNA vaccines have shown promising results in clinical trials. In one such study, Sahin et al. (2017) developed synthetic RNA vaccines, each encoding five linker-connected 27-mer neoantigens. Thirteen patients with stage III and IV melanoma were dosed with vaccine percutaneously into inguinal lymph nodes under ultrasound control. Each patient had received a maximum of 20 neoantigen vaccine doses by the end of the treatment, which were well tolerated without any serious adverse events. The vaccine enhanced preexisting weak responses against one-third of the immunogenic neoantigens, and there was a significant decrease in the cumulative rates of metastasis after vaccination, resulting in a sustained PFS. Eight (of the 13) patients mounted strong immune responses postvaccination and remained tumor-free throughout the follow-up period, and another patient developed a response to the vaccination when combined with PD-1 blockade therapy (Li et al., 2017; Sahin et al., 2017).

Cancer vaccine delivery systems

Viral vectors

Viral vector systems have advantages over other approaches because of their ability to deliver and express tumor antigen transgenes efficiently, both in vitro and in vivo (Larocca & Schlom, 2011; Lundstrom, 2016). Other features of viral vectors include low toxicity (Lundstrom, 2016) and the ability to enhance the presentation of tumor antigens encoded in the vaccine vector (i.e., recombinant virus), thus increasing immunogenicity (Larocca & Schlom, 2011). Several viral species including alphaviruses, adeno-associated viruses, adenoviruses, vesicular stomatitis virus, measles virus, poliovirus, and HBV have been assessed for use as recombinant vectors for vaccine delivery (Bolhassani & Zahedifard, 2012). Viral vector vaccines containing antigens (TAA, or neoantigens) with or without costimulatory molecules or cytokines, infect cells where the antigens are transcribed and translated, and then processed into epitopes and presented via MHC I on the surface of infected cells, or MHC II on infected APCs (Maeng & Berzofsky, 2019). Viral vectors stimulate cytotoxic T lymphocyte responses that are essential for antitumor immunity, and also induce antibody responses (Bolhassani & Zahedifard, 2012).

The delivery route (such as intramuscular, oral, intrarectal, or aerosol) for different viral vectors can determine the type of memory responses generated against the transgene (Bolhassani & Zahedifard, 2012). Other studies have demonstrated that antigens (peptides or soluble proteins) combined with adjuvants, or other immunotherapies (e.g., whole tumor cell and DC vaccines) are less immunogenic than transgenes expressed by viral vectors. Viral vectors designed for vaccine delivery are easier and cheaper to produce, purify, and store than vaccines complexed with adjuvant, or cellular therapies, particularly those that are personalized (Larocca & Schlom, 2011). The limitation of this delivery system is that some viral vectors can stimulate the patient's immune system to produce neutralizing antibodies to the vectors, thereby resulting in a reduction in efficacy (Larocca & Schlom, 2011).

Nanocarriers

Nanocarriers are technological innovations that are used for delivering antigens to induce antitumor immunity (Beg et al., 2020). Nanocarrier-based strategies help to boost immunogenicity and antigen presentation by APCs, elicit protective and specific T cell immune responses, and overcome immunosuppression in the TME, thereby ensuring that both primary tumor and metastasized cells are eliminated effectively (Yang & Zhang, 2020). Some of the nanocarriers for cancer vaccine delivery that have produced promising clinical results include nanoparticles, nanoemulsions, and liposomes. There are also indications that nanocarriers may be useful for localized delivery of cancer vaccines, with the help of targeting ligands (Beg et al., 2020).

Nanoparticles

Nanoparticles are globular particles with sizes ranging from 10 to 500 nm, with hollow or solid cores that can hold genetic material (Beg et al., 2020). Nanoparticles are often made of natural, semisynthetic, or synthetic polymers, which are biocompatible and biodegradable. The most common nanoparticles used in therapeutics that have been approved by US FDA are composed of poly(D,L-lactic-*co*-glycolic) acid (PLGA) and poly(lactic acid) (PLA) polymers. PLGA and PLA-based nanoparticles are safe, nonimmunogenic, and very efficient in delivering vaccine adjuvants into target immune cells (Beg et al., 2020). PLGA-based nanoparticles have been proposed as suitable delivery vehicles for antigens in DC-based cancer vaccines (Shen et al., 2006). PLGA nanoparticles protect the encapsulated antigenic peptide from degradation in vivo. The encapsulated antigens are then released in a controlled manner, to ensure availability in vivo for continuous immune stimulation (Jahan et al., 2018). A study in mice showed that PLGA nanoparticles were effective in the delivery of TAA and TLR-4 ligand, thereby stimulating sufficient CTL responses to inhibit tumor growth (Beg et al., 2020).

Liposomes

Liposomes are uni- or multilamellar spherical lipid vesicles with phospholipid bilayers and a hydrophilic core (Beg et al., 2020). They are particularly suitable for delivery in vivo because of the similarity between their membrane structure and that of cell membranes, as well as the ease of loading drugs into them. Furthermore, liposomes are biocompatible and biodegradable (like nanoparticles), and they can stabilize therapeutic compounds and enhance their biodistribution (Patra et al., 2018). They can be used to encapsulate and/or adsorb cancer antigens (in the form of proteins, peptides or RNA) for delivery to immune cells. Normally, liposomes have little or no intrinsic adjuvant properties, but they may be combined with adjuvants to boost antigenicity and strengthen the humoral immune response. Liposomes have recently been tested clinically as a vaccine carrier for delivery of HPV16 E7, a protein that is involved in causing cervical cancer (Beg et al., 2020). Another clinical study evaluated the safety and activity of tetravalent RNA-liposomal cancer vaccines containing four TAAs in patients with advanced melanoma. The study showed that the vaccines are safe and well tolerated (Sayour et al., 2018).

Nanoemulsions

Nanoemulsions are colloidal dispersions with a globule size of 20–200 nm, used in local or systemic delivery of antigens to boost the cellular immune response (Beg et al., 2020). Nanoemulsions of approximately 100 nM can be engulfed from the blood by macrophages of the reticuloendothelial system, and can cross hepatic endothelium or spread to the spleen and bone marrow through the lymphatic system. When the size of nanoemulsions is less than 50 nm, they can penetrate tumors following intravenous, intra-arterial, intraperitoneal, or oral administration (Shi et al., 2005). A nanoemulsion vaccine coencapsulated with a gastric cancer-specific antigen MG7 mimotope peptide and adjuvant CpG oligodeoxynucleotides 1645 was evaluated for its effects on gastric cancer in a murine model (Shi et al., 2005). The vaccine was reported to induce high levels of T cell and humoral immunogenicity, and a significant tumor growth inhibition rate of 82.5%, demonstrating the promise of this cancer vaccine approach, at least in animal models (Beg et al., 2020; Shi et al., 2005).

Nanocarrier vaccines are frequently deployed in place of viral vectors as delivery vehicles for genes and antigens because they are highly stable in vitro, strongly biocompatible, persistent in vivo, and biodegradable. Further clinical studies are needed to evaluate the safety and efficacy of nanocarriers for use in humans, as well as investigating their mechanism of action in inducing systematic immune responses (Beg et al., 2020).

Summary

There have been intensified research efforts in cancer vaccine design and development due to advances in the discovery of tumor antigens, approaches to boost their immunogenicity, and a better understanding of mechanisms involved in the presentation of these antigens by APCs and subsequently, induction of immune responses. Currently, various cancer vaccine strategies allow for optimal presentation of tumor antigens to immune cells, in combination with adjuvants, costimulatory molecules and/or with cytokines. These combinations elicit humoral and cell-mediated immune responses that aim to overcome tumor immunosuppression and destroy the target tumor. However, except for some neoantigen vaccines, clinical efficacy has been poor. Some vaccine approaches are restricted to a given cancer indication, a specific MHC allele, or a specific patient (e.g., personalized neoantigen vaccines), making deployment restrictive, complicated, and expensive. Several studies have shown that clinical outcomes may be greatly improved when vaccines are used in combination with other immunotherapeutic options, for example checkpoint inhibitor antibodies. Additionally, biodegradable vaccine delivery systems are under development to target antigens to APCs, ensuring long-lasting antitumor immune responses.

References

Aldous, A. R., & Dong, J. Z. (2018). Personalized neoantigen vaccines: A new approach to cancer immunotherapy. *Bioorganic & Medicinal Chemistry*, *26*(10), 2842–2849. Available from https://doi.org/10.1016/j.bmc.2017.10.021.

Ayob, A. Z., & Ramasamy, T. S. (2018). Cancer stem cells as key drivers of tumor progression. *Journal of Biomedical Science*, *25*(1), 20. Available from https://doi.org/10.1186/s12929-018-0426-4.

Beg, S., Alharbi, K. S., Alruwaili, N. K., Alotaibi, N. H., Almalki, W. H., Alenezi, S. K., Altowayan, W. M., Alshammari, M. S., & Rahman, M. (2020). Nanotherapeutic systems for delivering cancer vaccines: Recent advances. *Nanomedicine (London, England)*, *15*(15), 1527–1537. Available from https://doi.org/10.2217/nnm-2020-0046.

Biernacki, M. A., & Bleakley, M. (2020). Neoantigens in hematologic malignancies. *Frontiers in Immunology*, *11*, 121. Available from https://doi.org/10.3389/fimmu.2020.00121.

Bolhassani, A., & Zahedifard, F. (2012). Therapeutic live vaccines as a potential anticancer strategy. *International Journal of Cancer*, *131*(8), 1733–1743. Available from https://doi.org/10.1002/ijc.27640.

Bringmann, A., Held, S. A. E., Heine, A., & Brossart, P. (2010). RNA vaccines in cancer treatment. *Journal of Biomedicine & Biotechnology, 2010*, 623687. Available from https://doi.org/10.1155/2010/623687.

Chiang, C. L.-L., Benencia, F., & Coukos, G. (2010). Whole tumor antigen vaccines. *Seminars in Immunology, 22*(3), 132–143.

Chiang, C. L.-L., Coukos, G., & Kandalaft, L. E. (2015). Whole tumor antigen vaccines: Where are we? *Vaccines, 3*(2), 344–372. Available from https://doi.org/10.3390/vaccines3020344.

Cornelissen, R., Hegmans, J. P. J. J., Maat, A. P. W. M., Kaijen-Lambers, M. E. H., Bezemer, K., Hendriks, R. W., Hoogsteden, H. C., & Aerts, J. G. J. V. (2016). Extended tumor control after dendritic cell vaccination with low-dose cyclophosphamide as adjuvant treatment in patients with malignant pleural mesothelioma. *American Journal of Respiratory and Critical Care Medicine, 193*(9), 1023–1031. Available from https://doi.org/10.1164/rccm.201508-1573OC.

Dhodapkar, M. V., & Dhodapkar, K. M. (2011). Vaccines targeting cancer stem cells: Are they within reach? *Cancer Journal (Sudbury, Mass.), 17*(5), 397–402. Available from https://doi.org/10.1097/PPO.0b013e318233e730.

Esprit, A., De Mey, W., Bahadur Shahi, R., Thielemans, K., Franceschini, L., & Breckpot, K. (2020). Neo-antigen mRNA vaccines. *Vaccines, 8*(4). Available from https://doi.org/10.3390/vaccines8040776.

Fang, Y., Mo, F., Shou, J., Wang, H., Luo, K., Zhang, S., Han, N., Li, H., Ye, S., Zhou, Z., Chen, R., Chen, L., Liu, L., Wang, H., Pan, H., & Chen, S. (2020). A pan-cancer clinical study of personalized neoantigen vaccine monotherapy in treating patients with various types of advanced solid tumors. *Clinical Cancer Research: An Official Journal of the American Association for Cancer Research, 26*(17), 4511–4520. Available from https://doi.org/10.1158/1078-0432.CCR-19-2881.

Fry, T. J., Shand, J. L., Milliron, M., Tasian, S. K., & Mackall, C. L. (2009). Antigen loading of DCs with irradiated apoptotic tumor cells induces improved anti-tumor immunity compared to other approaches. *Cancer Immunology, Immunotherapy: CII, 58*(8), 1257–1264. Available from https://doi.org/10.1007/s00262-008-0638-7.

Ge, C., Li, R., Song, H., Geng, T., Yang, J., Tan, Q., Song, L., Wang, Y., Xue, Y., Li, Z., Dong, S., Zhang, Z., Zhang, N., Guo, J., Hua, L., Chen, S., & Song, X. (2017). Phase I clinical trial of a novel autologous modified-DC vaccine in patients with resected NSCLC. *BMC Cancer, 17*(1), 884. Available from https://doi.org/10.1186/s12885-017-3859-3.

Gu, Z., Da Silva, C. G., Van Der Maaden, K., Ossendorp, F., & Cruz, L. J. (2020). Liposome-based drug delivery systems in cancer immunotherapy. *Pharmaceutics, 12*(11). Available from https://doi.org/10.3390/pharmaceutics12111054.

Heine, A., Juranek, S., & Brossart, P. (2021). Clinical and immunological effects of mRNA vaccines in malignant diseases. *Molecular Cancer, 20*(1), 52. Available from https://doi.org/10.1186/s12943-021-01339-1.

Islam, F., Gopalan, V., & Lam, A.K.-Y. (2019). Cancer stem cells: Role in tumor progression and treatment resistance. In Oncogenomics: From *basic research to precision m*edicine (pp. 77–87). Academic Press. <https://doi.org/10.1016/B978-0-12-811785-9.00006-5>.

Jahan, S. T., Sadat, S. M., & Haddadi, A. (2018). Design and immunological evaluation of anti-CD205-tailored PLGA-based nanoparticulate cancer vaccine. *International Journal of Nanomedicine, 13*, 367–386. Available from https://doi.org/10.2147/IJN.S144266.

Kantoff, P. W., Higano, C. S., Shore, N. D., Berger, E. R., Small, E. J., Penson, D. F., Redfern, C. H., Ferrari, A. C., Dreicer, R., Sims, R. B., Xu, Y., Frohlich, M. W., & Schellhammer, P. F. (2010). Sipuleucel-T immunotherapy for castration-resistant prostate cancer. *The New England Journal of Medicine, 363*(5), 411–422. Available from https://doi.org/10.1056/NEJMoa1001294.

Kiyotani, K., Chan, H. T., & Nakamura, Y. (2018). Immunopharmacogenomics towards personalized cancer immunotherapy targeting neoantigens. *Cancer Science, 109*(3), 542–549. Available from https://doi.org/10.1111/cas.13498.

Labani-Motlagh, A., Ashja-Mahdavi, M., & Loskog, A. (2020). The tumor microenvironment: A milieu hindering and obstructing antitumor immune responses. *Frontiers in Immunology, 11*, 940. Available from https://doi.org/10.3389/fimmu.2020.00940.

Larocca, C., & Schlom, J. (2011). Viral vector-based therapeutic cancer vaccines. *Cancer Journal (Sudbury, Mass.), 17*(5), 359–371. Available from https://doi.org/10.1097/PPO.0b013e3182325e63.

Lee, K., Kim, M., Seo, Y., & Lee, H. (2018). Development of mRNA vaccines and their prophylactic and therapeutic applications. *Nano Research, 11*, 5173–5192. Available from https://doi.org/10.1007/s12274-018-2095-8.

Li, L., Goedegebuure, S. P., & Gillanders, W. E. (2017). Preclinical and clinical development of neoantigen vaccines. *Annals of Oncology: Official Journal of the European Society for Medical Oncology, 28*(Suppl. 12), xii11–xii17. Available from https://doi.org/10.1093/annonc/mdx681.

Lu, L., Tao, H., Chang, A. E., Hu, Y., Shu, G., Chen, Q., Egenti, M., Owen, J., Moyer, J. S., Prince, M. E., Huang, S., Wicha, M. S., Xia, J.-C., & Li, Q. (2015). Cancer stem cell vaccine inhibits metastases of primary tumors and induces humoral immune responses against cancer stem cells. *Oncoimmunology, 4*(3), e990767.

Lundstrom, K. (2016). Viral vector-based cancer immunotherapy. *Austin Immunology, 1*(2), 1008. Available from https://austinpublishinggroup.com/austin-immunology/fulltext/ai-v1-id1008.php.

Maeng, H. M., & Berzofsky, J. A. (2019). Strategies for developing and optimizing cancer vaccines. *F1000Research, 8*. Available from https://doi.org/10.12688/f1000research.18693.1.

Malonis, R. J., Lai, J. R., & Vergnolle, O. (2020). Peptide-based vaccines: Current progress and future challenges. *Chemical Reviews, 120*(6), 3210–3229. Available from https://doi.org/10.1021/acs.chemrev.9b00472.

Mastelic-Gavillet, B., Balint, K., Boudousquie, C., Gannon, P. O., & Kandalaft, L. E. (2019). Personalized dendritic cell vaccines-recent breakthroughs and encouraging clinical results. *Frontiers in Immunology, 10*, 766. Available from https://doi.org/10.3389/fimmu.2019.00766.

McNamara, M. A., Nair, S. K., & Holl, E. K. (2015). RNA-based vaccines in cancer immunotherapy. *Journal of Immunology Research, 2015*, 794528. Available from https://doi.org/10.1155/2015/794528.

Miao, L., Zhang, Y., & Huang, L. (2021). mRNA vaccine for cancer immunotherapy. *Molecular Cancer, 20*(1), 41. Available from https://doi.org/10.1186/s12943-021-01335-5.

Ott, P. A., Hu, Z., Keskin, D. B., Shukla, S. A., Sun, J., Bozym, D. J., Zhang, W., Luoma, A., Giobbie-Hurder, A., Peter, L., Chen, C., Olive, O., Carter, T. A., Li, S., Lieb, D. J., Eisenhaure, T., Gjini, E., Stevens, J., Lane, W. J., ... Wu, C. J. (2017). An immunogenic personal neoantigen vaccine for patients with melanoma. *Nature, 547*(7662), 217–221. Available from https://doi.org/10.1038/nature22991.

Ouyang, X., Telli, M. L., & Wu, J. C. (2019). Induced pluripotent stem cell-based cancer vaccines. *Frontiers in Immunology, 10*, 1510. Available from https://doi.org/10.3389/fimmu.2019.01510.

Pardi, N., Hogan, M. J., Porter, F. W., & Weissman, D. (2018). mRNA vaccines – A new era in vaccinology. *Nature Reviews. Drug Discovery, 17*(4), 261–279. Available from https://doi.org/10.1038/nrd.2017.243.

Patra, J. K., Das, G., Fraceto, L. F., Campos, E. V. R., Rodriguez-Torres, M. D. P., Acosta-Torres, L. S., Diaz-Torres, L. A., Grillo, R., Swamy, M. K., Sharma, S., Habtemariam, S., &

Shin, H.-S. (2018). Nano based drug delivery systems: Recent developments and future prospects. *Journal of Nanobiotechnology*, *16*(1), 71. Available from https://doi.org/10.1186/s12951-018-0392-8.

Peng, M., Mo, Y., Wang, Y., Wu, P., Zhang, Y., Xiong, F., Guo, C., Wu, X., Li, Y., Li, X., Li, G., Xiong, W., & Zeng, Z. (2019). Neoantigen vaccine: An emerging tumor immunotherapy. *Molecular Cancer*, *18*(1), 128. Available from https://doi.org/10.1186/s12943-019-1055-6.

Qin, H., Sheng, J., Zhang, D., Zhang, X., Liu, L., Li, B., Li, G., & Zhang, Z. (2019). New strategies for therapeutic cancer vaccines. *Anti-Cancer Agents in Medicinal Chemistry*, *19*(2), 213–221. Available from https://doi.org/10.2174/1871520618666181109151835.

Razi, S., Rezaei, N. (2019). Chapter 9 – Immune cell vaccine for cancer. In Vaccines for cancer immunotherapy an evidence-based review on current status and future perspectives (pp. 117–127). Academic Press. <https://doi.org/10.1016/B978-0-12-814039-0.00009-6>.

Sahin, U., Derhovanessian, E., Miller, M., Kloke, B.-P., Simon, P., Löwer, M., Bukur, V., Tadmor, A. D., Luxemburger, U., Schrörs, B., Omokoko, T., Vormehr, M., Albrecht, C., Paruzynski, A., Kuhn, A. N., Buck, J., Heesch, S., Schreeb, K. H., Müller, F., … Türeci, Ö. (2017). Personalized RNA mutanome vaccines mobilize poly-specific therapeutic immunity against cancer. *Nature*, *547*(7662), 222–226. Available from https://doi.org/10.1038/nature23003.

Sayour, E. J., Mendez-Gomez, H. R., & Mitchell, D. A. (2018). Cancer vaccine immunotherapy with RNA-loaded liposomes. *International Journal of Molecular Sciences*, *19*(10). Available from https://doi.org/10.3390/ijms19102890.

Schwartzentruber, D. J., Lawson, D. H., Richards, J. M., Conry, R. M., Miller, D. M., Treisman, J., Gailani, F., Riley, L., Conlon, K., Pockaj, B., Kendra, K. L., White, R. L., Gonzalez, R., Kuzel, T. M., Curti, B., Leming, P. D., Whitman, E. D., Balkissoon, J., Reintgen, D. S., … Hwu, P. (2011). gp100 peptide vaccine and interleukin-2 in patients with advanced melanoma. *The New England Journal of Medicine*, *364*(22), 2119–2127. Available from https://doi.org/10.1056/NEJMoa1012863.

Shen, H., Ackerman, A. L., Cody, V., Giodini, A., Hinson, E. R., Cresswell, P., Edelson, R. L., Saltzman, W. M., & Hanlon, D. J. (2006). Enhanced and prolonged cross-presentation following endosomal escape of exogenous antigens encapsulated in biodegradable nanoparticles. *Immunology*, *117*(1), 78–88.

Shi, R., Hong, L., Wu, D., Ning, X., Chen, Y., Lin, T., Fan, D., & Wu, K. (2005). Enhanced immune response to gastric cancer specific antigen peptide by coencapsulation with CpG oligodeoxynucleotides in nanoemulsion. *Cancer Biology & Therapy*, *4*(2), 218–224.

Shindo, Y., Hazama, S., & Nagano, H. (2019). [Cancer vaccine focused on neoantigens]. *Gan To Kagaku Ryoho. Cancer & Chemotherapy*, *46*(9), 1367–1371.

Slingluff, C. L. (2011). The present and future of peptide vaccines for cancer: Single or multiple, long or short, alone or in combination? *Cancer Journal (Sudbury, Mass.)*, *17*(5), 343–350. Available from https://doi.org/10.1097/PPO.0b013e318233e5b2.

Sondak, V. K., & Sosman, J. A. (2003). Results of clinical trials with an allogenic melanoma tumor cell lysate vaccine: Melacine. *Seminars in Cancer Biology*, *13*(6), 409–415.

Tanyi, J. L., Bobisse, S., Ophir, E., Tuyaerts, S., Roberti, A., Genolet, R., Baumgartner, P., Stevenson, B. J., Iseli, C., Dangaj, D., Czerniecki, B., Semilietof, A., Racle, J., Michel, A., Xenarios, I., Chiang, C., Monos, D. S., Torigian, D. A., Nisenbaum, H. L., … Kandalaft, L. E. (2018). Personalized cancer vaccine effectively mobilizes antitumor T cell immunity in ovarian cancer. *Science Translational Medicine*, *10*(436). Available from https://doi.org/10.1126/scitranslmed.aao5931.

Tsai, M.-J., Chang, W.-A., Huang, M.-S., & Kuo, P.-L. (2014). Tumor microenvironment: A new treatment target for cancer. *ISRN Biochemistry*, *2014*, 351959. Available from https://doi.org/10.1155/2014/351959.

Tsang, K. Y., Jochems, C., & Schlom, J. (2015). Insights on peptide vaccines in cancer immunotherapy. In P. A. Ascierto, D. F. Stroncek, & E. Wang (Eds.), Developments in T cell based cancer immunotherapies. Cancer *drug discovery and dev*elopment. (pp. 1–27). Humana Press, Cham. <https://doi.org/10.1007/978-3-319-21167-1_1>.

Vergati, M., Intrivici, C., Huen, N.-Y., Schlom, J., & Tsang, K. Y. (2010). Strategies for cancer vaccine development. *Journal of Biomedicine & Biotechnology*, *2010*. Available from https://doi.org/10.1155/2010/596432.

Wen, Y.-J., Min, R., Tricot, G., Barlogie, B., & Yi, Q. (2002). Tumor lysate-specific cytotoxic T lymphocytes in multiple myeloma: Promising effector cells for immunotherapy. *Blood*, *99*(9), 3280–3285.

Yaddanapudi, K., Mitchell, R. A., & Eaton, J. W. (2013). Cancer vaccines: Looking to the future. *Oncoimmunology*, *2*(3), e23403.

Yang, J., & Zhang, C. (2020). Regulation of cancer-immunity cycle and tumor microenvironment by nanobiomaterials to enhance tumor immunotherapy. *WIREs Nanomedicine and Nanobiotechnology*, *12*, e1612. Available from https://doi.org/10.1002/wnan.1612.

Yu, Z., Pestell, T. G., Lisanti, M. P., & Pestell, R. G. (2012). Cancer stem cells. *The International Journal of Biochemistry & Cell Biology*, *44*(12), 2144–2151. Available from https://doi.org/10.1016/j.biocel.2012.08.022.

Yu, C., & Liu, H. (2018). New strategies to improve therapeutic vaccines. In S. S. Athari (Ed.), In Immunoregulatory aspects of immunotherapy. IntechOpen. <https://doi.org/10.5772/intechopen.74900>.

Zhang, L., Huang, Y., Lindstrom, A. R., Lin, T. Y., Lam, K. S., & Li, Y. (2019). Peptide-based materials for cancer immunotherapy. *Theranostics*, *9*(25), 7807–7825. Available from https://doi.org/10.7150/thno.37194.

Zhao, W., Wu, J., Chen, S., & Zhou, Z. (2020). Shared neoantigens: Ideal targets for off-the-shelf cancer immunotherapy. *Pharmacogenomics*, *21*(9), 637–645. Available from https://doi.org/10.2217/pgs-2019-0184.

Zheng, F., Dang, J., Zhang, H., Xu, F., Ba, D., Zhang, B., Cheng, F., Chang, A. E., Wicha, M. S., & Li, Q. (2018). Cancer stem cell vaccination with PD-L1 and CTLA-4 blockades enhances the eradication of melanoma stem cells in a mouse tumor model. *Journal of Immunotherapy (Hagerstown, Md.: 1997)*, *41*(8), 361–368. Available from https://doi.org/10.1097/CJI.0000000000000242.

Chapter 8

Vaccines against SARS-CoV-2

Andrés Noé[1,2] and Katie Ewer[1]
[1]*Jenner Institute, University of Oxford, Oxford, United Kingdom,* [2]*Murdoch Children's Research Institute, Parkville, VIC, Australia*

The COVID-19 pandemic

In late December 2019, the World Health Organization was notified of a cluster of cases of pneumonia caused by an unknown infectious agent in Wuhan City, China. Two weeks later the genetic sequence of a virus isolated from samples collected from a number of cases implicated a novel coronavirus as the cause of the outbreak. In late January 2020, the virus had caused nearly 8000 cases and spread to 18 countries beyond China leading the Director-General of WHO, Dr. Tedros Adhanom Ghebreyesus, to declare this novel coronavirus outbreak a Public Health Emergency of International Concern (PHEIC). Cases were rising very rapidly in five of the six WHO regions, particularly in Italy, Iran, South Korea, and Japan and human-to-human transmission was strongly indicated. This was only the sixth PHEIC to have been declared since International Health Regulations were enacted in 2005 and early modeling of virus transmission indicated that self-sustaining outbreaks in major cities were inevitable necessitating large-scale public health interventions (Wu et al., 2020). In March 2020, WHO characterized the outbreak as a pandemic, based on the rapid spread of infections and severity of disease caused. By the first week of April 2020, some 3.9 billion people globally, more than half the world's population, were placed under some form of lockdown. By the end of 2020 this novel coronavirus and the disease it causes, now titled COVID-19, had caused more than 1.9 million deaths globally and more than 89 million confirmed infections, likely with a significantly higher number of unconfirmed infections.

Strategies to contain the pandemic have focused on well-established principles of infection control including promoting good hand hygiene, avoiding large gatherings, wearing suitable personal protective equipment (PPE) such as masks and face-coverings, isolating cases, and identifying close contacts through rigorous contact tracing. Other potential interventions to control the pandemic have included the evaluation of existing drugs and treatments to

Vaccinology and Methods in Vaccine Research. DOI: https://doi.org/10.1016/B978-0-323-91146-7.00001-9

determine efficacy against SARS-CoV-2 as well as the development of novel therapeutics. The careful evaluation of new and existing treatments through well-designed and executed clinical trials, such as the RECOVERY trial, led to the identification of existing medicines including dexamethasone that could substantially improve outcomes for COVID-19 patients on ventilators or receiving oxygen. It was also clear from the early stages of the pandemic that safe and effective vaccines that could be effectively deployed would also be a crucial component of controlling infection and reducing mortality. In April 2020, a target product profile for COVID 19 was published by WHO with the minimal target for vaccine efficacy of 50% and an array of candidate vaccine against SARS-CoV-2 entered clinical trials (WHO Target Product Profiles for COVID-19 Vaccines, 2020).

However, assessing the clinical efficacy of novel vaccines against SARS-CoV-2 in conventional Phase 3 randomized clinical trials (RCTs) presented a number of challenges due to the ongoing pandemic, including highly variable rates of transmission with waves of infection peaking and troughing unpredictably country by country. Another challenge was defining appropriate clinical endpoints, which ranged from preventing infection and disease through to preventing intensive care admission and death, with impact on virus shedding and transmission also likely to be a valuable endpoint in the elimination stage of the pandemic response (Hodgson et al., 2021). Furthermore, the clinical severity and mortality rates of COVID-19 disease vary widely with risk factors including age, sex, and ethnicity with older adults, males, and minority ethnic groups at greater risk of severe disease and death.

Prior to COVID-19, PHEICs had been declared five times: twice for Ebola outbreaks (2018 in Kivu and 2014 in West Africa), Zika virus in 2016, polio in 2014, and swine flu in 2009 (Mullen et al., 2020). The International Health Regulations, implemented in 2005 following the outbreak of SARS coronavirus, confer a legal responsibility on states to respond rapidly to the declaration of a PHEIC. Of previous PHEICs, the Zaire Ebola (ZEBOV) outbreak in West Africa that began in December 2013 in Guinea and rapidly spread to neighboring Liberia and Sierra Leone provided the best forewarning of what might occur in a modern global pandemic. This outbreak gave rise to 28,646 confirmed cases and 11,323 deaths with a mortality rate of 40%, with a substantial burden among frontline healthcare workers (HCWs) in the early months of the epidemic (WHO, 2016).

In contrast to the COVID-19 pandemic, several vaccines against Ebola virus disease (EVD) were already manufactured and in early-stage clinical development. Therefore the scientific response mainly focused on accelerating existing vaccine candidates through Phase 1 and 2 studies to facilitate large-scale Phase 3 efficacy trials with the aim of securing Emergency Use Authorization (EUA). This would allow these existing vaccines to be used to protect frontline HCWs in high-risk regions and to control new outbreaks at a local level with a strategy known as ring vaccination where all household and close contacts of known EVD cases were vaccinated to prevent further transmission (Merler et al., 2016). This ring vaccination approach was also incorporated into a novel clinical trial design for a Phase 3 efficacy trial of the recombinant Vesicular Stomatitis Virus (rVSV)-ZEBOV, which demonstrated up to 100% vaccine efficacy from 10 days after vaccination (Henao-Restrepo et al., 2017). The rVSV-ZEBOV vaccine and another vectored vaccine approach based on a heterologous prime-boost approach with an adenovirus-based priming vaccination (Ad26.ZEBOV) and a poxvirus booster dose (MVA BN Filo) were both subsequently licensed and deployed in later outbreaks of EVD in the Democratic Republic of Congo (Tomori & Kolawole, 2021).

In response to the EVD outbreak in West Africa, the WHO, with scientists and public health experts, produced the "Blueprint for R&D Preparedness" to accelerate research and development (R&D) for pathogens with epidemic potential. This generated a priority list of pathogens for which diagnostics, vaccines, and therapeutics were urgently needed to control potential future outbreaks. The priority list included EVD, Rift Valley fever virus (RVFV), Lassa fever, Middle Eastern respiratory syndrome (MERS), and severe acute respiratory syndrome (SARS) coronaviruses as well as a putative unidentified pathogen, named "Disease X." This acknowledgment that future pandemics would occur and require a rapid response provided substantial impetus for the research community to prioritize research toward interventions and for funders to support these programs of work. It led to the development of novel vaccine candidates against MERS and RVFV based on the same adenoviral-based vector technology that were the mainstay of vaccines against EVD, due to the large-scale manufacturing potential, cost-effectiveness, safety, and immunogenicity.

The spike protein of the SARS-CoV-2 virus

The SARS-CoV-2 virus is comprised of four structural proteins including envelope, membrane and, most relevant to vaccine design, spike (S). These are all present in the surface envelope of the virus. Nucleocapsid protein, the fourth structural protein, is within the ribonucleoprotein core. S protein mediates recognition of the host cell receptor, the dimeric human angiotensin-converting enzyme 2 (ACE2), to commence viral entry into cells. The homotrimeric S protein comprises an apical S1 subunit, which permits ACE2 binding via the receptor-binding domain (RBD), and a basal S2 subunit, responsible for membrane fusion. Upon fusion, S protein sheds the S1 subunit and exhibits a postfusion conformation and, when produced through recombinant expression, the S protein is metastable in a prefusion conformation. Indeed, due to the apical accessibility of the S1 subunit, it is immunodominant during SARS-CoV-2 infections and neutralizing epitopes are found mainly on the RBD. Stabilization and abundant expression of S protein in the prefusion state are thought to increase the quality and quantity of vaccine-induced antibodies. Most vaccines approaches have focused on using the prefusion S protein or RBD as antigenic targets.

COVID-19 vaccine development

Conventional vaccine platforms have had tremendous success in the control and elimination of several human infectious diseases. Since the beginning of this century alone, vaccine platform technologies have been advancing rapidly. A remarkable feature of COVID-19 vaccine development is the diversity of the platforms used for SARS-CoV-2 vaccines. These have included a similar technique to that applied centuries ago in smallpox vaccine development, using modified or killed whole organism, as well as protein-based vaccines and nanoparticle constructs. Newer approaches have included recombinant viral vectors and nucleic acids and, interestingly, it is these more novel technologies that have reached licensure and widespread use first. Each approach or platform has unique benefits and drawbacks comprising development time, scalability, cost, stability, as well as safety, immunogenicity, and efficacy profiles. This section will review the approaches that have been taken, their unique characteristics and how they went from concept to implementation, with a particular focus on nucleic acid-based and viral vector vaccines (Fig. 8.1).

Nucleic acid–based vaccines

Nucleic acid–based vaccines offer a cost-efficient and scalable approach to SARS-CoV-2 vaccine development. These vaccines can consist of DNA or RNA.

DNA vaccines

DNA vaccines contain a synthetic DNA sequence encoding the vaccine antigen. Bare DNA can be injected and taken up by antigen-presenting cells (APC) or APC can be transfected with the antigen DNA construct. Subsequent expression and presentation of the target antigen lead to antibody production and induction of antigen-specific CD4 + and CD8 + T cells.

DNA vaccines continue to face several challenges to become an effective vaccination platform. Following the initial proof of concept in the 1990s, most of the successes in preclinical studies have not yet persisted into the clinic phases of development. The main hurdle for this platform is low immunogenicity in larger animals and humans. This has resulted in various approaches to improve immunogenicity, for example, by (1) antigenic codon optimization, MHCI/II combination and fusion with invariant chain, (2)

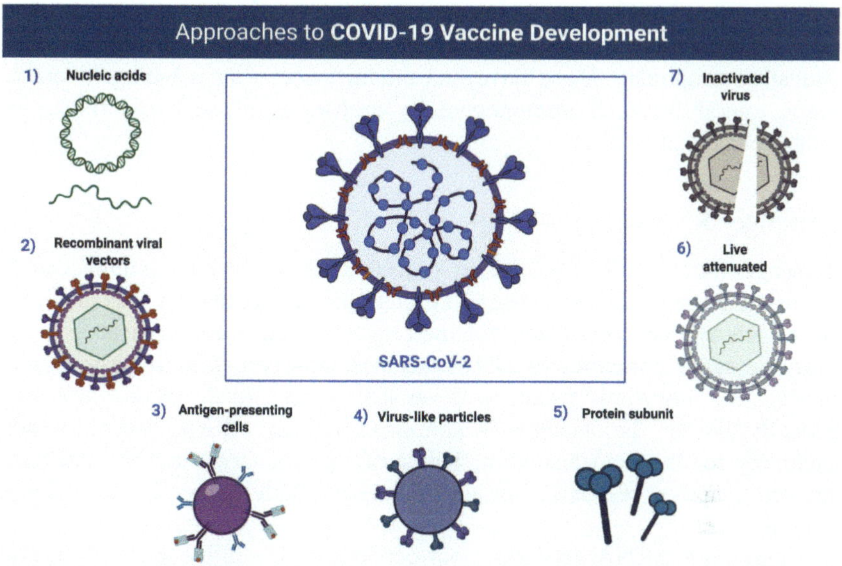

FIGURE 8.1 Approaches to COVID-19 vaccine development. An overview of different vaccine platforms for SARS-CoV-2 vaccine development. These vaccines types will be discussed further in this chapter. *Adapted from "Approaches to viral vaccine development," by BioRender.com (2021). Retrieved from https://app.biorender.com/biorender-templates*

encoding molecular adjuvants, (3) optimization of the vector backbone (the segment of the DNA not required for eukaryotic expression), and (4) the use of hybrid or specific promoters (Hobernik & Bros, 2018). A persisting concern regarding DNA vaccines has been the hypothetical potential for integration into host DNA resulting in dysregulated gene expression, although the risk of this is thought to be very low. While DNA uptake efficiency can be increased by using electroporation (Hobernik & Bros, 2018), this can also differentially shape the adaptive immune response (Haidari et al., 2017), therefore it remains unclear whether electroporation remains an effective strategy to improve immunogenicity directed against all pathogens (Wang et al., 2004). Prior to the pandemic, no DNA vaccine had been licensed for use in humans, but a DNA vaccine had been licensed for equine use against West Nile Virus in 2005 (Kaiser & Barrett, 2019). While the quality and safety considerations for human vaccines differ, regulators have therefore had at least some experience with this platform (Word Health Organization, 2007).

A number of studies showed the induction of nAb (neutralizing antibody) and cellular immunity using DNA vaccines against SARSCoV1 (Lundstrom, 2020; Roper & Rehm, 2009), though these approaches have limited utility in preventing significant disease upon challenge in animals. Several groups have progressed DNA vaccine candidates in the search for SARS-CoV-2 vaccines and this platform accounts for approximately 10% of candidates in clinical development. These have had modest success in inducing favorable safety, tolerability, and immunogenicity profiles in phase I and II clinical trials (Tebas et al., 2021).

Messenger RNA vaccines

Messenger RNA (mRNA) vaccine candidates have had an unprecedented role in the initial response to the COVID-19 pandemic. mRNA vaccines contain an RNA strand that codes for the target antigen. mRNA vaccines work using the same principles as DNA vaccines; however, they bypass nuclear translocation and transcription as these steps have already taken place synthetically. When the vaccine is administered, the mRNA uses host cell machinery for in vivo translation. This produces the target antigen and initiates an adaptive response, usually involving both humoral and cellular immunity.

Progressing mRNA vaccines from concept to feasibility required several technical challenges to be iteratively overcome before the pandemic that began in 2019. One of these was the immunogenic and resultant labile nature of in vitro–transcribed mRNA. The Weissman lab provided seminal contributions in this area by investigating the impact of nucleoside modification (Karikó et al., 2005, 2008). Unmodified RNA signals through human TLR3, TLR7, and TLR8, but this innate immune identification is abrogated if the RNA is constructed with modified nucleosides. These modifications provide mRNA with enhanced stability and translational capacity as well as diminished in vivo vector immunogenicity. A crucial step for the immunogenicity of an mRNA vaccine is the delivery of the construct into the cytoplasm of an APC. Liposomes complex with mRNA and can be endocytosed or directly fuse with a cell membrane, releasing mRNA into the cytoplasm. To this end, synthetic lipid nanoparticle formulation of mRNA demonstrates increased antigen production and immunogenicity in vivo (Geall et al., 2012). Two types of synthetic mRNA vaccines have emerged: conventional nonreplicating mRNA vaccines and self-amplifying mRNA vaccines (Fig. 8.2). Nonreplicating mRNA-based vaccines include an open reading frame (ORF) encoding the target protein antigen, flanking untranslated regions, a 5′ cap, and a poly(A) tail (Pardi et al., 2018). In self-amplifying vaccines, the ORF also encodes the replicative machinery derived from mRNA viruses allowing increased intracellular self-amplification and protein expression (Ljungberg & Liljeström, 2015). Self-amplification allows dose sparing and increased immunogenicity, compared to conventional mRNA vaccines (Vogel et al., 2018).

It was around 30 years between mRNA first being demonstrated as a vaccine technology and its worldwide deployment to improve human health

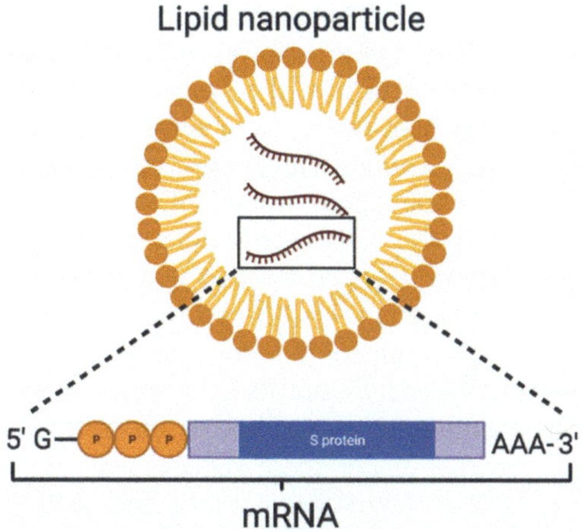

FIGURE 8.2 mRNA-based vaccines. This is an example of a lipid nanoparticle-encapsulated mRNA vaccine encoding the SARS-CoV-2 Spike (S) protein. The RNA component may be comprised or includes self-amplifying, uridine-containing mRNA or nucleoside-modified RNA regions. *Adapted from "COVID-19 vaccine candidate: mRNA-1273," by BioRender.com (2021). Retrieved from https://app.biorender.com/biorender-templates*

during the COVID-19 pandemic. In 1990, Wolff et al. showed that direct injection of mRNA into murine skeletal muscle could express the encoded protein (Wolff et al., 1990). Twenty-two years later, Petsch et al. were the first to report the application of mRNA vaccination to infectious disease, showing immunogenicity or protective effects against influenza A virus infection in mice, ferrets, and pigs (Petsch et al., 2012).

mRNA-1273, an mRNA vaccine developed by Moderna, Inc., was the first vaccine against SARS-CoV-2 to enter human trials, with the first volunteer receiving the vaccine 10 weeks after the pathogen sequence was published. Since then, approximately 15% of global SARS-CoV-2 clinical vaccine development efforts have centered on mRNA vaccines (World Health Organization, 2021). mRNA-1273 and BNT162b—another vaccines developed by BioNTech, Pfizer, and Fosun Pharma—are two mRNA candidates that have performed exceedingly well in safety, immunogenicity, and efficacy trials, as well as postlicensure evaluation. Separate clinical trials of over 30,000 individuals showed that both vaccine candidates conferred >90% efficacy at preventing COVID-19, including severe disease, and apart from short-lived local and systemic reactions, no safety concerns were identified (Baden et al., 2021; Polack et al., 2020). Following authorization for emergency use in various jurisdictions in late 2020 and early 2021, observational studies of vaccinated individuals and in mass-vaccination settings suggested that both mRNA vaccines were effective against a broad range of COVID-19-related outcomes, consistent with data from the RCTs (Dagan et al., 2021; Keehner et al., 2021; Thompson et al., 2021).

While the development of mRNA vaccines generally follows a structure, the process details mostly remain proprietary and not publicly available, posing regulatory and manufacturing challenges, particularly during the COVID-19 pandemic. The first step in the process is prioritization and selection of the antigen by sequencing and bioinformatic methods, followed by synthesis and cloning of a template plasmid DNA. mRNA is then transcribed from the template by a phage DNA-dependent RNA polymerase in vitro. Following this, the RNA may be complexed in lipid nanoparticles. mRNA vaccine thermal stability is a key challenge. The cold chain requirement is high—demanding ultra-low temperatures for long-term storage, conventional freezers for medium-term storage, and fridge storage in the short-term until administration. It may prove difficult to ensure transportation and storage of COVID-19 mRNA vaccines under these requirements in low- and middle-income countries (LMICs). Rapid, inexpensive, and scalable manufacturing due to the high yields of in vitro transcription reactions has been touted as an advantage of mRNA vaccines (Pardi et al., 2018). The antigen, however, is not the only component of the vaccine and the antigen delivery molecules and purification processes, currently require proprietary, bespoke, and novel compounds and components.

Prior to the pandemic, many regulatory jurisdictions did not have guidelines that specifically addressed mRNA vaccines (Hinz et al., 2017). As regulatory experience with mRNA vaccines increases, there will likely be harmonization and more consistency among regulators regarding the quality and considerations of the mRNA vaccine development process. Specific regulatory considerations include the purity of the mRNA, the fraction of full-length transcripts, short transcripts, and double-stranded mRNA, and the proportion of capped and polyadenylated mRNA (Knezevic et al., 2021).

Viral vector vaccines

Viral vectored vaccines can infect host cells and deliver transgene products that encode antigenic proteins for production inside the cell. The vectors are usually genetically adapted to be replication deficient and for reduced virulence. Nonreplicating vectors enter cells and produce the antigen of interest, but, importantly, no new viral particles develop. The most commonly used nonreplicating viral vectors include adenoviruses and poxviruses. Replicating viral vectors can also be used in SARS-CoV-2 vaccine design as they simulate the natural process of infection, have self-adjuvanting properties, and are potentially more potent than nonreplicating viral vectors (Bezbaruah et al., 2021).

Nonreplicating viral vectors constitute about 15% of the COVID-19 vaccine candidates in development in early 2021. Most of these use adenoviruses as the vector and the majority encode the SARS-CoV-2 spike protein. Human adenoviruses have a good safety profile and are highly immunogenic; however, preexisting immunity ("antivector immunity") can hamper the induction of immunity against the encoded antigen. To overcome this issue, animal adenoviruses can be used as the vaccine vector. This approach has been used by the AstraZeneca/University of Oxford COVID-19 vaccine program, using the vector ChAdOx1 derived from chimpanzee adenovirus Y25 (Fig. 8.3). A single dose of nonhuman adenoviruses can still induce antivector immunity, however, potentially affecting the potency of booster doses. Notwithstanding apprehensions about antivector immunity, many of the adenovirus-based SARS-CoV-2 vaccines were initially trialed and planned to be given as single dose vaccines (Flanagan et al., 2020). In early 2021, multiple adenoviral-based vaccines had been either approved for general use or authorized for emergency use in various constituencies. The replicating viral vectors that have been developed clinically against COVID-19 have been based on measles and recombinant vesicular stomatitis viruses and contribute a small proportion of vaccine candidates progressing into development.

Previous vaccine development against SARS-CoV-1 and other coronaviruses provided a foundation on which to progress and deploy viral vector vaccines against SARS-CoV-2. In animal challenge studies, high titers of neutralizing antibodies to SARS-CoV-1 and MERS virus correlated with protection against disease (Flanagan et al., 2020). Vesicular stomatitis virus, measles virus, and adeno-associated virus-vectored vaccines had been shown

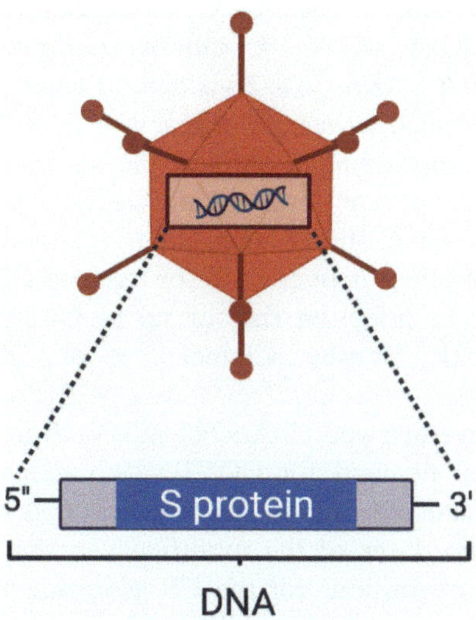

FIGURE 8.3 Viral vector-based vaccines. This is an example of an engineered adenovirus capable of producing the spike (S) protein of SARS-CoV-2. The vector (virus transporting the transgene of interest) may be an adenovirus or other immunogenic replicating or nonreplicating viruses. *Adapted from "COVID-19 vaccine candidate: AZD1222," by BioRender.com (2021). Retrieved from https://app.biorender.com/biorender-templates*

to be immunogenic and protect against challenge with SARS-CoV-1 in animal models (Roper & Rehm, 2009); however, none of these were trialed clinically. The only SARS-CoV-1 vaccines to reach clinical trials were based on inactivated virus, protein subunit, and DNA approaches (Padron-Regalado, 2020). By contrast, MERS vaccines trialed clinically were based on DNA and viral vector platforms (Folegatti et al., 2020; Koch et al., 2020).

In an example of the versatility of vaccine platform technologies, after the publication of the SARS-CoV-2 genome, a promising candidate directed against MERS, ChAdOx1 MERS, was repurposed to encode the full-length codon-optimized spike protein of SARS-CoV-2. ChAdOx1 is a recombinant E1 E3-deleted vector with comparable immunogenicity, but lower human seroprevalence, relative to other simian adenoviruses (Dicks et al., 2012). Preclinical studies demonstrated ChAdOx1 MERS to be highly immunogenic, to induce both polyfunctional CD8 + T cells and neutralizing antibodies, and to be protective in transgenic mice, nonhuman primates and in natural transmission settings among dromedary camels (Alharbi et al., 2017, 2019; Munster et al., 2017; Van Doremalen et al., 2020). Data from a phase 1 clinical trial showed that ChAdOx1 MERS was safe and well tolerated at the doses the ChAdOx1 COVID-19 vaccine has been given, and immunogenic against MERS virus peptides within 1 month of vaccination (Folegatti et al., 2020).

The laboratory, clinical, and regulatory experience with ChAdOx1 MERS allowed this vector to be quickly reengineered and trialed against SARS-CoV-2. ChAdOx1 nCoV-19, otherwise known as AZD1222 or Vaxzevria, was the first SARS-CoV-2 vaccine to enter phase 3 clinical trials worldwide. It demonstrated a vaccine efficacy of over 70% against COVID-19. Similar to other front-runner vaccines, the vaccine was shown to have 100% efficacy against severe disease and death. Large-scale prospective observational studies have shown that after a single dose of ChAdOx1 nCoV-19, the vaccine effect is over 85% for reduced COVID-19 hospitalization and a reduction in infection risk of up to 60% (Menni et al., 2021; Vasileiou et al., 2021; Voysey, Clemens, et al., 2021; Voysey, Costa Clemens, et al., 2021).

Several factors have allowed ChAdOx1 nCoV-19 to be a key component in meeting the global demand for COVID-19 vaccines. Prior to the pandemic, large-scale manufacturing and distribution had been performed for a viral vectored vaccine; doses of the modified vaccinia Ankara vector were produced in the tens of millions for the US government's smallpox vaccine program, thus demonstrating the feasibility of upscaling production. The process of ChAdOx1 manufacture builds on well-established methods involving replication in mammalian cells, and cultivation and purification of the virus, allowing yields of up to 5×10^{13} purified virus particles per liter of culture (Fedosyuk et al., 2019; Kallel & Kamen, 2015).

Early on in the pandemic AstraZeneca established partnerships and technology transfer agreements with large-scale vaccine producers experienced in cell-based medicines, GMP manufacturing, and aseptic filling of sterile products. In particular, a partnership with the Serum Institute of India, the world's largest vaccine producer at that time, enabled very large-scale production at low cost to ensure availability of vaccine to LMICs. These factors have allowed the vaccine to be produced in vast quantities at a low price. As the vaccine can be refrigerated rather than frozen for medium-term storage (up to 6 months), it is suitable for deployment in LMICs where distribution of vaccines at fridge temperature is well established. The regulatory environment for approval of COVID-19 vaccines has also facilitated rapid access to viral vector platforms. Despite the first viral vector vaccine being licensed in 2019, for prophylaxis against Dengue virus disease, there has been internationally accepted and specific regulatory guidance on the quality, nonclinical and clinical aspects of viral vector vaccine development since 2010 (Nakayama & Aruga, 2015; Sakurai et al., 2019).

Antigen-presenting cell-based vaccines

To mount an effective adaptive immune response, APCs are required to process and display fragmented antigens on their cell surface. APC-based vaccines are designed with this in mind and bypass all the in vivo steps leading up to this point. Conventionally, an individual's dendritic cells are harvested, expanded, and altered to present the target antigen, and infused back into the same individual. There are a handful of APC-based vaccines being developed clinically against COVID-19, although all of these also require coadministration of, or modification by, recombinant viral vectors (Word Health Organization, 2021). Thus far, this relatively nascent platform technology has been prohibitively cost- and labor-intensive for large-scale deployment. The cold-chain and IV infusion requirements for such a cell-based vaccine also encumber the deployment of these vaccines on a large scale, particularly given that multiple doses are often required for potent immunogenicity. As a result, no vaccines for infectious diseases have been fully or partially licensed based on this novel platform (Van Riel & De Wit, 2020).

Protein-based vaccines

Protein-based vaccines encompass vaccines derived from protein purified from the SARS-CoV-2 virus or infected cells, recombinant protein or virus-like particles (VLP). Protein-based vaccines often need to be formulated with an adjuvant to evoke a sufficiently strong immune response for lasting protection.

A major consideration when manufacturing protein-based vaccines is the recombinant protein production platform. Several systems are available, including mammalian, yeast, bacterial, insect and plant-based expression, which can be differentiated by varying yields, safety, speed, scalability, regulatory compliance and cost (Shin et al., 2020). Conventionally, bioreactors and mammalian, bacterial or yeast cell cultures have been used in large-scale manufacturing processes but these platforms are expensive and are vulnerable to human pathogen contamination (Schneemann & Young, 2003; Vieira Gomes et al., 2018). New manufacturing platforms have been proposed, such as plant-based protein expression, and it has been suggested these have the potential to rapidly meet global demand requirements in response to pandemics (Shin et al., 2020). A successful example of this was during the 2014 Ebola outbreak; an antibody cocktail named ZMapp was produced by molecular farming of a plant-based expression system (Davey et al., 2016). However, during the COVID-19 pandemic, successfully licensed vaccines have relied on either mammalian or bacterial systems.

Virus-like particles

VLPs are multimeric molecules that can self-assemble into viral structures but are noninfectious because they contain no viral genome and nonstructural proteins. This platform has generated increased interest in contemporary vaccine design owing to its good safety profile and the capacity to target the vaccine to dendritic cells (Zhao et al., 2014). VLPs, being composed of viral structural proteins, mimic pathogen-associated molecular patterns, and their interior surface elements can be packaged with TLR ligands (Mohsen et al., 2017). Therefore, these vaccines can act as self-adjuvants, allowing efficient activation and amplification of antigen-specific immune response. VLP vaccines also provide enhanced stability and a lower propensity for degradation compared to constructs in DNA and RNA, protein subunit, peptide vaccines. A further advantage of the VLP display platform is its modularity and scalability. Peptides can be adapted as new information about SARS-CoV-2 and further variants emerge.

The developmental track record for VLP-based vaccines provides optimism for their implementation as COVID-19 vaccines. Numerous VLP-based vaccines are already commercially available such as vaccines against Human Papilloma Virus and Hepatitis B Virus (HBV). In addition, the first licensed malaria vaccine, RTS,S/AS01, is based on the HBV VLP and recently entered an implementation pilot study across Africa. Several VLP-based SARS-CoV-2 vaccines are in preclinical development and this platform accounts for 5% of clinical stage candidates, all in early phase development.

Protein subunit

The most prevalent approach in developing vaccines against COVID-19 is based on protein (or peptide) vaccines. More than 30% of all candidates in clinical trials are protein subunit-based. Compared to nucleic acid, viral vector and VLP-based vaccines, they are simple and cheap to make, and they are safer than whole organism-based methods (Flanagan et al., 2020). Protein subunit-based vaccines, however, often lack sufficient immunogenicity and require adjuvants and multiple doses to induce robust responses (Richmond et al., 2021). Protein-based vaccines have the advantage of a broader immune response, compared to peptide approaches, but can lose immunogenicity if the vaccine is based purely on conformational epitopes. Peptide-based vaccines include important T and B cell immunogens, often of 15–30 amino acids in length, and are designed to specifically target immunity. However, these types of vaccines are hampered by HLA restriction for recognition of T cell epitopes, absence of receptor cross-linking in the case of B cell epitopes, absence of cross-reactive or heterosubtypic immunity and difficulties in inducing long-lasting immunity. Previous experimental experience with peptide-based vaccines is considerable, particularly against viral pathogens, but none have been licensed for use in humans (Di Natale et al., 2020).

Whole organism-based vaccines

Whole organism-based vaccines were crucial to the widespread improvement in human health throughout the 20th century and continue to be critical for protection against disease and death. Whole organism-based approaches can contain noninfectious inactivated virus, or live-attenuated virus, which can stimulate a robust immune response but does not cause disease.

It is difficult to develop whole virus-based vaccines against outbreak pathogens, such as SARS-CoV-2, quickly, because of the time, knowledge and stringent biosecurity requirements. Lengthy and wide-ranging safety testing would be needed to confirm that live attenuated viruses are safe and do not readily revert to wild type (leading to the situation where vaccine-related SARS-CoV-2 infection may reintroduce COVID-19 to a region that is not sufficiently vaccinated). There would also be a requirement for very large amounts of virus to be grown in biosafety level 3 (BSL3) conditions for these approaches. BSL3 would also be required throughout quality control processes as well, including potency, sterility and purity testing.

Live attenuated virus vaccines

Live attenuated vaccines mimic the natural infection process and, consequently, they can be sufficiently immunogenic with a single nonadjuvanted administration. Attenuation would traditionally involve SARS-CoV-2 to be altered through passage such that it loses enough virulence factors so as to not cause pathology. A more novel approach has been to attenuate virus through reverse genetics by inserting site-directed mutations into genes associated with virulence. This has shown promise in SARS-CoV-1 and MERS-CoV vaccine development (Almazán et al., 2013; Regla-Nava et al., 2015). Such reverse genetics techniques require sufficient knowledge of the genes that are important for viral replication and confirmation that the inserted mutations are stable. A key aspect of this vaccine approach is to balance attenuation with replication. Virus that is excessively attenuated may not replicate enough to be immunogenic (Chen & Chen, 2020). Unfortunately, these vaccines are also often contraindicated in special risk groups, for example individuals who are pregnant or immuno-compromised. Live attenuated virus vaccines against SARS-CoV-2 currently represent less than 2% of the total candidates in clinical development.

Inactivated virus vaccines

Isolation, inactivation and administration of a virus are a well-established method for vaccination. The whole virus can be used after inactivation or alternatively, antigenic sections can be purified to generate a subunit vaccine. Generally, these vaccines are considered safer than live-attenuated vaccines however they are also less immunogenic and may need an adjuvant. Less than 15% of candidates currently in clinical development are based on inactivated SARS-CoV-2.

Inactivation, by chemicals (e.g., formaldehyde), heat or radiation, have all been trialed in vaccine development against other coronaviruses. Unfortunately, vaccine-induced disease enhancement has been a cause for apprehension in SARS-CoV-1 and MERS-CoV-inactivated vaccine development. Such effects were first seen in the 1960s, when a formalin-inactivated respiratory syncytial virus (RSV) vaccine induced predominantly nonneutralizing antibodies, and previously seronegative children had worse RSV outcomes and hospitalization rates compared to controls (Kim et al., 1969). Comparably, immunization with a gamma radiation inactivated MERS vaccine and, separately, a UV radiation inactivated SARS-CoV-1 vaccine led to eosinophilic pro-inflammatory lung pathology on challenge with live virus (Agrawal et al., 2016; Bolles et al., 2011). It has recently been shown that the choice of adjuvant and inactivation method may alter the propensity for a vaccine to cause eosinophil-mediated vaccine-related pathology (Deng et al., 2018). Due to the sequence homology between SARS-CoV-2, SARS-CoV and MERS-CoV, early in the pandemic antibody-dependent enhancement (ADE) and vaccine-associated enhanced respiratory disease (VAERD) were seen as potential safety issues (Lambert et al., 2020; Zellweger et al., 2020). Thus far, definitive VAERD and ADE have not been described in SARS-CoV-2 vaccine animal studies, phase 3 trials or postlicensure surveillance (Kim et al., 1969). More data are needed on whether waning antibody titers could increase the risk of ADE after subsequent virus exposure.

Future vaccine requirements and challenges

Despite widespread acknowledgment after the Ebola outbreak in 2014 that large scale disease outbreaks have the potential to pose a severe threat to human health and economic security (Piot et al., 2019), most countries were poorly prepared to deal with the COVID-19 pandemic. The combination of globalization with environmental, political and socioeconomic causes is increasing the conditions in which novel infectious diseases can emerge. Investment in public health systems and infrastructure in individual countries, combined with effective global surveillance and communication are key to the response to new emerging pathogens. However, significant and sustained investment in R&D on pathogens with pandemic potential is also crucial, particularly for Disease X situations. To facilitate this, WHO produced an R&D blueprint for action to prevent epidemics to improve emergency preparedness, highlight the importance of research in the response to epidemics and to promote coordinated research and product development activities.

Another initiative established in 2016 was the Coalition for Epidemic Preparedness Innovation (CEPI), a global partnership focused on developing vaccines and diagnostics for emerging pathogens and to ensure equitable access to those vaccines. CEPI's approach was to accelerate vaccine development against pathogens with known threats as well as funding the development of technologies to facilitate rapid development of vaccines against previously unknown pathogens with a target of 16 weeks from identification of antigen to release of vaccine for clinical trials. However, responding to an outbreak of a novel pathogen with a vaccine requires much more than just generation of material for clinical trials, and significant obstacles must be overcome after acceptable safety and efficacy of any candidate vaccine has been demonstrated. Deployment of any new medicinal product requires either licensure or use under EUA regulations on a country-by-country basis meaning that each national medicine regulatory authority must undertake its own review process prior to grant EUA. This is extremely time-consuming and results in substantial duplication of effort. While it is essential that each country's national regulator gives assurance that a new product is suitable for its own citizens, some elements of the review process could be shared, such as audits of manufacturing facilities to accelerate the approval process.

Other obstacles to deployment of new vaccines under EUA include the availability of large-scale manufacturing capacity, with much of the available capacity centered in a few mostly high-income areas, such as the UK, Europe, North America and India. As a result, during the first six months of the COVID-19 vaccine rollout, only 1 dose in LMIC was given for every 73 doses administered in the developed world, as high-income countries prioritized vaccination of their own citizens over supporting global vaccine distribution. COVAX was established by CEPI, WHO and GAVI to both accelerate vaccine manufacturing and ensure equitable access to COVID-19 vaccines for every country in the world, potentially providing doses for up to 20% of each country's population. However, COVAX was dependent on individual countries and manufacturers providing doses for distribution to LMIC and delays with supply reduced the ability of COVAX to meet the enormous demand for doses in the initial stages of the vaccine rollout. Only 1.3 billion doses of vaccine were pledged to COVAX in 2021, well short of the estimated 2 billion doses required. Democratization of vaccine supply through the establishment of more manufacturing capacity globally, particularly in LMIC, would ensure equitable access to vaccines in future pandemics.

In addition to manufacturing capacity, specialist reagents and supplies are also required for large-scale manufacturing, many of which are only produced by a few suppliers, creating bottlenecks. Furthermore, supplies for administering vaccines and related equipment, such as fridges, freezers and thermally insulated transport packaging are also required at enormous scale for vaccines to be distributed to populations. All this infrastructure and logistic capacity is essential, requiring significant national planning and investment by governments, and for this to be maintained and renewed between pandemics. In the case of COVID-19, the clinical development of safe and effective vaccines was relatively easily achieved, however, much needs to be improved to ensure successful large-scale manufacturing, deployment and equitable access to those vaccines once approved for use.

References

Agrawal, A. S., Tao, X., Algaissi, A., Garron, T., Narayanan, K., Peng, B.-H., Couch, R. B., & Tseng, C.-T. K. (2016). Immunization with inactivated Middle East Respiratory Syndrome coronavirus vaccine leads to lung immunopathology on challenge with live virus. *Human Vaccines & Immunotherapeutics*, *12*(9), 2351–2356. Available from https://doi.org/10.1080/21645515.2016.1177688.

Alharbi, N. K., Padron-Regalado, E., Thompson, C. P., Kupke, A., Wells, D., Sloan, M. A., Grehan, K., Temperton, N., Lambe, T., Warimwe, G., Becker, S., Hill, A. V. S., & Gilbert, S. C. (2017). ChAdOx1 and MVA based vaccine candidates against MERS-CoV elicit neutralising antibodies and cellular immune responses in mice. *Vaccine*, *35*(30), 3780–3788. Available from https://doi.org/10.1016/j.vaccine.2017.05.032.

Alharbi, N. K., Qasim, I., Almasoud, A., Aljami, H. A., Alenazi, M. W., Alhafufi, A., Aldibasi, O. S., Hashem, A. M., Kasem, S., Albrahim, R., Aldubaib, M., Almansour, A., Temperton, N. J., Kupke, A., Becker, S., Abu-Obaidah, A., Alkarar, A., Yoon, I.-K., Azhar, E., ... Balkhy, H. H. (2019). Humoral immunogenicity and efficacy of a single dose of ChAdOx1 MERS vaccine candidate in dromedary camels. *Scientific Reports*, *9*(1), 16292. Available from https://doi.org/10.1038/s41598-019-52730-4.

Almazán, F., DeDiego, M. L., Sola, I., Zuñiga, S., Nieto-Torres, J. L., Marquez-Jurado, S., Andrés, G., & Enjuanes, L. (2013). Engineering a replication-competent, propagation-defective Middle East respiratory syndrome coronavirus as a vaccine candidate. *MBio, 4*(5). Available from https://doi.org/10.1128/mBio.00650-13, e00650−13.

Baden, L. R., El Sahly, H. M., Essink, B., Kotloff, K., Frey, S., Novak, R., Diemert, D., Spector, S. A., Rouphael, N., Creech, C. B., McGettigan, J., Khetan, S., Segall, N., Solis, J., Brosz, A., Fierro, C., Schwartz, H., Neuzil, K., Corey, L., ... Zaks, T. (2021). Efficacy and Safety of the mRNA-1273 SARS-CoV-2 Vaccine. *The New England Journal of Medicine, 384*(5), 403−416. Available from https://doi.org/10.1056/NEJMoa2035389.

Bezbaruah, R., Borah, P., Kakoti, B. B., Al-Shar'I, N. A., Chandrasekaran, B., Jaradat, D. M. M., Al-Zeer, M. A., & Abu-Romman, S. (2021). Developmental landscape of potential vaccine candidates based on viral vector for prophylaxis of COVID-19. *Frontiers in Molecular Biosciences, 8*635337. Available from https://doi.org/10.3389/fmolb.2021.635337.

Bolles, M., Deming, D., Long, K., Agnihothram, S., Whitmore, A., Ferris, M., Funkhouser, W., Gralinski, L., Totura, A., Heise, M., & Baric, R. S. (2011). A double-inactivated severe acute respiratory syndrome coronavirus vaccine provides incomplete protection in mice and induces increased eosinophilic proinflammatory pulmonary response upon challenge. *Journal of Virology, 85*(23), 12201−12215. Available from https://doi.org/10.1128/JVI.06048-11.

Chen, J.-W., & Chen, J.-M. (2020). Potential of live pathogen vaccines for defeating the COVID-19 pandemic: History and mechanism. *Journal of Medical Virology, 92*(9), 1469−1474. Available from https://doi.org/10.1002/jmv.25920.

Dagan, N., Barda, N., Kepten, E., Miron, O., Perchik, S., Katz, M. A., Hernán, M. A., Lipsitch, M., Reis, B., & Balicer, R. D. (2021). BNT162b2 mRNA Covid-19 vaccine in a nationwide mass vaccination setting. *The New England Journal of Medicine, 384*(15), 1412−1423. Available from https://doi.org/10.1056/NEJMoa2101765.

Davey, R. T., Dodd, L., Proschan, M. A., Neaton, J., Neuhaus Nordwall, J., Koopmeiners, J. S., Beigel, J., Tierney, J., Lane, H. C., Fauci, A. S., Massaquoi, M. B. F., Sahr, F., & Malvy, D. (2016). A randomized, controlled trial of ZMapp for Ebola virus infection. *The New England Journal of Medicine, 375*(15), 1448−1456.

Deng, Y., Lan, J., Bao, L., Huang, B., Ye, F., Chen, Y., Yao, Y., Wang, W., Qin, C., & Tan, W. (2018). Enhanced protection in mice induced by immunization with inactivated whole viruses compare to spike protein of middle east respiratory syndrome coronavirus. *Emerging Microbes & Infections, 7*(1), 60. Available from https://doi.org/10.1038/s41426-018-0056-7.

Di Natale, C., La Manna, S., De Benedictis, I., Brandi, P., & Marasco, D. (2020). Perspectives in peptide-based vaccination strategies for syndrome Coronavirus 2 pandemic. *Frontiers in Pharmacology, 11*578382. Available from https://doi.org/10.3389/fphar.2020.578382.

Dicks, M. D. J., Spencer, A. J., Edwards, N. J., Wadell, G., Bojang, K., Gilbert, S. C., Hill, A. V. S., & Cottingham, M. G. (2012). A novel chimpanzee adenovirus vector with low human seroprevalence: Improved systems for vector derivation and comparative immunogenicity. *PLoS One, 7*(7)e40385. Available from https://doi.org/10.1371/journal.pone.0040385.

Fedosyuk, S., Merritt, T., Peralta-Alvarez, M. P., Morris, S. J., Lam, A., Laroudie, N., Kangokar, A., Wright, D., Warimwe, G. M., Angell-Manning, P., Ritchie, A. J., Gilbert, S. C., Xenopoulos, A., Boumlic, A., & Douglas, A. D. (2019). Simian adenovirus vector production for early-phase clinical trials: A simple method applicable to multiple serotypes and using entirely disposable product-contact components. *Vaccine, 37*(47), 6951−6961. Available from https://doi.org/10.1016/j.vaccine.2019.04.056.

Flanagan, K. L., Best, E., Crawford, N. W., Giles, M., Koirala, A., Macartney, K., Russell, F., Teh, B. W., & Wen, S. C. (2020). Progress and pitfalls in the quest for effective SARS-

CoV-2 (COVID-19) vaccines. *Frontiers in Immunology*, *11*579250. Available from https://doi.org/10.3389/fimmu.2020.579250.

Folegatti, P. M., Bittaye, M., Flaxman, A., Lopez, F. R., Bellamy, D., Kupke, A., Mair, C., Makinson, R., Sheridan, J., Rohde, C., Halwe, S., Jeong, Y., Park, Y.-S., Kim, J.-O., Song, M., Boyd, A., Tran, N., Silman, D., Poulton, I., . . . Gilbert, S. (2020). Safety and immunogenicity of a candidate Middle East respiratory syndrome coronavirus viral-vectored vaccine: A dose-escalation, open-label, non-randomised, uncontrolled, phase 1 trial. *The Lancet Infectious Diseases*, *20*(7), 816–826. Available from https://doi.org/10.1016/S1473-3099(20)30160-2.

Geall, A. J., Verma, A., Otten, G. R., Shaw, C. A., Hekele, A., Banerjee, K., Cu, Y., Beard, C. W., Brito, L. A., Krucker, T., O'Hagan, D. T., Singh, M., Mason, P. W., Valiante, N. M., Dormitzer, P. R., Barnett, S. W., Rappuoli, R., Ulmer, J. B., & Mandl, C. W. (2012). Nonviral delivery of self-amplifying RNA vaccines. *Proceedings of the National Academy of Sciences of the United States of America*, *109*(36), 14604–14609. Available from https://doi.org/10.1073/pnas.1209367109.

Haidari, G., Cope, A., Miller, A., Venables, S., Yan, C., Ridgers, H., Reijonen, K., Hannaman, D., Spentzou, A., Hayes, P., Bouliotis, G., Vogt, A., Joseph, S., Combadiere, B., McCormack, S., & Shattock, R. J. (2017). Combined skin and muscle vaccination differentially impact the quality of effector T cell functions: The CUTHIVAC-001 randomized trial. *Scientific Reports*, *7*(1), 13011. Available from https://doi.org/10.1038/s41598-017-13331-1.

Henao-Restrepo, A. M., Camacho, A., Longini, I. M., Watson, C. H., Edmunds, W. J., Egger, M., Carroll, M. W., Dean, N. E., Diatta, I., Doumbia, M., Draguez, B., Duraffour, S., Enwere, G., Grais, R., Gunther, S., Gsell, P.-S., Hossmann, S., Watle, S. V., Kondé, M. K., . . . Kieny, M.-P. (2017). Efficacy and effectiveness of an rVSV-vectored vaccine in preventing Ebola virus disease: Results from the Guinea ring vaccination, open-label, cluster-randomised trial (Ebola ÇaSuffit!). *Lancet (London, England)*, *389*(10068), 505–518. Available from https://doi.org/10.1016/S0140-6736(16)32621-6.

Hinz, T., Kallen, K., Britten, C. M., Flamion, B., Granzer, U., Hoos, A., Huber, C., Khleif, S., Kreiter, S., Rammensee, H.-G., Sahin, U., Singh-Jasuja, H., Türeci, Ö., & &Kalinke, U. (2017). The European regulatory environment of RNA-based vaccines. *Methods in Molecular Biology (Clifton, N.J.)*, *1499*, 203–222.

Hobernik, D., & Bros, M. (2018). DNA vaccines-how far from clinical use? *International Journal of Molecular Sciences*, *19*(11). Available from https://doi.org/10.3390/ijms19113605.

Hodgson, S. H., Mansatta, K., Mallett, G., Harris, V., Emary, K. R. W., & Pollard, A. J. (2021). What defines an efficacious COVID-19 vaccine? A review of the challenges assessing the clinical efficacy of vaccines against SARS-CoV-2. *The Lancet Infectious Diseases*, *21*(2), e26–e35. Available from https://doi.org/10.1016/S1473-3099(20)30773-8.

Kaiser, J. A., & Barrett, A. D. T. (2019). Twenty years of progress toward West Nile virus vaccine development. *Viruses*, *11*(9). Available from https://doi.org/10.3390/v11090823.

Kallel, H., & Kamen, A. A. (2015). Large-scale adenovirus and poxvirus-vectored vaccine manufacturing to enable clinical trials. *Biotechnology Journal*, *10*(5), 741–747. Available from https://doi.org/10.1002/biot.201400390.

Karikó, K., Buckstein, M., Ni, H., & Weissman, D. (2005). Suppression of RNA recognition by Toll-like receptors: The impact of nucleoside modification and the evolutionary origin of RNA. *Immunity*, *23*(2), 165–175.

Karikó, K., Muramatsu, H., Welsh, F. A., Ludwig, J., Kato, H., Akira, S., & Weissman, D. (2008). Incorporation of pseudouridine into mRNA yields superior nonimmunogenic vector with increased translational capacity and biological stability. *Molecular Therapy: The Journal of the American Society of Gene Therapy*, *16*(11), 1833–1840. Available from https://doi.org/10.1038/mt.2008.200.

Keehner, J., Horton, L. E., Pfeffer, M. A., Longhurst, C. A., Schooley, R. T., Currier, J. S., Abeles, S. R., & Torriani, F. J. (2021). SARS-CoV-2 infection after vaccination in health care workers in California. *The New England Journal of Medicine*, *384*(18), 1774–1775. Available from https://doi.org/10.1056/NEJMc2101927.

Kim, H. W., Canchola, J. G., Brandt, C. D., Pyles, G., Chanock, R. M., Jensen, K., & Parrott, R. H. (1969). Respiratory syncytial virus disease in infants despite prior administration of antigenic inactivated vaccine. *American Journal of Epidemiology*, *89*(4), 422–434.

Knezevic, I., Liu, M. A., Peden, K., Zhou, T., & Kang, H.-N. (2021). Development of mRNA vaccines: Scientific and regulatory issues. *Vaccines*, *9*(2). Available from https://doi.org/10.3390/vaccines9020081.

Koch, T., Dahlke, C., Fathi, A., Kupke, A., Krähling, V., Okba, N. M. A., Halwe, S., Rohde, C., Eickmann, M., Volz, A., Hesterkamp, T., Jambrecina, A., Borregaard, S., Ly, M. L., Zinser, M. E., Bartels, E., Poetsch, J. S. H., Neumann, R., Fux, R., ... Addo, M. M. (2020). Safety and immunogenicity of a modified vaccinia virus Ankara vector vaccine candidate for Middle East respiratory syndrome: An open-label, phase 1 trial. *The Lancet Infectious Diseases*, *20*(7), 827–838. Available from https://doi.org/10.1016/S1473-3099(20)30248-6.

Lambert, P.-H., Ambrosino, D. M., Andersen, S. R., Baric, R. S., Black, S. B., Chen, R. T., Dekker, C. L., Didierlaurent, A. M., Graham, B. S., Martin, S. D., Molrine, D. C., Perlman, S., Picard-Fraser, P. A., Pollard, A. J., Qin, C., Subbarao, K., & Cramer, J. P. (2020). Consensus summary report for CEPI/BC March 12-13, 2020 meeting: Assessment of risk of disease enhancement with COVID-19 vaccines. *Vaccine*, *38*(31), 4783–4791. Available from https://doi.org/10.1016/j.vaccine.2020.05.064.

Ljungberg, K., & Liljeström, P. (2015). Self-replicating alphavirus RNA vaccines. *Expert Review of Vaccines*, *14*(2), 177–194. Available from https://doi.org/10.1586/14760584.2015.965690.

Lundstrom, K. (2020). Coronavirus pandemic-therapy and vaccines. *Biomedicines*, *8*(5). Available from https://doi.org/10.3390/biomedicines8050109.

Menni, C., Klaser, K., May, A., Polidori, L., Capdevila, J., Louca, P., Sudre, C. H., Nguyen, L. H., Drew, D. A., Merino, J., Hu, C., Selvachandran, S., Antonelli, M., Murray, B., Canas, L. S., Molteni, E., Graham, M. S., Modat, M., Joshi, A. D., ... Spector, T. D. (2021). Vaccine side-effects and SARS-CoV-2 infection after vaccination in users of the COVID Symptom Study app in the UK: A prospective observational study. *The Lancet Infectious Diseases*, *21*(7), 939–949. Available from https://doi.org/10.1016/S1473-3099(21)00224-3.

Merler, S., Ajelli, M., Fumanelli, L., Parlamento, S., Pastore, Y., Piontti, A., Dean, N. E., Putoto, G., Carraro, D., Longini, I. M., Halloran, M. E., & Vespignani, A. (2016). Containing Ebola at the source with ring vaccination. *PLoS Neglected Tropical Diseases*, *10*(11), e0005093. Available from https://doi.org/10.1371/journal.pntd.0005093.

Mohsen, M. O., Zha, L., Cabral-Miranda, G., & Bachmann, M. F. (2017). Major findings and recent advances in virus-like particle (VLP)-based vaccines. *Seminars in Immunology*, *34*, 123–132. Available from https://doi.org/10.1016/j.smim.2017.08.014.

Mullen, L., Potter, C., Gostin, L. O., Cicero, A., & Nuzzo, J. B. (2020). An analysis of International Health Regulations Emergency Committees and Public Health Emergency of International Concern Designations. *BMJ Global Health*, *5*(6). Available from https://doi.org/10.1136/bmjgh-2020-002502.

Munster, V. J., Wells, D., Lambe, T., Wright, D., Fischer, R. J., Bushmaker, T., Saturday, G., Van Doremalen, N., Gilbert, S. C., De Wit, E., & Warimwe, G. M. (2017). Protective efficacy of a novel simian adenovirus vaccine against lethal MERS-CoV challenge in a transgenic human DPP4 mouse model. *NPJ Vaccines*, *2*, 28. Available from https://doi.org/10.1038/s41541-017-0029-1.

Nakayama, Y., & Aruga, A. (2015). Comparison of current regulatory status for gene-based vaccines in the U.S., Europe and Japan. *Vaccines*, *3*(1), 186–202. Available from https://doi.org/10.3390/vaccines3010186.

Padron-Regalado, E. (2020). Vaccines for SARS-CoV-2: Lessons from other coronavirus strains. *Infectious Diseases and Therapy*, 1–20. Available from https://doi.org/10.1007/s40121-020-00300-x.

Pardi, N., Hogan, M. J., Porter, F. W., & Weissman, D. (2018). mRNA vaccines—A new era in vaccinology. *Nature Reviews Drug Discovery*, *17*(4), 261–279. Available from https://doi.org/10.1038/nrd.2017.243.

Petsch, B., Schnee, M., Vogel, A. B., Lange, E., Hoffmann, B., Voss, D., Schlake, T., Thess, A., Kallen, K.-J., Stitz, L., & Kramps, T. (2012). Protective efficacy of in vitro synthesized, specific mRNA vaccines against influenza A virus infection. *Nature Biotechnology*, *30*(12), 1210–1216. Available from https://doi.org/10.1038/nbt.2436.

Piot, P., Soka, M. J., & Spencer, J. (2019). Emergent threats: Lessons learnt from Ebola. *International Health*, *11*(5), 334–337. Available from https://doi.org/10.1093/inthealth/ihz062.

Polack, F. P., Thomas, S. J., Kitchin, N., Absalon, J., Gurtman, A., Lockhart, S., Perez, J. L., Pérez Marc, G., Moreira, E. D., Zerbini, C., Bailey, R., Swanson, K. A., Roychoudhury, S., Koury, K., Li, P., Kalina, W. V., Cooper, D., Frenck, R. W., Hammitt, L. L., … Gruber, W. C. (2020). Safety and efficacy of the BNT162b2 mRNA Covid-19 vaccine. *The New England Journal of Medicine*, *383*(27), 2603–2615. Available from https://doi.org/10.1056/NEJMoa2034577.

Regla-Nava, J. A., Nieto-Torres, J. L., Jimenez-Guardeño, J. M., Fernandez-Delgado, R., Fett, C., Castaño-Rodríguez, C., Perlman, S., Enjuanes, L., & DeDiego, M. L. (2015). Severe acute respiratory syndrome coronaviruses with mutations in the E protein are attenuated and promising vaccine candidates. *Journal of Virology*, *89*(7), 3870–3887. Available from https://doi.org/10.1128/JVI.03566-14.

Richmond, P., Hatchuel, L., Dong, M., Ma, B., Hu, B., Smolenov, I., Li, P., Liang, P., Han, H. H., Liang, J., & Clemens, R. (2021). Safety and immunogenicity of S-Trimer (SCB-2019), a protein subunit vaccine candidate for COVID-19 in healthy adults: A phase 1, randomised, double-blind, placebo-controlled trial. *Lancet (London, England)*, *397*(10275), 682–694. Available from https://doi.org/10.1016/S0140-6736(21)00241-5.

Roper, R. L., & Rehm, K. E. (2009). SARS vaccines: Where are we? *Expert Review of Vaccines*, *8*(7), 887–898. Available from https://doi.org/10.1586/erv.09.43.

Sakurai, A., Ogawa, T., Matsumoto, J., Kihira, T., Fukushima, S., Miyata, I., Shimizu, H., Itamura, S., Ouchi, K., Hamada, A., Tani, K., Okabe, N., & Yamaguchi, T. (2019). Regulatory aspects of quality and safety for live recombinant viral vaccines against infectious diseases in Japan. *Vaccine*, *37*(43), 6573–6579. Available from https://doi.org/10.1016/j.vaccine.2019.08.031.

Schneemann, A., & Young, M. J. (2003). Viral assembly using heterologous expression systems and cell extracts. *Advances in Protein Chemistry*, *64*, 1–36.

Shin, M. D., Shukla, S., Chung, Y. H., Beiss, V., Chan, S. K., Ortega-Rivera, O. A., Wirth, D. M., Chen, A., Sack, M., Pokorski, J. K., & Steinmetz, N. F. (2020). COVID-19 vaccine development and a potential nanomaterial path forward. *Nature Nanotechnology*, *15*(8), 646–655. Available from https://doi.org/10.1038/s41565-020-0737-y.

Tebas, P., Yang, S., Boyer, J. D., Reuschel, E. L., Patel, A., Christensen-Quick, A., Andrade, V. M., Morrow, M. P., Kraynyak, K., Agnes, J., Purwar, M., Sylvester, A., Pawlicki, J., Gillespie, E., Maricic, I., Zaidi, F. I., Kim, K. Y., Dia, Y., Frase, D., … Humeau, L. M. (2021). Safety and immunogenicity of INO-4800 DNA vaccine against SARS-CoV-2: A preliminary report of an open-label, Phase 1 clinical trial. *EClinicalMedicine*, *31*100689. Available from https://doi.org/10.1016/j.eclinm.2020.100689.

Thompson, M. G., Burgess, J. L., Naleway, A. L., Tyner, H. L., Yoon, S. K., Meece, J., Olsho, L. E. W., Caban-Martinez, A. J., Fowlkes, A., Lutrick, K., Kuntz, J. L., Dunnigan, K., Odean, M. J., Hegmann, K. T., Stefanski, E., Edwards, L. J., Schaefer-Solle, N., Grant, L., Ellingson, K., … Gaglani, M. (2021). Interim estimates of vaccine effectiveness of BNT162b2 and mRNA-1273 COVID-19 vaccines in preventing SARS-CoV-2 infection among health care personnel, first responders, and other essential and frontline workers—Eight U.S. Locations, December 2020-March 2021. *Morbidity and Mortality Weekly Report*, *70*(13), 495–500. Available from https://doi.org/10.15585/mmwr.mm7013e3.

Tomori, O., & Kolawole, M. O. (2021). Ebola virus disease: Current vaccine solutions. *Current Opinion in Immunology*, *71*, 27–33. Available from https://doi.org/10.1016/j.coi.2021.03.008.

Van Doremalen, N., Haddock, E., Feldmann, F., Meade-White, K., Bushmaker, T., Fischer, R. J., Okumura, A., Hanley, P. W., Saturday, G., Edwards, N. J., Clark, M. H. A., Lambe, T., Gilbert, S. C., & Munster, V. J. (2020). A single dose of ChAdOx1 MERS provides protective immunity in rhesus macaques. *Science Advances*, *6*(24). Available from https://doi.org/10.1126/sciadv.aba8399, eaba8399.

Van Riel, D., & De Wit, E. (2020). Next-generation vaccine platforms for COVID-19. *Nature Materials*, *19*(8), 810–812. Available from https://doi.org/10.1038/s41563-020-0746-0.

Vasileiou, E., Simpson, C. R., Shi, T., Kerr, S., Agrawal, U., Akbari, A., Bedston, S., Beggs, J., Bradley, D., Chuter, A., De Lusignan, S., Docherty, A. B., Ford, D., Hobbs, F. R., Joy, M., Katikireddi, S. V., Marple, J., McCowan, C., McGagh, D., … Sheikh, A. (2021). Interim findings from first-dose mass COVID-19 vaccination roll-out and COVID-19 hospital admissions in Scotland: A national prospective cohort study. *Lancet (London, England)*, *397*(10285), 1646–1657. Available from https://doi.org/10.1016/S0140-6736(21)00677-2.

Vieira Gomes, A. M., Souza Carmo, T., Silva Carvalho, L., Mendonça Bahia, F., & Parachin, N. S. (2018). Comparison of yeasts as hosts for recombinant protein production. *Microorganisms*, *6*(2). Available from https://doi.org/10.3390/microorganisms6020038.

Vogel, A. B., Lambert, L., Kinnear, E., Busse, D., Erbar, S., Reuter, K. C., Wicke, L., Perkovic, M., Beissert, T., Haas, H., Reece, S. T., Sahin, U., & Tregoning, J. S. (2018). Self-amplifying RNA vaccines give equivalent protection against influenza to mRNA vaccines but at much lower doses. *Molecular Therapy: The Journal of the American Society of Gene Therapy*, *26*(2), 446–455. Available from https://doi.org/10.1016/j.ymthe.2017.11.017.

Voysey, M., Clemens, S. A. C., Madhi, S. A., Weckx, L. Y., Folegatti, P. M., Aley, P. K., Angus, B., Baillie, V. L., Barnabas, S. L., Bhorat, Q. E., Bibi, S., Briner, C., Cicconi, P., Collins, A. M., Colin-Jones, R., Cutland, C. L., Darton, T. C., Dheda, K., Duncan, C. J. A., … Pollard, A. J. (2021). Safety and efficacy of the ChAdOx1 nCoV-19 vaccine (AZD1222) against SARS-CoV-2: An interim analysis of four randomised controlled trials in Brazil, South Africa, and the UK. *Lancet (London, England)*, *397*(10269), 99–111. Available from https://doi.org/10.1016/S0140-6736(20)32661-1.

Voysey, M., Costa Clemens, S. A., Madhi, S. A., Weckx, L. Y., Folegatti, P. M., Aley, P. K., Angus, B., Baillie, V. L., Barnabas, S. L., Bhorat, Q. E., Bibi, S., Briner, C., Cicconi, P., Clutterbuck, E. A., Collins, A. M., Cutland, C. L., Darton, T. C., Dheda, K., Dold, C., ... Pollard, A. J. (2021). Single-dose administration and the influence of the timing of the booster dose on immunogenicity and efficacy of ChAdOx1 nCoV-19 (AZD1222) vaccine: A pooled analysis of four randomised trials. *Lancet (London, England)*, *397*(10277), 881–891. Available from https://doi.org/10.1016/S0140-6736(21)00432-3.

Wang, Z., Troilo, P. J., Wang, X., Griffiths, T. G., Pacchione, S. J., Barnum, A. B., Harper, L. B., Pauley, C. J., Niu, Z., Denisova, L., Follmer, T. T., Rizzuto, G., Ciliberto, G., Fattori, E., Monica, N. L., Manam, S., & Ledwith, B. J. (2004). Detection of integration of plasmid DNA into host genomic DNA following intramuscular injection and electroporation. *Gene Therapy*, *11*(8), 711–721.

Word Health Organization. (2007). Guidelines for assuring the quality and non-clinical safety evaluation of DNA vaccines. WHO norms and standards for biological products. https://www.who.int/publications/m/item/guidelines-for-assuring-the-quality-and-non-clinical-safety-evaluation-of-dna-vaccines.

World Health Organization. (2016). WHO Ebola situation report. https://apps.who.int/ebola/ebola-situation-reports?m = 20141224.

World Health Organization. (2020). WHO target product profiles for COVID-19 vaccines. https://www.who.int/docs/default-source/blue-print/who-target-product-profiles-for-covid-19-vaccines.pdf.

World Health Organization. (2021). COVID-19 vaccine tracker and landscape. https://www.who.int/publications/m/item/draft-landscape-of-covid-19-candidate-vaccines.

Wolff, J. A., Malone, R. W., Williams, P., Chong, W., Acsadi, G., Jani, A., & Felgner, P. L. (1990). Direct gene transfer into mouse muscle in vivo. *Science (New York, N.Y.)*, *247*(4949 Pt 1), 1465–1468.

Wu, J. T., Leung, K., & Leung, G. M. (2020). Nowcasting and forecasting the potential domestic and international spread of the 2019-nCoV outbreak originating in Wuhan, China: A modelling study. *Lancet (London, England)*, *395*(10225), 689–697. Available from https://doi.org/10.1016/S0140-6736(20)30260-9.

Zellweger, R. M., Wartel, T. A., Marks, F., Song, M., & Kim, J. H. (2020). Vaccination against SARS-CoV-2 and disease enhancement—Knowns and unknowns. *Expert Review of Vaccines*, *19*(8), 691–698. Available from https://doi.org/10.1080/14760584.2020.1800463.

Zhao, L., Seth, A., Wibowo, N., Zhao, C.-X., Mitter, N., Yu, C., & Middelberg, A. P. J. (2014). Nanoparticle vaccines. *Vaccine*, *32*(3), 327–337. Available from https://doi.org/10.1016/j.vaccine.2013.11.069.

Chapter 9

Adjuvants, immunomodulators, and adaptogens

Anita Milicic[1], Sören Reinke[1], Joannah Fergusson[1], Erik B. Lindblad[2], Aneesh Thakur[3], George Corby[4], Stephanie Longet[5], Sabina Górska[6], Agnieszka Razim[6], Kefei Hu[2,7], Bror Morein[8], Marina Luchner[9], Dennis Christensen[10], Slavcho Mrenoshki[11], Sebnem Ercelen Ceylan[12], Sveinbjorn Gizurarson[13,14] and Malachy Chigozie Ugwu[15]

[1]Jenner Institute, University of Oxford, Oxford, United Kingdom, [2]R&D, Inari Consult ApS, Copenhagen, Denmark, [3]University of Copenhagen, Copenhagen, Denmark, [4]University of Oxford, Oxford, United Kingdom, [5]Wellcome Centre for Human Genetics, University of Oxford, Oxford, United Kingdom, [6]Hirszfeld Institute of Immunology and Experimental Therapy Polish Academy of Sciences, Wroclaw, Poland, [7]R&D Unit, Croda Denmark, Frederikssund, Denmark, [8]DVM, Savacc AB, Uppsala, Sweden, [9]Technical University of Munich, Munich, Germany, [10]Statens Serum Institut, Copenhagen, Denmark, [11]Ss. Cyril and Methodius University, Faculty of Veterinary Medicine, Skopje, North Macedonia, [12]Genetic Engineering and Biotechnology Institute, Kocaeli, Turkey, [13]Faculty of Pharmaceutical Sciences, University of Iceland, Reykjavik, Iceland, [14]Pharmacy Department, Kamuzu University of Health Sciences, Chichiri, Malawi, [15]Department of Pharmaceutical Microbiology and Biotechnology, Faculty of Pharmaceutical Sciences, Nnamdi Azikiwe University, Awka, Nigeria

Introduction

Vaccine adjuvants (derived from the Latin "adiuvare"—to help) were first described by Ramon as "substances used in combination with a specific antigen that produced a more robust immune response than the antigen alone" (Ramon, 1924). From the serendipitous observation that horses with an inflammation at the site of toxoid injection had greater vaccine efficacy, Ramon linked acute inflammation with a more potent immune response to a vaccine, and proceeded to test a range of foodstuffs and other substances as additives to veterinarian vaccines (Ramon, 1925).

Since these seminal discoveries, a plethora of immunological agents have been trialed as adjuvants, especially in a veterinary context. In terms of their function and role in inducing the immune response, adjuvants are best described as diverse compounds that fall on a spectrum of functions, ranging

Vaccinology and Methods in Vaccine Research. DOI: https://doi.org/10.1016/B978-0-323-91146-7.00009-3

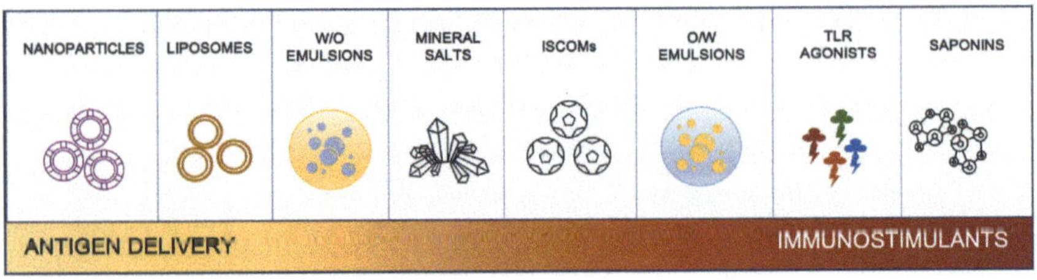

FIGURE 9.1 Adjuvants can be described within a spectrum of functions ranging from antigen delivery to immune stimulation.

from predominantly enhancing antigen presentation and delivery to acting as immunostimulants, with the precise immunological mechanisms still poorly understood (Fig. 9.1).

Today, there are a few adjuvants in use in human vaccines, and many more in vaccines for animals. The few approved for people are the aluminum-based salts (longest established and the only adjuvant type licensed for use as a separate medicinal product), MF59, developed by Novartis and licensed as part of flu (influenza) vaccines, and adjuvant systems AS01, AS03, and AS04, developed by GlaxoSmithKline (GSK) and licensed as part of vaccines against shingles (AS01), flu (AS03), hepatitis B (HepB, AS04), and Human Papilloma virus (HPV, AS04). More recently, vaccines have been approved that contain the CpG adjuvant (hepatitis B vaccine Heplisav) and Matrix M adjuvant (Covid-19 vaccine NVX-CoV2373 by Novavax).

Antigen presentation

When antigens are delivered as a vaccine, for example, subcutaneously, the host starts to eliminate the "foreign body." The elimination is especially rapid for vaccines delivered to mucosal surfaces (e.g., intranasally), as mucociliary clearance may result in a half-life of 15–20 minutes. It is essential that this rate of removal should not exceed the time it takes for antigen to encounter immune cells if a successful immune response is to be initiated. Adjuvants may preferentially augment lymphatic uptake of antigens instead of uptake into the plasma, ensuring an adaptive immune response is initiated in secondary lymphoid organs. This may be assisted by using larger molecules and nanoparticles, which are too large to enter the circulation and can slowly be taken up by the lymphatic system.

Antigen stabilization

Following delivery, antigens may rapidly degrade, and adjuvants have a key role in preventing this. Antigens can be complex molecules comprising proteins, peptides, toxoids, polysaccharides, conjugates, or other compounds.

Thus lipophilicity, ionic strength, pH, and other factors postimmunization need to be considered in inducing an effective immune response against the desired epitopes. Vaccines comprising messenger RNA (mRNA) are especially fragile, and require appropriate formulation to prevent destruction.

Right place, right time

Adjuvants should promote an environment at the injection site (or mucosal surface) that ensures sufficient recruitment of antigen presenting cells (APCs). Injectable vaccines will generate IgG responses whereas mucosal vaccination will also induce IgA antibodies—acting as a security system to prevent microbe entry into the mucosa.

Immunogenicity

Improving the immunogenicity of weak antigens has enabled the design of safer vaccines. The earliest vaccines were live (including live attenuated) pathogens that did not require adjuvants (Singh & O'Hagan, 2003). However, many of these are not appropriate for immuno-compromised individuals, in whom vaccination may cause actual pathological infection (Troy et al., 2013), as has been reported for the Zostavax live attenuated shingles vaccine (Bharucha et al., 2017). The development of the AS01 adjuvant by GSK has led to the Shingrix vaccine based on a recombinant subunit antigen, which is safe in immuno-compromised people and has 97.2% efficacy in adults over 50 (Cunningham et al., 2016). It should be noted, however, that the inflammatory response induced by adjuvants may itself be pathological—and so the balance between adjuvanticity and toxicity must be carefully assessed.

Dose sparing

As early as the 1930s, Volk reported on the superiority of two doses of an alum-adjuvanted tetanus vaccine over three doses of the nonadjuvanted vaccine (Volk & Bunney, 1939). In addition, each adjuvanted dose can achieve the same effect with less antigen. This reduces the production cost, as well as improving the speed of production—which is critical in a pandemic situation. The reduced requirement for booster shots additionally increases vaccination coverage and compliance.

Adjuvant potency is strongly influenced by the immune response required to confer protection against a given pathogen, for example, disease susceptibility to humoral versus cell-mediated immunity, and the degree to which the adjuvant promotes these immune mechanisms. Apart from aluminum-based compounds, new adjuvants are licensed as part of a specific vaccine formulation, rather than as an independent component.

Understanding the mode of action has allowed selection of adjuvants that are likely to promote the most effective immune response for a particular disease—for instance preferentially inducing either a Th1 or Th2 response. The first immune response to antigen may direct subsequent immune responses. Optimization of adjuvant-induced immune responses may be determined in early trials (Kester et al., 2009), or predicted based on existing understanding of immune response in a disease and adjuvant mechanisms likely to augment them. A notable example of this is AS03, observed to promote the activation of naïve B cells and broaden their specificity (Ellebedy et al., 2020)—which is of particular importance for a pandemic influenza vaccine (Henry et al., 2018; Pulendran et al., 2021).

Both historically and today, vaccine development is focused on the antigen and its immunogenicity. Though there have been noteworthy achievements in improving existing vaccines and delivery systems (Cervantes et al., 2012; Greenwood, 2014), when studies using a "standard" adjuvant fail to show the desired response, researchers often look to modify the antigen rather than search for an alternative adjuvant that may provide the right subtype of immune response. This has resulted in few adjuvants being licensed and may represent a good opportunity for improvements in vaccination through optimization of adjuvant systems.

Application

The choice of an adjuvant in any application depends on a number of factors: (1) vaccine antigen characteristics, such as type of antigen, size, and electric charge; (2) purpose—therapeutic vaccines (e.g., for cancer) generally have a higher reactogenicity threshold than prophylactic vaccines, especially for pediatric use; (3) route of immunization—not all adjuvants are suitable for administration via specific routes; (4) nature of the desired immune response, for example, antibody or T-cell response, Th1 or Th2.

Adjuvant testing and down-selection, preclinically as well as clinically, is carried out in association with a chosen antigen. The immunogenicity and potential to enhance vaccine efficacy of a given adjuvant will depend on both the vaccine platform deployed and on the specific antigen of interest. Methods used for adjuvant testing are the same as those applied to vaccine assessment, described in Chapter 4, Vaccine development: from the laboratory to the field.

Here we describe a variety of commonly used vaccine adjuvants and compounds with the ability to stimulate the immune system, or to enhance vaccine efficacy in some other way.

Aluminum-based adjuvants

Origin and chemical composition

In 1926 Alexander Glenny was investigating the immune response to diphtheriae antigen precipitated with salts (Glenny et al., 1926). Glenny worked with potassium alum ($KAl(SO_4)_2$:12 H_20) that can precipitate proteins when mixed under the right set of chemical conditions. This approach is referred to as "alum-precipitated vaccines." When injected into guinea pigs, 1.0 c.c. of precipitated diphtheriae antigen resulted in the same magnitude of antitoxin titers as 5.0 c.c. of soluble toxoid without alum.

This first approach, however, had several disadvantages. For example, since the precipitate was formed in the presence of the antigen, the aluminum adjuvant could not be characterized before mixing with antigen. In addition, the resulting precipitate may vary depending on the protein formulation, including buffer ions that could coprecipitate with the antigen. Finally, the approach required working aseptically, since sterilization by autoclaving would destroy heat sensitive antigens. Although this technology is now widely considered obsolete, there are still a few examples of vaccines prepared using precipitation.

The shortcomings of alum-precipitation were overcome when it was found that it was possible to adsorb protein antigens onto preformed "gels" of both aluminum hydroxide and aluminum phosphate in the form of hydrated colloid suspensions (Lindblad & Duroux, 2016). Both aluminum hydroxide and aluminum phosphate can be precipitated using different aluminum salts as starting materials; the resulting physico-chemical characteristics of the formulations reflect both the chemical starting material and the chemical process (Hem & White, 1984). For example, the electrical charge of aluminum hydroxide adjuvant particles is influenced by the presence of di- or trivalent anions in the process. Most importantly, with this advance it was now possible to standardize and characterize aluminum adjuvants before mixing them with the more costly antigen, and the adjuvant can be autoclaved to achieve sterility before mixing. The approach is known as "aluminum-adsorbed vaccines" in contrast to "alum-precipitated vaccines" mentioned earlier.

Relevant parameters for comprehensive characterization of aluminum adjuvants comprise the following, although no commercial supplier offers the entire range as standard: (1) chemical composition, (2) pH, (3) sterility, (4) protein adsorption capacity, with a model antigen in a standardized set-up, (5) particle size distribution, (6) zeta-potential, (7) point of zero charge measured in a standardized set-up, (8) absence of pyrogenicity/endotoxin.

Application

The use of aluminum adjuvants has been reviewed in more detail (Lindblad, 2004). Aluminum adjuvants have a track record of more than 80 years in commercial production of both human and veterinary vaccines. In addition to this, the protein adsorption capacity of aluminum hydroxide has also been used for purification of blood products, such as blood clotting Factor VIII for the treatment of hemophilia, although the aluminum hydroxide itself is removed during the process and is not part of the final product.

TABLE 9.1 Examples of human formulations containing aluminum adjuvants (alphabetical listing of pathogens).

Prophylactic vaccines		Therapeutic vaccines
Bacterial antigens	**Viral antigens**	**Adsorbed allergens**
Bacillus anthracis (Anthrax)	COVID-19 Corona virus	*Alternaria alternate*
Bordetella pertussis (Whooping cough)	Hepatitis A virus	*Artemisia vulgaris*
*Clostridium botulinum** (Botulism)	Human Papilloma virus	*Canis familiaris*
Clostridium tetani (Tetanus)	Inactivated poliovirus	*Dermatophagoides farina*
Corynebacterium diphtheriae (Diphtheria)	Influenza A /Vietnam (H5N1)	*Dermatophagoides pteronyssinus*
Hemophilus influenzae type B (HiB infections)	Japanese Encephalitis virus	*Equus caballus*
Streptococcus pneumoniae (pneumococcal disease)		*Felis domesticus*
		Phleum pratense
		Secale cereal

*Typically used for high-risk laboratory and military personnel.

Aluminum adjuvants are generally not used in live-attenuated vaccines, only in killed/inactivated vaccines and subunit/recombinant vaccines. Aluminum adjuvanted vaccines are frost sensitive and must not be exposed to temperatures that introduce a risk of freezing (Kurzatkowski et al., 2013).

Tables 9.1 and 9.2 present an overview of both human and veterinary vaccines adjuvanted with aluminum adjuvants.

In addition to these examples, aluminum adjuvants are also part of several novel vaccines in clinical testing, for example, the Takeda vaccine TAK 426 against Zika virus (https://clinicaltrials.gov/ct2/show/NCT03343626).

Mechanism of action

The mechanism of action of aluminum adjuvants is overly complex. Originally it was suggested that they delayed clearing of antigen from the injection site by forming an antigen depot. This explanation was, however,

TABLE 9.2 Examples of veterinary vaccine formulations containing aluminum adjuvants (alphabetical listing of pathogens).

Prophylactic vaccines	
Bacterial antigens	**Viral antigens**
Avibacterium paragallinarum	Bovine coronavirus
Clostridium chauvoei	Bovine rotavirus
Clostridium haemolyticum	Bluetongue virus
Clostridium novyi	Duck virus hepatitis
Clostridium perfringens	Feline calicivirus
Clostridium septicum	Feline leukemia virus
Clostridium tetani	Feline rhinotracheitis virus
Eschericia coli	Parainfluenza-3 virus
Erysipelothrix rhusiopathiae	Rabies virus
Hemophilus somnus	
Leptospira interrogans	
Pasteurella multocida	
Pasteurella trehalosi	
Pseudomonas aeruginosa	
Salmonella Dublin	
Salmonella enteritidis	
Salmonella typhimurium	

soon challenged and is now considered inadequate. Instead, two other mechanisms, both supported by experimental data, have been put forward; for a more in-depth discussion, please see Lindblad and Duroux (2016).

Back in 1985, Mannhalter and co-workers convincingly showed that aluminum-adsorbed tetanus toxoid was more efficiently taken up by APCs compared to the corresponding soluble protein (Mannhalter et al., 1985) resulting in a positive impact on the T-cell proliferative response. A likely explanation is that APCs take up antigen more effectively by phagocytosis than pinocytosis. After uptake by dendritic cells (DCs), aluminum adjuvanted vaccines have been shown to stimulate the formation of the NALP3 inflammasome and secretion of the pro-inflammatory cytokines IL-1β and IL-18 (Li et al., 2007, 2008).

Type of immune response induced

The general perception is that aluminum adjuvants elicit a predominantly Th2 response. Grun and Maurer demonstrated in 1989 (Grun & Maurer, 1989) that anti-IL-1α or anti-IL-4 was able to inhibit an antigen-specific T-cell proliferative response after immunization with aluminum adjuvant (but not if the mice were immunized with Freund's Complete Adjuvant). In addition, the proliferative responses were inhibited by anti-CD4 antibody, regardless of the adjuvant used. These data indicated that proliferating CD4 + T-cells from mice immunized with aluminum adjuvant were of the Th2 subset. It is worth noting recent work that indicates that by reducing the particle size into the nanoparticle range, aluminum hydroxide may in fact also stimulate a Th1 response (Orr et al., 2019). Until 1997, aluminum hydroxide and aluminium phosphate, together with the calcium phosphate adjuvant, remained the only adjuvant licensed for use in humans across Europe and North America, until the licensing of the MF59-adjuvanted flu vaccine (O'Hagan et al., 2013).

Emulsion-based adjuvants

Origin and chemical composition

The first oil-rich adjuvant was Freund's adjuvant, introduced in 1937 and used in both human and animal vaccines (Stills, 2005). However, due to numerous postimmunization side effects it was withdrawn from clinical use and new emulsion-based adjuvants with a better safety profile were developed.

Emulsions are defined as mixtures of two immiscible liquids, typically composed of oil droplets suspended in an aqueous dispersion medium, creating an oil-in-water system (O/W), or aqueous droplets suspended in an oil medium, forming a water-in-oil system (W/O). Another more complex system is a double-emulsion, that is, water-in-oil-in-water (W/O/W) that consists of water droplets suspended in larger oil droplets, which again are suspended in an aqueous dispersion medium. Stable emulsions require the addition of an appropriate surfactant (Tween 80, Kolliphor, Labrasol), which reduces the interfacial tension and prevents spontaneous aggregation (Xi et al., 2009). Small droplet size (<30 nm) gives transparent or translucent optical properties, whereas larger particles result in a more opaque/white appearance. Emulsion can be manufactured by methods that require low-energy (spontaneous emulsification, phase inversion method), high-energy (high pressure homogenizers, sonicators), or a combination of the two (Kotta et al., 2015).

Application

Emulsions can be supplied as liquids, aerosols, or sprays, and delivered by different routes (e.g., topical, oral, intranasal, or pulmonary), offering a broad and diverse potential in vaccine development (Lemoine et al., 2020; Razim et al., 2021).

Emulsion-based adjuvants used in licensed human and veterinary vaccines as well as those studied in clinical trials are summarized in Table 9.3.

MF59 and AS03 are the only emulsion-based adjuvants licensed for human use, both based on a squalene oil in water (O/W) emulsion. In addition, MF59 contains two nonionic surfactants, Tween 80 and Span 85, and a citrate buffer as an aqueous dispersion medium. AS03 comprises two oils: squalene and α-tocopherol (bioavailable form of vitamin E) and polysorbate 80 surfactant in phosphate-buffered saline. MF59 is licensed for use in seasonal influenza vaccines, and has also been tested clinically with HIV, RSV, and COVID-19 vaccine candidates (Table 9.3).

Many more emulsion-based adjuvants are approved for use in animal vaccines, as the veterinary market is more competitive and economically oriented with less strict regulations. Montanide ISA are a series of adjuvants approved for use in animals, differing in the ratio of ingredients and the size of emulsion droplets, suitable for use with inactivated or live antigens in a wide range of species (Ibrahim et al., 2015).

Mechanism of action

Emulsion adjuvants induce a strong humoral response and the mechanism of action is dependent mostly on the type of emulsion. Activity of O/W emulsions is based on localized chemokine and cytokine induction leading to enhanced recruitment of APCs. W/O emulsions, due to a slow and sustained release of the antigen (depot effect), enhance antigen engulfment and presentation by APCs. W/O/W emulsions combine fast and extended release that allow both instant and prolonged stimulation of the immune cells (Bozkir & Hayta, 2004) (Fig. 12.2) and offer a significant dose sparing effect (Caillet et al., 2010) (Fig. 9.2).

The complex mechanism of action of emulsion-based adjuvants is exemplified by MF59, one of the most versatile and promising adjuvant of this type. MF59 increases vaccine effectiveness among vulnerable populations such as the elderly (Domnich et al., 2017) and young children (McGuire et al., 2018; Zedda et al., 2015), by multiple mechanisms including induction of robust humoral responses, pro-inflammatory cytokines, and chemokines, as well as promotion of antigen uptake, processing, and presentation. MF59 creates an immuno-competent environment and recruits immune cells to the site of injection (SOI) where it is taken up by neutrophils, monocytes, DCs, and B cells, and transported to the draining lymph nodes (dLN) (O'Hagan et al., 2012). Monocyte recruitment leads to their differentiation into DCs, which represent the major source of antigen-loaded and activated APCs after vaccination with MF59 in combination with antigen (Cioncada et al., 2017). MF59 is of particular interest because it elicits a persistent response to vaccination, by localizing in the subcapsular sinus and promoting the accumulation of unprocessed antigen, simulating a depot effect (Cantisani et al., 2015). It increases the

TABLE 9.3 Summary of emulsion-based adjuvants used in licensed human and veterinary vaccines.

No.	Name	Type	Mechanism of action/type of induced response	Vaccines/examples of tested antigens
Emulsion adjuvants in human licensed vaccines				
1	MF59	O/W	Induction of cytokines and chemokines, enhancing antigen uptake, processing, and presentation/humoral response	Fluad, Focetria, Celtura, Fluad Quadrivalent. Also, Aflunov, used in veterinary vaccines.
2	AS03	O/W	APCs recruitment and activation, induction of cytokines and chemokines/cellular and humoral response	Pandemrix, Arepanrix
Emulsion adjuvants in veterinary licensed vaccines				
1	Montanide ISA	W/O, O/W, W/OW	Humoral and cellular response, Th1-biased	Newcastle disease in chickens, *Mycoplasma hyopneumoniae* in pigs, FMD in swine and cattle, various infections in farmed fish
2	Emulsigen	O/W	Immune cells recruitment, induction of cytokines at the site of injection, depot effect/humoral and cellular response	Aptovac, Biofel PCH/FMD and flu in pigs, A2HV-1, *Mycoplasma bovis*, *Staphylococcus aureus*, and *Staphylococcus chromogenes* in cattle
3	MetaStim	O/W	Induction of cytokines and expression of APC surface antigens/humoral and cellular response, Th1-biased	Pyramid, Fostera PCV MetaStim/Equine influenza virus in horses, BRD

Emulsion adjuvants in clinical trails				
1	Montanide ISA 51	W/O	Depot effect/humoral and CTL response	Flu (phase II, NCT03180801), malaria (phase II, NCT02083068), SARS-CoV-2 (phase I NCT04546841)
2	AS02	O/W	Release of endogenous danger signal, maturation of DCs/cellular and humoral immunity	Malaria (many studies in all phases), tuberculosis (phase II, CT00146744, NCT00397943), HIV (phase I, NCT00027365), hepatitis B (phase III, many studies), pneumonia (phase II, NCT00307528)
3	GLA-SE	O/W	TLR4 agonist/Th1, IgG2-skewed antibody response	Schistosomiasis (phase II, NCT03041766, NCT03799510); HIV (phase I, NCT03409276, NCT03856996); malaria (phase I, NCT01540474, NCT02647489); tuberculosis (phase II, NCT03806686); flu (phase II, NCT01991561, NCT01147068); Leishmania (phase I, NCT01751048); RSV (phase I, NCT02289820)
4	MAS-1 (MER4101)	O/W	Th2-type and humoral response, T cell–based immune regulatory control	Flu (phase I, NCT02500680)
5	Nanoemulsion	O/W	Dendritic cell (DC) maturation, TLR2 and TLR4 activation/humoral response, systemic and mucosal Th1 and Th17 responses	Anthrax (phase I, NCT04148118), flu (phase I, NCT01354379)

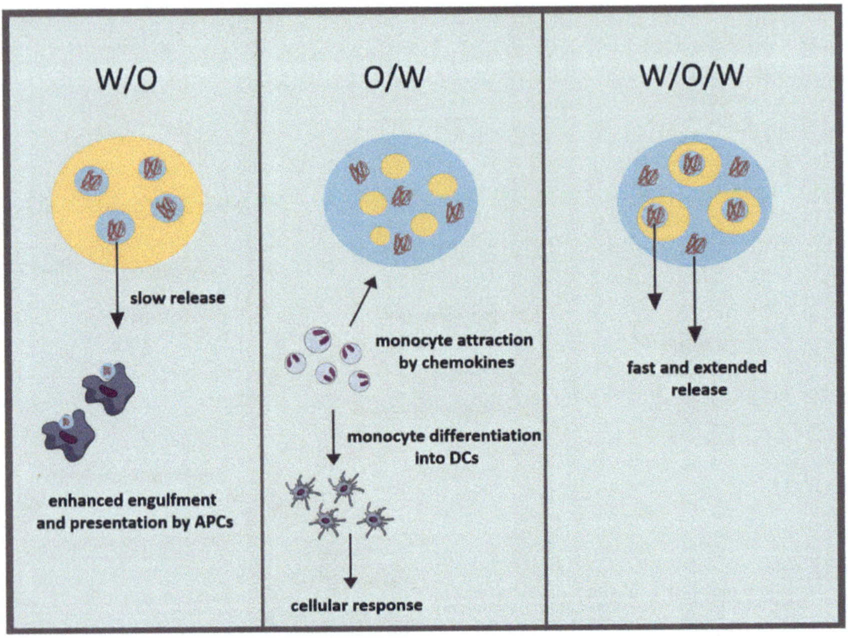

FIGURE 9.2 Mechanism of action of emulsion-based adjuvants.

differentiation of germinal center B cells in secondary lymph nodes, along with enhancing the frequency of T follicular helper cells (Lofano et al., 2015). Using systems serology, Boudreau et al. showed that MF59 induced antibody responses that strongly stimulated neutrophil-phagocytosis and complement activity, and a weak recruitment of NK cells and monocyte-phagocytosis, compromised by the lack of FCGR3A receptor-binding capacity by MF59-induced antibodies (Boudreau et al., 2020).

Saponins

Origin and chemical composition

Saponins are naturally found glycosides with a wide range of properties including the ability to modulate mammalian immune responses at low doses. Extracts from the bark of the South American tree *Quillaja saponaria* Molina have been known to possess potent adjuvant activity since as early as 1925 (Ramon, 1925; Warshakoon et al., 2009), with the immunomodulatory components subsequently purified to an enriched mixture, known as Quil A, in 1974 (Dalsgaard, 1974). Further purification by reverse-phase chromatography identified fractions with potent adjuvant activity and low toxicity, in particular QS-21 that has since become one of the most widely studied saponin adjuvants (Kensil et al., 1991).

Saponins consist of one or more hydrophilic sugar chains attached via ester bonds to a hydrophobic aglycone backbone. This amphipathic structure is considered to be essential for adjuvant activity (Oda et al., 2000, 2003). Due to the inherent lability of ester linkages, these saponins are susceptible to spontaneous hydrolysis and the formation of inactive or toxic byproducts, creating challenges in formulation and storage of saponin adjuvants (Cleland et al., 1996). A further challenge arises from the affinity of the aglycone for cholesterol in cell membranes (Bangham, 1972), which, while important for adjuvanticity, can also result in hemolytic toxicity (Sun et al., 2009). However, this property also enables their formulation into secondary structures in the presence of cholesterol, which can reduce hemolytic activity while retaining adjuvanticity, activities that are structurally independent (Oda et al., 2000). These drawbacks may also be overcome by the chemical synthesis of optimized analogs (Fernández-Tejada et al., 2014, 2017). Examples include liposomes, such as the AS01 adjuvant system, and immune stimulating complexes (ISCOMs).

Application

Quil A, a heterogeneous mixture of approximately 25 different saponins, was initially used in a commercial veterinary foot-and-mouth disease vaccine (Dalsgaard, 1978) and has since been used in a variety of other veterinary vaccines (Macy, 1997; Santos et al., 2007). However, its use in humans is limited by its toxicity. In contrast, the more purified fraction QS-21 has significantly reduced toxicity enabling its use in clinical trials (Kashala et al., 2002; Schijns & O'Hagan, 2006), although it is associated with injection site pain, sometimes accompanied by mild flu-like symptoms, which may be modulated by vaccine formulation (Bigaeva et al., 2016; Petrovsky, 2015; Waite et al., 2001).

In preclinical animal models, QS-21 has demonstrated the ability to augment responses to weakly immunogenic antigens including protein (Ma et al., 1994), glycoprotein (Newman et al., 1992; Powell et al., 1994), peptide (Kim et al., 1999), polysaccharide (Coughlin et al., 1995), and DNA (Sasaki et al., 1998). In combination with conjugate antigens, QS21 has shown superior capacity to induce IgG antibodies in cancer vaccine trials including melanoma (Slingluff et al., 2001), breast (Gilewski et al., 2000, 2007), and prostate (Gin & Slovin, 2011), and has proven well-tolerated and immunogenic, although the vaccines themselves have often been insufficiently efficacious (Eggermont et al., 2010).

QS-21 has also been used as an adjuvant in experimental prophylactic vaccines against infectious diseases including bacterial (Coughlin et al., 1995; Ma et al., 1994), parasitic (Kashala et al., 2002), and viral (Newman et al., 1992; Powell et al., 1994) pathogens, demonstrating a dose sparing effect of up to 100-fold (Wyde et al., 2001) and reduction in the number of doses required for effective immunization in baboons (Powell et al., 1994) and humans (Evans et al., 2001). However, to-date QS-21 has only been licensed for use in human vaccines as part of the adjuvant system AS01 (see Section "GSK adjuvant systems").

Mechanism of action and type of immune response induced

The mechanism of action is not yet completely understood, and it is unlikely that a universal mechanism will be applicable to all saponins (Marciani, 2018). When studied as part of adjuvant systems, QS-21 has been shown to be a potent activator of the inflammasome (Detienne et al., 2016; Marty-Roix et al., 2016), a crucial step in activating innate immunity and inducing cell recruitment. However, individual saponins have been noted to differ in their ability to induce either Th1 or Th2 adaptive immunity (Marciani, 2018; Press et al., 2000), likely reflecting the specific chemical composition (Oda et al., 2003).

The aldehyde moiety of triterpene aglycone saponins is thought to play a role in inducing Th1 responses through receptors such as CD2 on T cells. This provides costimulatory signals to the T cell favoring high IL-2 and IFN-γ production indicative of Th1 immunity, and can substitute conventional CD28 costimulation that is susceptible to CTLA4-mediated downregulation (Marciani, 2003). However, this costimulation only activates T cells in the presence of TCR ligation provided by APCs such as DCs. In DCs, exogenous antigen and saponins such as QS-21 enter via cholesterol-dependent endocytosis, mediated by the saponin's affinity for cholesterol, where QS-21 is then thought to disrupt the endosomal/lysosomal membrane allowing escape of the antigen into the cytosol and cross-presentation on MHC class I (Den Brok et al., 2016). Consequently, this activates CD8 T cell-mediated immunity important for protection against intracellular pathogens, and which is often lacking in responses to subunit vaccines.

Delivery of antigen to MHC class II is thought to be restricted by acylation of fucospyranosyl residues in the oligosaccharide chain of QS-21, which sterically hinders binding to receptors such as DC-SIGN on DCs, and prevents receptor-mediated endocytosis from delivering exogenous antigen to the conventional MHC II presentation pathway (Marciani, 2018). Deacylation of these chains, such as in the semisynthetic QS-21 analog QT-0101 or following QS-21 degradation, enables delivery of antigen to MHC class II and resultant Th2-mediated humoral responses (Marciani, 2014; Marciani et al., 2001). These Th2 responses may be more effective in combating neurodegenerative or autoimmune disorders with Th1-mediated pathologies. However, this phenomenon also illustrates the danger of saponin instability that can result in changes to the nature of immunity generated (Marciani et al., 2001).

Immune stimulating complexes

Origin and chemical composition

ISCOMs are spherical open cage-like structures, typically 40 nm in diameter (Fig. 9.3), that are spontaneously formed when cholesterol, phospholipids, and *Quillaja* saponins are combined with a specific stoichiometry (Morein et al., 1984). By being bound to cholesterol and phospholipid, the saponin is blocked from interacting with the cell surface cholesterol, preventing its cytolytic activity. Today, *Quillaja* saponins formulated into ISCOMs are virtually free of side effects when combined with vaccines in clinical trials.

The active component in the nanoparticles is the *Quillaja* saponin, either in a semipurified (Quil A), or highly purified form such as QS21. *Quillaja* saponin has also been separated by HPLC into subfractions, named QH-A (most hydrophilic fractions) and QH-C (most hydrophobic fractions, including QS21). Fractions QH-A and QH-C are used separately to form ISCOMs with distinct properties, such as the ISCOM-MATRIX M1 now owned by Novavax and used in their COVID-19 vaccine.

Cholesterol is not only used as a structural component of the ISCOM cages, but most importantly also to block the cytolytic effect of the saponin.

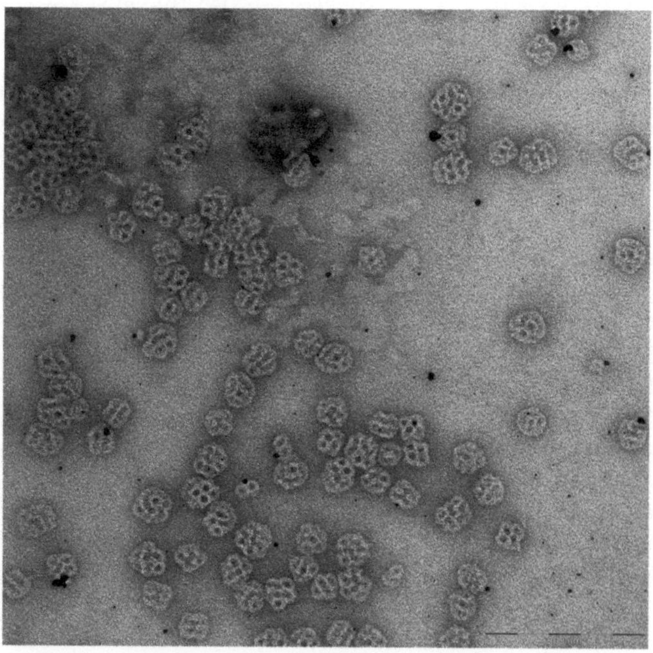

FIGURE 9.3 Electron micrograph of 40 nm spherical ISCOM particles. Cholesterol and *Quillaja* saponins form hexagonal rings glued together by lipid, for example, phosphatidylcholine, to form the structure.

It should be noted that, due to the potential contamination by BSE, animal cholesterol is avoided and cholesterol of plant origin is used. Phospholipids act as a glue in forming the three-dimensional, cage-like structures. Recent study shows although it is possible to form ISCOMs without the phospholipids, exceedingly high energy must be introduced into the process that would hamper large-scale production (Hu et al., unpublished).

Two types of ISCOMs are described in the literature. The first is classic ISCOM, formed by the combination of cholesterol, saponin, phospholipid, and protein (initially viral envelope proteins, such as influenza). The second is known as ISCOM matrix, also called empty ISCOMs or ISCOMATRIX without the inbuilt viral proteins, which has the advantage of versatility and ease of use by simple admixture with the antigen of choice (Sjölander et al., 1998).

Application

Foot and mouth disease virus was a huge global burden in bovine and porcine species, causing suffering in animals and enormous economic losses until the introduction of mass vaccination in the 1950s. The vaccine deployed was adjuvanted with alum and *Quillaja* saponin.

Later, *Quillaja* saponin in particulate form, that is, ISCOMs became available, which are almost free from side effects (Morein et al., 1984).

ISCOM-based vaccines have been shown to be effective against a substantial number of pathogens in several species. In laboratory animals, ISCOMs containing antigens from a variety of pathogens have been tested for immunogenicity and immune protection, including picornavirus in dogs (Osterhaus et al., 1985), feline leukemia virus (Osterhaus et al., 1989), CDV in seals (Putkonen et al., 1994), *Mycoplasma gallisepticu* (Sundquist et al., 1996), and avian influenza H5N1 (Rimmelzwaan et al., 1999) in chickens. Most significantly, several experimental ISCOM vaccines have been tested in primates against lentivirus infections, that is, HIV-1, HIV-2, and simian immune deficiency virus with protective efficacy varying between 40% and 100% among eight studies (Hu et al., 1998; Osterhaus et al., 1992). Antigens from bacterial species and parasites such as malaria have also been shown to evoke protective responses when combined with ISCOMs, with a highly promising malaria vaccine R21 in Matrix M (Novavax) adjuvant showing 77% efficacy in the field and currently entering large Phase III trials in children in Africa (Datoo et al., 2021). Most recently, a COVID-19 vaccine containing Matrix M adjuvant has shown high (90%) protective efficacy in Phase III trials and is likely to become the first ISCOM-based vaccine approved for use in humans (Mahase, 2021).

Mechanism of action and type of immune response induced

ISCOMs can deliver antigens via both the cytosol and the endosomal pathways, leading to, respectively, MHC class I and II antigen presentation and resulting in the induction of both cell- and antibody-mediated immune responses (Hu et al., 1998; Sjölander et al., 1998). It is thought that the membrane-disrupting properties of the saponin component of the ISCOM/ISCOMATRIX adjuvant might aid endosomal escape, thereby releasing antigen to the cytosol and aiding cross-presentation (Pearse & Drane, 2005; Villacres et al., 1998). Several cytokines are stimulated by ISCOM adjuvanted experimental vaccines encompassing a mixed Th1/Th2 profile, as well as upregulation of cell surface makers including MHC class II molecules, which was observed both in animals and man (Sjölander et al., 2001).

TLR agonists

Origin and chemical composition

At the 1989 Cold Spring Harbor Symposium, Charles Janeway predicted that pattern-recognition receptors (PRRs) represent the link between innate and adaptive immunity (Janeway, 1989). In the subsequent years, several innate signaling molecules, designated as Toll-like receptors (TLRs), were discovered. In 1998, it was shown that TLR4 recognizes, and signals in response to, bacterial lipopolysaccharide (LPS) (Poltorak et al., 1998). This was followed by the identification of cytidine-phospho-guanosine (CpG) DNA as a TLR9 agonist (Hemmi et al., 2002) and imidazoquinolines as TLR7 agonists through a MyD88-dependent signaling pathway (Hemmi et al., 2002). After another decade, it was discovered that TLR8 can sense bacterial RNA and single-stranded viral RNA within the phagosomal vacuoles (Cervantes et al., 2012).

TLRs are transmembrane receptors predominantly expressed by innate immune cells and can be classified into cell surface and intracellular TLRs expressed on the surface of endosomes (described in more detail in Chapter 3, "Dendritic Cells as Vaccine Targets" Activation of the TLR signaling cascade links innate with adaptive immunity, giving TLR agonists the potential to stimulate an effective immune response against pathogens, vaccines, or tumor cells. Importantly, mRNA vaccines act as self-adjuvating agents that can stimulate innate immunity through TLR3, TLR7, TL BuR8, RIG-I, and MDA5, eventually inducing antibody secretion by adaptive mechanisms (Zhang et al., 2019).

Application, mechanism of action, and type of immune response induced

Each TLR agonist family has a different application and mode of response; these are described later for each subset of agonists.

TLR3 agonists

Poly-IC$_{12}$U (Ampligen, Hemispherx) and poly-ICLC (Hiltonol, Oncovir) represent modified variants of the synthetic dsRNA molecule poly-IC. Poly-IC$_{12}$U promotes a dose-sparing effect when coadministered with the antiretroviral drug zidovudine, as a therapeutic treatment for HIV-1 (Martins et al., 2015). In vitro studies documented the ability of poly-$IC_{12}U$ to activate DCs (Martins et al., 2015). Subcutaneous administration of a sister compound, poly-ICLC, was found to markedly increase the immunogenicity of a peptide vaccine against ovarian cancer in a clinical trial (Sabbatini et al., 2012). In vivo, poly-ICLC induces a clear Th1 phenotype characterized by the induction of IL-12, TNF-α, IFN-γ, IL-6, and type I IFN, as well as the chemokines KC, MCP1, MIP1-α, and MIP1-β (Martins et al., 2015).

TLR4 agonists

Monophosphoryl lipid A (MPL), an LPS-derivative, was the first FDA-approved TLR agonist for use in humans, as part of the papillomavirus vaccine Cervarix, and the EU-authorized hepatitis B vaccine Fendrix (Reinke et al., 2020). Clinical trials of the herpes zoster vaccine Shingrix illustrate the benefits from MPL as an adjuvant, showing superior protection in older adults and efficacy in immuno-compromised individuals (Heineman et al., 2019). While LPS-induced signaling is predominantly transmitted via MyD88, MPL is a TIR-domain-containing adapter-inducing interferon-β (TRIF)-biased agonist of TLR4, which might explain its reduced toxicity (Luchner et al., 2021). To expand the beneficial effects of MPL, a synthetic compound GLA in stable emulsion (GLA-SE) was developed. A GLA-SE adjuvanted H1N1 influenza vaccine demonstrates dose-sparing capacity, which would be of particular benefit in a pandemic scenario (Pillet et al., 2018). GLA-SE also showed promising results as an adjuvant in the ID93 tuberculosis vaccine (Coler et al., 2018) and in the immunotherapeutic treatment of cancer (Somaiah et al., 2017). GLA increases humoral and cellular immunogenicity through APC-induced production of inflammatory cytokines such as IFN-γ, TNF, IL-2, and IL-1β (Reinke et al., 2020).

TLR5 agonist

Mobilan, an adenovirus-based vector for intratumor delivery that encodes human TLR5 and its ligand flagellin, induces strong adaptive antitumor immune responses (Li et al., 2007, 2008). Entolimod, a pharmacologically optimized flagellin derivative, has shown potential for immunotherapy against cancer in combination with radiotherapy in mouse tumor models (Li et al., 2017). In the first completed clinical study in prostate cancer patients, mobilan-induced stronger NF-κB signaling than entolimod, and led to the recruitment of innate immune cells, including neutrophils and NK cells (Eremina et al., 2020).

TLR7/8 agonists

Clinical use of topical imiquimod (Aldara Imiquimod 5% cream) is EU authorized for external genital warts, superficial basal cell carcinoma, and actinic keratosis. Topical imiquimod was tested in clinical trials as an adjuvant for testis-cancer antigen NY-ESO-1 immunotherapy in patients with high-risk melanoma and for treatment of breast cancer skin metastases (Li et al., 2017), demonstrating tolerability and systemic immune responses (Adams et al., 2012, 2008). TLR7/8 agonists such as imiquimod have been shown to activate APCs, induce humoral and cellular immunity, and especially enhance Th1 responses (Luchner et al., 2021). Originally thought to be nonfunctional, TLR8 is now known to have a complex signaling mechanism that includes cross-talk with other endosomal TLRs (Cervantes et al., 2012). It is also implicated in the generation of autoimmunity (Guiducci et al., 2013).

TLR9 agonists

Synthetic oligodeoxynucleotides (ODN) are immuno-stimulatory through unmethylated CpG motifs (Scheiermann & Klinman, 2014). CpG 1018 is included in the FDA-approved hepatitis B vaccine HEPLISAV-B (Fox, 2017). In clinical studies against anthrax, malaria, and influenza, CpG-adjuvanted vaccines demonstrated enhanced immunogenicity with no apparent adverse reactions, even in immuno-compromised individuals, as well as improved sero-protection and dose-sparing effects. CpG induces the expression of NLRP3, ASC, caspase-1, and IL-1β in a TLR9-dependent fashion (Reinke et al., 2020). Combining the immuno-stimulatory effect of CpG-ODN and poly-IC has led to the design of IC31, which has shown promising results in the first clinical trial of the tuberculosis vaccine candidate H56: IC31 (Luabeya et al., 2015).

Nanoparticles

Origin and chemical composition

Delivery of an antigen in a particulate form, either encapsulated or surface adsorbed to polymeric nanoparticles, facilitates sensing and uptake by APCs. Polymers are attractive materials for formulating nanoparticulate adjuvants because of their biocompatibility, biodegradability, and low toxicity (Peres et al., 2017). Polymeric nanoparticles enable sustained release of antigen and improve the stability of labile antigens from in vivo enzymatic degradation (Wibowo et al., 2021). Surface modification of nanoparticles with ligands or antibodies targeting various PRRs can enhance active targeting of antigen to specific APCs. In addition, nanoparticles enable codelivery of vaccine antigens with immunostimulatory components to induce a synergistic effect. A variety of natural and synthetic biodegradable polymer-based nanoparticles have been developed and tested to enhance immunogenicity for vaccine delivery (Table 9.4). Importantly, vaccine antigens encapsulated in polymer-based nanoparticles and administered through the mucosal route are protected from degradation and released locally to stimulate potent mucosal

TABLE 9.4 Examples of natural and synthetic polymer-based nanoparticulate adjuvants.

Polymer type	Name	Advantages	Disadvantages	Vaccine antigen cargo
Natural polymers	Chitosan	Nontoxic, cationic, biocompatible, biodegradable, mucoadhesive, easily modified, humoral, and cell-mediated immunity	Low water soluble, weakly immunogenic, poor strength	Protein, peptide, DNA, RNA
	Starch	Nontoxic, degradable, abundant, cheap	Water resistant, poor mechanical properties	Protein, peptide
	Dextran	Hydrophilic, strong humoral immunity	Rare adverse reactions	Protein, peptide
	Alginate	Biodegradable, mucoadhesive	Anionic (limited delivery system properties)	Protein, peptide
	Cellulose	Abundant, cheap, sustainable	Rare adverse reactions	Protein, peptide
	Inulin	Nontoxic, biocompatible, biodegradable	–	Protein, peptide
	Pullulan	Nontoxic, biocompatible, biodegradable	Weakly immunogenic	Protein, peptide, DNA, RNA
	Hyaluronic acid	Hydrophilic, biocompatible, biodegradable, transdermal immunization	–	Protein, peptide, DNA

Synthetic polymers	Poly(lactic-glycolic acid) (PLGA)	Biocompatible, biodegradable, low toxicity, induces IgG responses	Low antigen loading	Protein, peptide, DNA
	Polylactic acid (PLA)	Biocompatible, biodegradable, low toxicity, induces $CD8^+$ T cells	Low antigen loading	Protein, peptide, DNA
	Polymethyl methacrylate resin (PMMA)	Biocompatible	Monomer is cytotoxic	Protein, peptide
	Polyethylene glycol (PEG)	Flexible, hydrophilic	Poor antigen loading	Protein, peptide
	Polyethyleneimine (PEI)	Intrinsic immunomodulation	Low transfection	Protein, peptide, DNA
	Polycaprolactone (PCL)	Biocompatible, biodegradable, low toxicity	Not intrinsically immunogenic	Protein, peptide
	Polyphosphazenes	Biodegradable	Unstable	Protein, peptide
	Polyanhydrides	Biodegradable, low toxicity	Poor antigen loading	Protein, peptide
	Dendrimers	Biocompatible, multivalent	Low cytotoxicity	Protein, peptide, DNA, RNA

immunity (Rose et al., 2018). Natural polymers are a renewable resource that can be degraded into water, carbon dioxide, and inorganic small molecules, and are classified as nontoxic, biocompatible, and biodegradable materials (Wibowo et al., 2021).

Application

Chitosan, starch, dextran, alginate, cellulose, inulin, pullulan, and hyaluronic acid are some examples of natural polymers that have been explored as adjuvants in vaccine formulations (Han et al., 2018; Rose et al., 2018; Wibowo et al., 2021). Synthetic polymers include poly(lactic-glycolic acid) (PLGA), poly(lactic acid) (PLA), polymethyl methacrylate resin (PMMA), poly(ethylene glycol) (PEG), polyethylenimine (PEI), poly(ε-caprolactone) (PCL), polyphosphazenes, polyanhydride, and dendrimers. Synthetic polymers are produced through chemical synthesis and serve as excellent vaccine adjuvants due to their biocompatibility, biodegradability, and low cost (Han et al., 2018). Despite several advantages, natural polymers provide restricted immune modulation compared to the synthetic polymers and are often combined with synthetic polymers or inorganic materials to develop hybrid systems with improved properties (Rose et al., 2018). Nanoparticles prepared from cationic polymers have been shown to deliver nucleic acids including mRNA. Examples of polymers used for in vivo delivery of mRNA include derivatives of PEI, poly(amido-amine), poly-lysine, chitosan, and poly-beta amino-esters (Chahal et al., 2016; Kaczmarek et al., 2016; Kauffman et al., 2016). Recently, lipid polymer hybrid nanoparticles formed by integrating lipid and polymeric nanomaterials are being developed for the delivery of mRNA vaccines (Zhao et al., 2018).

Mechanism of action and type of immune response induced

The immunomodulatory properties of nanoparticulate adjuvants have been attributed to several mechanisms: a depot effect, inflammasome activation, complement activation, immune cell recruitment, PRR-dependent signal transduction, repetitive antigen display, MHC class I or II antigen presentation, T-cell differentiation, and release of cytokines or chemokines (Pulendran et al., 2021). Several methods are available for the formulation of polymer-based nanoparticles such as solvent evaporation, emulsification-solvent diffusion, and nanoprecipitation (Vauthier & Bouchemal, 2009). However, irrespective of the preparation method the shape, size, and chemistry (surface charge, hydrophobicity, stability) of nanoparticles govern their ability to trigger or block immune pathways and consequently their adjuvant function (Bookstaver et al., 2018). Polymer nanoparticulate adjuvants have shown great promise in potentiating the immunogenicity of vaccines (Han et al., 2018; Wibowo et al., 2021). However, further research efforts are required to confirm their safety profile, exact molecular and immunological mechanism of adjuvant potentiation, and development of novel nanoparticulate adjuvants for specific disease antigens.

Liposomes

Origin and chemical composition

The term liposome is a broad definition for vesicles composed of lipid bilayers enclosing an aqueous compartment. They were first described by Bangham et al. in 1965 (Bangham et al., 1965) with the discovery that phospholipids dispersed in water form closed vesicles. Liposomes, formed by self-assembly upon dispersion of certain amphiphilic lipids in aqueous buffer, were initially utilized as model membranes (Bangham, 1972; Lasic, 1998). Shortly after their discovery, the application of liposomes as drug delivery systems was suggested by Gregoriadis and Ryman (Gregoriadis et al., 1971; Gregoriadis & Ryman, 1972), which led to extensive research programs within the pharmaceutical industry and academia.

Liposomes consist of amphiphilic lipid bilayers whose hydrophobic regions will orient toward each other, and the hydrophilic regions will orient toward the aqueous phase in which the lipids are suspended; the adjacent lipid molecules are held together by van der Waals forces. This results in the spontaneous formation of lipid spherules that surround internal aqueous spaces. The polar region typically consists of a phosphate group, an acidic group and/or tertiary or quaternary ammonium salts, and can either be charged or neutral, depending on the composition of the lipid head-groups. The apolar region typically consists of one or more long fatty acid chains with >8 carbons with/without cholesterol (Apolinário et al., 2021).

Application

The application of liposomes as vaccine delivery systems was first investigated in the early 1970s, with the discovery of improved antibody responses to diphtheria toxoid formulated with liposomes (Allison & Gregoriadis, 1974). Numerous vaccines formulated with liposomes have since reached clinical trials and even been approved for human use, namely, Mosquirix and Shingrix, both containing different versions of the liposome-based AS01 adjuvant (Didierlaurent et al., 2017). Currently, there is a growing number of drug products approved for human use, not limited to vaccines (Doxil/Caeylx, AmBisome/Abelcet, DepoDur, Mosquirix, Shingrix), with the liposomes improving the delivery of drugs and excipients to their target-sites, or forming a drug depot and subsequently reducing systemic toxicity.

The liposome composition and preparation method influence important physicochemical characteristics that will affect their functionality. These include particle size, membrane fluidity, hydrophobicity, and surface charge (Blakney et al., 2019; De Serrano & Burkhart, 2017; Hamborg et al., 2014; Henriksen-Lacey et al., 2010, 2011; Hussain et al., 2014; Perrie et al., 2013). The versatility of liposomes allows incorporation of different types of molecules into the same liposome dispersion, for example, lipid- and/or nucleic acid–based immuno-stimulators and protein-based antigen.

One good example of this is the AS01 adjuvant, which is based on phosphatidylcholine/cholesterol liposomes containing the two immunostimulants MPL (TLR4 agonist) and saponin QS-21. MPL has highly apolar regions and formulation in liposomes will improve its stability as well as immunogenicity. QS-21 on the other hand is a water-soluble amphiphilic triterpene glycoside with hemolytic activity, which can be eliminated by formulation into liposomal membranes (Didierlaurent et al., 2017). Several the liposome-based adjuvants in clinical development, including ALFQ (Alving et al., 2020) and GLA-LSQ (Baldwin et al., 2021; Friedman-Klabanoff et al., 2021), are remarkably like AS01 containing both MPL and QS-21 incorporated into phosphatidylcholine/cholesterol-based liposomes.

Adjuvants based on other liposome backbones in clinical development include the CAF adjuvants. CAF01 and CAF09b are both based on cationic liposomes incorporating different immuno-stimulators (Pedersen et al., 2018). Here the liposomes incorporate apolar glycolipid monomycoloyl glycerol (MMG, a Minkle agonist) in the apolar region while adsorbing the water soluble negatively charged double-stranded RNA compound polyinosinic: polycytidylic acid (poly I:C; TLR3/MDA5/RIG-I agonist) on the cationic surface (Korsholm et al., 2014; Schmidt et al., 2016, 2019).

Mechanism of action

The function of liposomes as delivery systems is primarily to act as a depot for the vaccine antigens and to present/deliver the antigens and immuno-stimulators to APCs. In addition, some liposome formulations have intrinsic immuno-stimulatory activity, with the biological function highly dependent on their physicochemical properties. In particular, the surface charge has a major influence on the adjuvant effect and mechanism of action. Most in vivo studies demonstrate that cationic liposomes are superior compared to both anionic and neutral liposomes (Ma et al., 1994), due to an efficient interaction between the cationic entities and the negatively charged cell surface. Thus anionic liposomes interact with an extremely low percentage of human and murine DCs in vitro, whereas the opposite is true for cationic liposomes, leading to their localization in the endosomal compartment or the cytoplasm (Foged et al., 2004) of the DCs. This has brought cationic liposomes into focus as vaccine delivery systems.

Biodistribution of the vaccine might play an even bigger role mechanistically. The physicochemical properties of the liposomes very much define whether the liposomes form a depot at the injection site, localize in the local dLNs, or distribute more systemically in the body. This factor is controlled by the surface charge and particle size of the liposomes, and will determine which type of immune cells the vaccine contacts, and thus the type of immune response elicited.

Type of immune response induced

The addition of immuno-stimulators influences the resulting immune response, with different responses (e.g., Thl/Th2/Th17) possible with a certain combination of antigen, immuno-stimulators, and liposomal carrier. Most liposome-based adjuvants have been associated with strong T-cell responses, especially Th1 and Th17; however, this function is more dependent on the incorporated immuno-stimulators, such as MPL, QS21, MMG, poly I:C, etc., than on the intrinsic function of the liposomes themselves.

Exosomes

Origin and chemical composition

Exosomes are naturally occurring small membrane vesicles around 30–100 nm in diameter, first described in the 1970s (Couzin, 2005). They are released from different types of immune cell, such as DCs, T and B cells, mast cells, and macrophages, and found in most biological fluids as well as cell culture supernatants (Keller et al., 2011). Exosomes have important functional roles in transferring proteins, nucleic acids, and lipids between cells, and thereby modulate cell–cell communication and immune regulation.

Application

Exosomes can also be engineered as potential delivery systems for proteins and nucleic acids, due to their intrinsic nature as endogenous delivery systems (Arima et al., 2019), for example, delivering a protein antigen to APCs (André et al., 2002). However, this is an emerging field with a limited number of studies to date. Anticoli et al. engineered exosomes that can form in vivo and proposed an exosome-based vaccine platform that can activate CTL against viral antigens (Anticoli et al., 2018; Di Bonito et al., 2017). The approach involves an intramuscular injection of a DNA vector expressing Human Papilloma Virus E7 protein (HPV-E7) fused to the C-terminus of an exosome-anchoring protein, Nefmut. The E7 protein is known to induce a strong antigen-specific CTL response. Different viral nucleoproteins and other products from Ebola, flu, Crimean–Congo hemorrhagic fever, West Nile virus, and hepatitis C virus were screened and found to incorporate into exosomes efficiently, independent from their size (Di Bonito et al., 2017).

An exosome-based vaccine containing the S (spike) protein of SARS-CoV-1 has been explored: Kuate et al. engineered exosomes by replacing the transmembrane and cytoplasmic domains of the S protein by those of glycoprotein of the vesicular stomatitis virus (VSV-G). The immunogenicity and efficacy of the S-containing exosomes were tested in mice alongside an adenoviral vector vaccine expressing the S protein, and both were found to induce neutralizing antibody titers exceeding those observed in the convalescent serum of a SARS patient (Kuate et al., 2007).

Exosomes could find an application in tumor vaccines and immunotherapy. Mashioto et al. obtained exosomes from murine melanoma cells and modified them with a CpG adjuvant (Morishita et al., 2016), observing an enhanced antitumor effect following immunization. There are currently two phase I and one phase II clinical trial in malignant melanoma and nonsmall-cell lung carcinoma exploring the use of DC-derived exosomes (DEX) as therapeutic antitumor vaccines (Nikfarjam et al., 2020).

Mechanism of action and type of immune response induced

Exosomes have an active role in inducing an immune response by antigen transfer to DCs, T-lymphocyte activation via antigen presentation, and natural killer (NK) cell activation by binding NKG2D ligand binding (Théry et al., 2009). The studies described earlier demonstrate that DEX-based vaccines are safe and have the capability to trigger both the adaptive (T cell) and the innate (NK cell) immune responses.

GSK adjuvant systems

Origin and chemical composition

In the late 1980s, GSK created the concept of so-called Adjuvant Systems (AS) to facilitate the design of vaccines against difficult pathogens such as HIV and malaria (Garçon & Di Pasquale, 2017). Immuno-modulating molecules MPL-A (TLR4 agonist), QS-21 (saponin), and CpG ODN (TLR9 agonist) were combined with aluminum salts, liposomes, or oil-in-water emulsion with the aim of inducing additive, complementary, or even synergistic effects on the immune response. Several AS resulted, of which AS01, AS03, and AS04, are clinically relevant today (Table 9.5) (Garçon & Di Pasquale, 2017; Laupèze et al., 2019).

Application, mechanism of action, and type of immune response induced

Each AS has a different application, mode of action, and type of immune response, and these are individually described below.

TABLE 9.5 GSK Adjuvant Systems.

Adjuvant system	Composition	Application (examples)
AS01	QS-21 + MPL + liposomes	Malaria* Herpes zoster* Malaria next generation# Tuberculosis vaccine# HIV vaccine#
AS02	QS-21 + MPL + O/W emulsion	HIV vaccine† Tuberculosis vaccine† Therapeutic melanoma vaccine† Malaria vaccine†
AS03	Squalene O/W emulsion + vitamin E	Prepandemic H5N1 vaccine* Pandemic H1N1 vaccine* SARS-CoV-2 vaccine#
AS04	MPL adsorbed onto $Al(OH)_3$ or $AlPO_4$	HPV vaccine* Hepatitis B vaccine* Herpes simplex vaccine†
AS15	CpG 7909 + QS-21 + MPL + liposomes	MAGE-A3 cancer immunotherapy†

*, Vaccines licensed or in Phase III trials; #, vaccines in Phase I or Phase II trials; †, vaccine development discontinued.
Source: Adapted from (Garcon & Di Pasquale 2017).

AS01

AS01 is part of the approved herpes zoster (RZV/AS01, Shingrix), and malaria (RTS,S/AS01, Mosquirix) vaccines.

It is a liposome-based formulation that contains the synthetic TLR4 agonist MPL and the triterpene saponin QS-21. Mosquirix targets the *P. falciparum* circumsporozoite protein. It initially showed good efficacy in young children in Africa, although the protection waned with time (Bejon et al., 2008; Guerra Mendoza et al., 2019; Stoute et al., 1997). It is currently undergoing a WHO implementation program in several sub-Saharan countries.

The licensed Shingrix contains the varicella zoster virus glycoprotein E in AS01. It has demonstrated high efficacy and good safety in older adults (Cunningham et al., 2016; Lal et al., 2015; Laupèze et al., 2019). AS01 has also been clinically assessed in candidate vaccines against tuberculosis and HIV (Harrer et al., 2014; Penn-Nicholson et al., 2015).

A key feature of AS01 is that immunostimulants MPL and QS-21 synergistically activate novel immunological pathways that are not triggered by the individual components alone (Coccia et al., 2017; Del Giudice et al., 2018): MPL activates TLR4 and induces NFκB signaling mainly via the adapter TRIF (Luchner et al., 2021), whereas QS-21 is suggested to promote adjuvanticity through activation of MyD88 and Caspase-1 in resident macrophages of the dLN (Detienne et al., 2016; Marty-Roix et al., 2016).

When combined in AS01, MPL and QS-21 synergistically induce the release of IFNγ by NK cells in the dLN of immunized mice (Coccia et al., 2017; Del Giudice et al., 2018). This mechanism is mediated by macrophages, IL-12, and IL-18, and eventually promotes a Th1-skewed immune response characterized by the robust generation of antigen-specific polyfunctional CD4 + T cells. Like the effects observed in mice, AS01 or RTS, S/AS01 induced IFNγ production in dLN of nonhuman primates (NHPs) and in the serum of healthy human individuals, respectively (Coccia et al., 2017). Interestingly, a systems biology approach detected a negative correlation between NK cell numbers in blood and protection against malaria after RTS,S/AS01 vaccination, potentially due to enhanced recruitment of NK cells to the SOI (Kazmin et al., 2017).

AS03

AS03 is an oil-in-water emulsion that contains alpha-tocopherol (vitamin E) and is part of the approved prepandemic vaccine H5N1/AS03 and pandemic H1N1/AS03 influenza vaccines (Arepanrix, Pandemrix) (Garçon & Di Pasquale, 2017). AS03 was the adjuvant of choice as it induced elevated levels of neutralizing antibodies, allowed antigen sparing, induced cross-clade protection, and is suited for rapid large-scale manufacturing, all key features in a pandemic situation (Chu et al., 2009; Garçon et al., 2012; Laupèze et al., 2019). However, postlicensure real-world data indicated an association between Pandemrix and narcolepsy, subsequently ascribed to antigen mimicry rather than AS03 (Cohet et al., 2019). More recently AS03 has been used in the development of COVID-19 vaccine candidates (Francica et al., 2021; Liang et al., 2021; Richmond et al., 2021).

In mice, AS03 activates the ER stress sensor IRE1α, contributing to the robust release of pro-inflammatory cytokines such as IL-6 and MCP-1, eventually supporting the production of high-affinity Ag-specific antibodies (Givord et al., 2018). It also modifies cytokine expression at the SOI, increases antigen loading of monocytes, and induces recruitment of granulocytes to the dLN (Morel et al., 2011). Furthermore, high-throughput B-cell receptor sequencing in immunized people showed increased activation of naïve B cells and enhanced memory recall (Galson et al., 2016).

AS04

In AS04, the TLR4 agonist MPL is adsorbed onto either aluminum hydroxide or aluminum phosphate, depending on the vaccine antigen. AS04 is part of licensed vaccines HBV/AS04 (Fendrix) and AS04-HPV-16/18 (Cervarix) against hepatitis B and human papillomavirus, respectively (Laupèze et al., 2019). HBV/AS04 was designed to circumvent the reduced immunogenicity and efficacy of aluminum salt-adjuvanted HBV vaccines in immuno-compromised individuals such as patients with renal insufficiency (Garçon & Di Pasquale, 2017). Importantly, in prehemodialysis and hemodialysis patients HBV/AS04 provided more rapid and durable protection, reducing the number of boosters required and allowing dose sparing (Kong et al., 2008; Tong et al., 2005). AS04-HPV-16/18 promotes long-lasting protection against HPV-16 and HPV-18, which are the predominant cervical cancer inducing HPV strains (Naud et al., 2014; Paavonen et al., 2009), and cross protection against other oncogenic HPV strains (Paavonen et al., 2009). Both licensed AS04-adjuvanted vaccines have an acceptable reactogenicity profile (Kong et al., 2008; Paavonen et al., 2009; Tong et al., 2005).

In mice, AS04 mediates NFκB activation and cytokine secretion, leading to increased numbers of Ag-loaded DCs and monocytes in the dLN, which eventually enhances activation of T and B cells (Didierlaurent et al., 2009). The driver of the innate cell activation is MPL-1, whereas the aluminum hydroxide prolongs the cytokine response at the SOI, which may promote a more effective adaptive immunity. Indeed, AS04-HPV-16/18 induced higher antibody responses and memory B cells in mice, monkeys, and humans, when compared to an HPV vaccine formulated in aluminum hydroxide only (Giannini et al., 2006).

Using HBsAg as a model antigen, AS01, AS03, AS04, and aluminum salts were compared in a clinical trial (Burny et al., 2017; Leroux-Roels et al., 2016). Interestingly, AS01 induced the highest number of HBs-specific CD4 + T cells (Leroux-Roels et al., 2016), whereas the frequency of HBs-specific antibodies and memory B cells was highest in AS01 and AS03 groups (Leroux-Roels et al., 2016). Analysis of innate responses showed that the different AS activate common pathways, with a higher magnitude response observed for AS01 and AS03, which is an explanation for their enhanced potential to generate HBs-specific adaptive responses (Burny et al., 2017; Leroux-Roels et al., 2016).

Cholera toxin

Origin and chemical composition

Cholera toxin (CT) is produced by the bacterium *Vibrio cholerae*. Although its existence was first postulated by Koch in 1886, CT was discovered in 1959 with a report that cell-free *V. cholerae* culture filtrate can elicit a massive accumulation of fluid in ligated ileal loops of rabbits (De, 1959).

The active substance is a holotoxin composed of the central subunit A ("Active") with five subunits B ("Binding") located peripherally (Bharati & Ganguly, 2011). It was also demonstrated in 1968 that *Escherichia coli* can produce enterotoxins (T. G. Evans et al., 2001). Heat-labile toxin (LT) composed of LTA and LTB subunits is structurally and functionally closely related to CT (Duan et al., 2019).

Application

The activity of CT and LT has been examined in prophylactic vaccines against bacterial, viral, and fungal pathogens for several decades in animal models (Freytag & Clements, 2005). Both CT and LT were reported to be potent mucosal adjuvants in animal models. However, they induced severe diarrhea in humans when administered orally (Fukuyama et al., 2015) and might lead to facial paralysis when administered by the nasal route (Mutsch et al., 2004). To address the problem of toxicity, derivatives of toxins have been developed and evaluated (Holmgren et al., 1993), showing similar adjuvanticity to native CT in mice (Ågren et al., 1997; Eriksson et al., 2004). In humans, there are several Phase I and II clinical trials completed or ongoing with double mutant LT in the context of killed or attenuated enterotoxigenic *E. coli* (ETEC) vaccines (Clements & Norton, 2018). Sublingual or oral vaccinations with ETEC vaccines administered with double mutant LT were reported to be safe and immunogenic in Phase I clinical trials in adults (Bernstein et al., 2019; Lundgren et al., 2014) and children (Qadri et al., 2020). Recombinant CTB (the B subunit of CT) is used in the commercialized oral cholera vaccine Dukoral and induces mucosal and systemic antibody responses following mucosal administration in humans (Bergquist et al., 1997; Jertborn et al., 2001; Kozlowski et al., 1997).

Mechanism of action

The pentameric B subunit of CT (CTB) binds to the GM1 ganglioside receptor found on a variety of cells including intestinal epithelial cells (Stratmann, 2015). CT is internalized into the endosomes before being transported as the holotoxin to the Golgi and then to the endoplasmic reticulum. At this stage, CTA (the A subunit of CT) ADP ribosylates the protein Gs leading to its activation, which stimulates cyclic adenosine monophosphate (cAMP) production. The increase of intracellular cAMP induces an osmotic movement of water into the intestinal lumen causing diarrhea (Datta et al., 2010; Häse et al., 1994; Vanden Broeck et al., 2007).

Type of immune response induced

CT has been shown to induce antigen-specific Th2 cytokine responses and IgG1, IgE, and mucosal IgA secretion, while LT promotes mixed Th1 and Th2 cytokine responses and IgG1, IgG2a, IgG2b, and mucosal-secreted IgA (SIgA) responses in mouse models (Bowman & Clements, 2001). CT and LT have been shown to induce IL-17-producing CD4 T cells in mice (Brereton et al., 2011; Datta et al., 2010), influencing their ability to enhance the SIgA response (Hirota et al., 2013).

In vitro, CT can enhance antigen presentation and promote the proliferation/differentiation of B and T cells (Holmgren et al., 1993). LT was shown to induce cell recruitment (Pizza et al., 2001; Ryan et al., 2000), DC maturation, antigen presentation, and B/T cell activation (Petrovska et al., 2003; Pitcovski et al., 2006).

Freund's adjuvant

Origin and chemical composition

Incomplete Freund's adjuvant (IFA) is composed of paraffin oil containing mannide mono-oleate. Complete Freund's adjuvant (CFA) additionally contains killed mycobacteria (Billiau & Matthys, 2001). Originally, Jules Freund observed that guinea pigs infected with *Mycobacterium tuberculosis* produced higher antibody responses. Similar results were observed with killed *M. tuberculosis* in a water-in-paraffin emulsion (Freund, 1951; Freund et al., 1937). Localized reactogenicity was attributed to impure emulsifiers and today's emulsion adjuvants contain highly purified and defined light mineral oils with low toxicity (Chang et al., 1998).

Application

IFA was evaluated in nonhuman primates and humans in the context of an influenza vaccine, demonstrating protection and immunogenicity without major side effects (Davenport & Hennessy, 1956; Salk et al., 1953). In other human studies with the tetanus toxoid vaccine (MacLennan et al., 1965) or *V. cholerae* formulation (Ogonuki et al., 1967), the use of IFA was associated with SOI reactions. However, IFA is still evaluated in humans particularly in the context of cancer therapeutic vaccines (Chang et al., 1998; Pollack et al., 2020). Due to significant localized and systemic side effects, CFA is considered too toxic for human use and it is limited to research studies in animal models (Stills, 2005).

Mechanism of action

IFA and CFA have been shown to act by a depot effect at the SOI leading to a slow antigen release (Herbert, 1968), and enhanced antigen uptake (Lind, 1968; McDevitt et al., 1966). It has recently been demonstrated that IFA could enhance Th1 responses, TLR signaling, and CD40 ligand expression in the tumor microenvironment using melanoma peptides and a TLR3 agonist tested on human skin biopsies (Pollack et al., 2020).

α-GalCer

Origin and chemical composition

Adjuvants targeting unconventional T cells such as invariant natural killer T (iNKT) cells have been considered, especially in the context of mucosal vaccine development. iNKT cells have the potential to bridge innate and adaptive immunity (Abautret-Daly et al., 2014) as they recognize glycolipids presented by CD1d-expressing APCs. The antigen can be bacterial or vegetal lipids such as α-GalactosylCeramide (α-GalCer), originally discovered in a marine sponge more than 25 years ago (Cerundolo et al., 2010).

Application

Upon activation, iNKT cells can modulate DC- and B-cell activity and induce activation of NK cells, leading to enhancement of adaptive immune responses (Fujii et al., 2003; Hermans et al., 2003). Consequently, iNKT cell agonists have been investigated as adjuvants to enhance immune responses in immunotherapy and vaccination strategies. Some studies have demonstrated the potential of the iNKT-cell activator α-GalCer as an oral adjuvant to enhance intestinal and systemic immune responses induced by experimental whole-cell killed ETEC (Davitt et al., 2016), *V. cholerae* (Davitt et al., 2019), and *Helicobacter pylori* (Longet et al., 2019) in mice. Even though α-GalCer has been used in over 30 antitumor clinical trials (Zhang et al., 2019), it has not yet been tested in prophylactic vaccinations in humans (Cerundolo et al., 2010; Lindqvist et al., 2009).

Mechanism of action

T-cell receptors of iNKT cells can recognize lipid antigens presented by MHC class I-like CD1d molecules expressed on DCs or B cells. It was shown that iNKT cell activation could be mediated by direct recognition of α-GalCer presented by CD1d molecules, or indirectly by stimulatory cytokines (e.g., IL-12, IFNs) released by DCs following TLR signaling (Paget et al., 2007; Salio et al., 2007).

Adjuvant properties of the microbiome

The *microbiome* is a complex community of diverse microorganisms (viruses, including bacteriophage, archaea, bacteria, and fungi), protozoa, and helminths, known collectively as *microbiota*, that inhabit a particular environment in animals and humans (Desselberger, 2018). Most of the microbiome is found in the intestinal tract (gut microbiota), although other body areas have their own specific colonizers. Microecological balance, with rich and diverse microbiota, is termed *eubiosis*, comprising a healthy state, that is, homeostasis between the host and microbiome (Li et al., 2019). Ecological disturbance due to antibiotic usage, diet alteration or other reasons, with reduced diversity, microbiota instability and opportunistic pathogen accumulation, is called *dysbiosis*, and can lead to pathological changes including altered immune responses (Li et al., 2019).

The microbiome has a significant effect on the host's innate and adaptive immunity (Zheng et al., 2020), with a potential impact on responses to vaccines (De Jong et al., 2020). Localized immune interactions, via APCs, B and T cells that detect microbe-associated molecular patterns (MAMPs) or microbial metabolites, are particularly intensive in the gastrointestinal tract (De Jong et al., 2020). Microbiota can also influence immune reactions at distal locations, such as lymph nodes, bone marrow, spleen, liver, and peritoneum. These interactions could be mediated by dissemination of bacterial products (e.g., LPS), cell signaling molecules, or microbiota metabolites (De Jong et al., 2020).

Various microbiota MAMPs, such as commensal (gut flora) DNA (Hall et al., 2008), flagellin (Oh et al., 2014), Gram-negative bacteria LPS (Harris et al., 2018), or conserved peptidoglycan motif muramyl dipeptide (Kim et al., 1999), have been reported as vaccine potentiators (Magwira & Taylor, 2018). Many different bacterial species, in distinct locations within the gut, have been implicated in immune modulation (Oh et al., 2014). Several findings have associated phyla Actinobacteria and Firmicutes with higher humoral and cellular responses, and phyla Proteobacteria and Bacteroidetes with lower responses to vaccination (Zimmermann & Curtis, 2018).

Microbiota disbalance is most frequently and severely provoked by antibiotic exposure, promoting a dysbiosis, where normal microbiota subsets are diminished or lost. This in turn reduces epithelial exposure to microbiota-derived MAMP and PRR signaling, which leads to changes in downstream regulation of innate defenses (Hosomi & Kunisawa, 2020), and reduction in vaccine responses (Ciabattini et al., 2019; Hagan et al., 2019; Lynn & Pulendran, 2018). Restoration of microbial homeostasis as a strategy to improve vaccine responses can be achieved through: (1) increasing "good" intestinal bacteria through use of prebiotics (Benyacoub et al., 2008) or probiotics (Zimmermann & Curtis, 2018); (2) selective depletion of harmful species [use of narrow spectrum antibiotics (Harris et al., 2018) or bacteriophages]; (3) fecal microbiota transplantation (Twitchell et al., 2016), or a combination of these (Yitbarek et al., 2019). Achieving a healthy gut microbiome could have a significant positive impact particularly in newborns and the elderly, who often present with a decreased microbial diversity, making them (among other factors) less responsive to vaccination (Cianci et al., 2020; De Jong et al., 2020).

Plant-derived adaptogens

Medicinal plants present a rich sources of phytochemicals (bioactive nutrient plant chemicals or nutraceuticals) characterized by diverse and pleiotropic properties including the ability to provide medical or health benefits (Di Sotto et al., 2020).

Immunomodulatory properties are linked to several phytochemicals, including polysaccharides, saponins, lectins, alkaloids, phenolics, glycoproteins, terpenoids, and fatty acids (Di Sotto et al., 2020; Reyna-Margarita et al., 2019). Many plant polysaccharides can stimulate the immune system by inducing T cells, cytokine production, and/or antibody responses. Astragalus polysaccharides provide adjuvant capacity to various veterinary (reviewed in Sun et al., 2009) and human (Du et al., 2012) viral vaccines. A delta inulin derivative Advax has shown an adjuvanting effect with influenza and hepatitis B surface antigens (Honda-Okubo et al., 2012; Saade et al., 2013). In addition, plant polysaccharides demonstrate lower toxicity than bacterial polysaccharides or synthetic compounds (Sun et al., 2018) and can be administered orally or parenterally. *Quillaja* saponins are a potent plant-derived adjuvant and are discussed earlier in this chapter. Lectins are proteins that bind to carbohydrates present on the cell surface, resulting in signal transduction and production of cytokines, primarily Th1, manifested by elevated levels of IFN-γ (Souza et al., 2013). There are studies indicating the possible utility of plant lectins as vaccine adjuvants for parasites of veterinary (*Neospora caninum*, *Leishmania*) or human (*Trypanosoma cruzi*, *Paracoccidioides brasiliensis*) relevance (reviewed in Reyna-Margarita et al., 2019).

Adaptogens are a unique class of diverse compounds found in plants, which may provide health benefits including modulation of immune responses (Panossian, 2017; Ratan et al., 2021). One of the most researched adaptogenic plants is *Panax ginsen* (Asian or Korean ginseng). Its main active components, saponins (also known as *ginsenosides*), exhibit immuno-modulatory effects when combined with different veterinary viral (Ginseng Leaf-Stem Saponins and ginsenoside Rd) or protozoal (ginsenoside Rg1) vaccines (reviewed in Ratan et al., 2021). An adjuvanting effect of purified ginsenoside Rb1 has also been demonstrated in dairy cattle in immunization with *Staphylococcus aureus* used for prevention of bovine mastitis (Hu et al., 1998; Osterhaus et al., 1992), and for human vaccines, for example, a ginsenoside Rg1 nano-emulsion in combination with hepatitis B surface antigen (HBsAg).

Comparison of the licensed Alhydrogel-adjuvanted vaccines against porcine parvovirus and *Erysipelothrix rhusiopathiae* with those supplemented with ginseng extract (Rivera, Daggfeldt, & Hu, 2003; Rivera, Hu, & Concha, 2003) shows that (1) the ginseng fractions Rb1 and Rg1 are potent adjuvants inducing higher or similar antibody titers than the Alhydrogel-adjuvanted vaccines, and (2) the ginseng-supplemented vaccines favored IgG2, suggesting ginseng as a cheap and safe alternative for improving the potency of Alhydrogel-adjuvanted vaccines especially if the IgG2 response is needed. Ginseng can stimulate basal NK cell activity and can activate macrophages to produce reactive nitrogen intermediates and become tumoricidal, as well as increasing oxygen radicals scavenging and decreasing lipid peroxidation (Liao et al., 2018).

Withania somnifera (L.) Dunal (Ashwagandha, Indian ginseng) contains steroidal lactones and has shown potential as an oral immuno-potentiating agent: mice given oral *W. somnifera* and immunized with Diphtheria-Pertussis-Tetanus (DPT) vaccine, followed by challenge with live *Bordetella pertussis* had a higher antibody titer and reduced mortality (compared to control animals with immunization and challenge alone) (Gautam et al., 2004).

The medicinal potential of *Asparagus racemosus* (Shatavari) is attributed to the presence of steroidal saponins (or shatavarin) as well as sapogenins. This traditional Ayurvedic plant may possess immuno-potentiating properties, based on an in vivo immunization studies with diphtheria, tetanus, pertussis (DTP) vaccine (Gautam et al., 2004) and hepatitis B surface antigen (Sidiq et al., 2011).

The modulatory effects of *curcumin*, the principal curcuminoid of turmeric (Curcuma longa) on the TLR4/MyD88/NF-κB signaling pathway, have been demonstrated in different pathological conditions (summarized in Catanzaro et al., 2018). TLR4 agonist-like properties have been observed for the polysaccharide (1,4)-α-D-glucan, isolated from *Tinospora cordifolia* (Guduchi or Giloy) (Raghu et al., 2009). Salidroside, a major component of *Rhodiola rosea* L., was shown to possess l adjuvant activity by regulating humoral and cellular immune responses to ovalbumin in mice (Guan et al., 2011).

The organo-sulfur compounds of *Allium sativum* (garlic) extract have been shown to inhibit growth of tumors in animals, via the activation of NK cells, stimulation of T lymphocytes and enhanced IL-2 production (Moutia et al., 2018).

Zingiber officinale (ginger) is a rich source of volatile oils, mostly zingiberol, zingiberene, phellandrene, and linalool. Along with garlic extract, ginger has been shown to modulate the innate immune response and enhance clearance of multidrug resistant *E. coli* in chickens (Elmowalid et al., 2019). Ginger essential oil was found to enhance the humoral immune response in immuno-suppressed mice (Carrasco et al., 2009). Ginger has also shown as promise in the prevention and treatment of inflammatory bowel disease and colitis-associated cancer (Zhang et al., 2016) and immuno-modulatory properties in multiple sclerosis (Jafarzadeh & Nemati, 2018).

References

Abautret-Daly, A. E., Davitt, C. J. H., & Lavelle, E. C. (2014). Harnessing the antibacterial and immunological properties of mucosal-associated invariant T cells in the development of novel oral vaccines against enteric infections. *Biochemical Pharmacology, 92*(2), 173–183. Available from https://doi.org/10.1016/j.bcp.2014.08.017.

Adams, S., Kozhaya, L., Martiniuk, F., Meng, T.-C., Chiriboga, L., Liebes, L., Hochman, T., Shuman, N., Axelrod, D., Speyer, J., Novik, Y., Tiersten, A., Goldberg, J. D., Formenti, S. C., Bhardwaj, N., Unutmaz, D., & Demaria, S. (2012). Topical TLR7 agonist imiquimod can induce immune-mediated rejection of skin metastases in patients with breast cancer. *Clinical Cancer Research, 18*(24), 6748–6757. Available from https://doi.org/10.1158/1078-0432.CCR-12-1149.

Adams, S., O'Neill, D. W., Nonaka, D., Hardin, E., Chiriboga, L., Siu, K., Cruz, C. M., Angiulli, A., Angiulli, F., Ritter, E., Holman, R. M., Shapiro, R. L., Berman, R. S., Berner, N., Shao, Y., Manches, O., Pan, L., Venhaus, R. R., Hoffman, E. W., & Bhardwaj, N. (2008). Immunization of malignant melanoma patients with full-length NY-ESO-1 protein using TLR7 agonist imiquimod as vaccine adjuvant. *Journal of Immunology, 181*(1), 776–784. Available from https://doi.org/10.4049/jimmunol.181.1.776.

Ågren, L. C., Ekman, L., Löwenadler, B., & Lycke, N. Y. (1997). Genetically engineered non-toxic vaccine adjuvant that combines B cell targeting with immunomodulation by cholera toxin A1 subunit. *Journal of Immunology, 158*(8), 3936–3946.

Allison, A. C., & Gregoriadis, G. (1974). Liposomes as immunological adjuvants. *Nature, 252* (5480), 252. Available from https://doi.org/10.1038/252252a0.

Alving, C. R., Peachman, K. K., Matyas, G. R., Rao, M., & Beck, Z. (2020). Army liposome formulation (ALF) family of vaccine adjuvants. *Expert Review of Vaccines, 19*(3), 279–292. Available from https://doi.org/10.1080/14760584.2020.1745636.

André, F., Schartz, N. E. C., Chaput, N., Flament, C., Raposo, G., Amigorena, S., Angevin, E., & Zitvogel, L. (2002). Tumor-derived exosomes: A new source of tumor rejection antigens. *Vaccine*, A28–A31. Available from https://doi.org/10.1016/s0264-410x(02)00384-5.

Anticoli, S., Manfredi, F., Chiozzini, C., Arenaccio, C., Olivetta, E., Ferrantelli, F., Capocefalo, A., Falcone, E., Ruggieri, A., & Federico, M. (2018). An exosome-based vaccine platform imparts cytotoxic T lymphocyte immunity against viral antigens. *Biotechnology Journal, 13* (4), 1700443. Available from https://doi.org/10.1002/biot.201700443.

Apolinário, A. C., Hauschke, L., Nunes, J. R., & Lopes, L. B. (2021). Lipid nanovesicles for biomedical applications: 'What is in a name'? *Progress in Lipid Research, 82*. Available from https://doi.org/10.1016/j.plipres.2021.101096.

Arima, Y., Liu, W., Takahashi, Y., Nishikawa, M., & Takakura, Y. (2019). Effects of localization of antigen proteins in antigen-loaded exosomes on efficiency of antigen presentation. *Molecular Pharmaceutics, 16*(6), 2309–2314. Available from https://doi.org/10.1021/acs.molpharmaceut.8b01093.

Baldwin, S. L., Reese, V. A., Larsen, S. E., Beebe, E., Guderian, J., Orr, M. T., Fox, C. B., Reed, S. G., Coler, R. N., & Izzo, A. A. (2021). Prophylactic efficacy against *Mycobacterium tuberculosis* using ID93 and lipid-based adjuvant formulations in the mouse model. *PLoS One, 16* (3), e0247990. Available from https://doi.org/10.1371/journal.pone.0247990.

Bangham, A. D. (1972). Lipid bilayers and biomembranes. *Annual Review of Biochemistry, 41*, 753–776. Available from https://doi.org/10.1146/annurev.bi.41.070172.003541.

Bangham, A. D., Standish, M. M., & Watkins, J. C. (1965). Diffusion of univalent ions across the lamellae of swollen phospholipids. *Journal of Molecular Biology, 13*(1), 238–252. Available from https://doi.org/10.1016/S0022-2836(65)80093-6.

Bejon, P., Lusingu, J., Olotu, A., Leach, A., Lievens, M., Vekemans, J., Mshamu, S., Lang, T., Gould, J., Dubois, M. C., Demoitié, M. A., Stallaert, J. F., Vansadia, P., Carter, T., Njuguna, P., Awuondo, K. O., Malabeja, A., Abdul, O., Gesase, S., & Von Seidlein, L. (2008). Efficacy of RTS,S/AS01E vaccine against malaria in children 5 to 17 months of age. *New England Journal of Medicine, 359*(24), 2521–2532. Available from https://doi.org/10.1056/NEJMoa0807381.

Benyacoub, J., Rochat, F., Saudan, K. Y., Rochat, I., Antille, N., Cherbut, C., Von Der Weid, T., Schiffrin, E. J., & Blum, S. (2008). Feeding a diet containing a fructooligosaccharide mix can enhance *Salmonella* vaccine efficacy in mice. *Journal of Nutrition*, *138*(1), 123–129. Available from https://doi.org/10.1093/jn/138.1.123.

Bergquist, C., Johansson, E. L., Lagergård, T., Holmgren, J., & Rudin, A. (1997). Intranasal vaccination of humans with recombinant cholera toxin B subunit induces systemic and local antibody responses in the upper respiratory tract and the vagina. *Infection and Immunity*, *65* (7), 2676–2684. Available from https://doi.org/10.1128/iai.65.7.2676-2684.1997.

Bernstein, D. I., Pasetti, M. F., Brady, R., Buskirk, A. D., Wahid, R., Dickey, M., Cohen, M., Baughman, H., El-Khorazaty, J., Maier, N., Sztein, M. B., Baqar, S., & Bourgeois, A. L. (2019). A Phase 1 dose escalating study of double mutant heat-labile toxin LTR192G/L211A (dmLT) from Enterotoxigenic *Escherichia coli* (ETEC) by sublingual or oral immunization. *Vaccine*, *37*(4), 602–611. Available from https://doi.org/10.1016/j.vaccine.2018.12.011.

Bharati, K., & Ganguly, N. K. (2011). Cholera toxin: A paradigm of a multifunctional protein. *Indian Journal of Medical Research*, *133*(2), 179–187. Available from http://icmr.nic.in/ijmr/2011/february/0207.pdf.

Bharucha, T., Ming, D., & Breuer, J. (2017). A critical appraisal of 'Shingrix,' a novel herpes zoster subunit vaccine (HZ/Su or GSK1437173A) for varicella zoster virus. *Human Vaccines & Immunotherapeutics*, 1789–1797. Available from https://doi.org/10.1080/21645515.2017.1317410.

Bigaeva, E., Doorn, E., Van, Liu, H., Hak, E., & Sambhara, S. (2016). *Meta*-analysis on randomized controlled trials of vaccines with QS-21 or ISCOMATRIX adjuvant: Safety and tolerability. *PLoS One*, *11*(5), e0154757. Available from https://doi.org/10.1371/journal.pone.0154757.

Billiau, A., & Matthys, P. (2001). Modes of action of Freund's adjuvants in experimental models of autoimmune diseases. *Journal of Leukocyte Biology*, *70*(6), 849–860.

Blakney, A. K., McKay, P. F., Christensen, D., Yus, B. I., Aldon, Y., Follmann, F., & Shattock, R. J. (2019). Effects of cationic adjuvant formulation particle type, fluidity and immunomodulators on delivery and immunogenicity of saRNA. *Journal of Controlled Release*, *304*, 65–74. Available from https://doi.org/10.1016/j.jconrel.2019.04.043.

Bookstaver, M. L., Tsai, S. J., Bromberg, J. S., & Jewell, C. M. (2018). Improving vaccine and immunotherapy design using biomaterials. *Trends in Immunology*, *39*(2), 135–150. Available from https://doi.org/10.1016/j.it.2017.10.002.

Boudreau, C. M., Yu, W. H., Suscovich, T. J., Talbot, H. K., Edwards, K. M., & Alter, G. (2020). Selective induction of antibody effector functional responses using MF59-adjuvanted vaccination. *Journal of Clinical Investigation*, *130*(2), 662–672. Available from https://doi.org/10.1172/JCI129520.

Bowman, C. C., & Clements, J. D. (2001). Differential biological and adjuvant activities of cholera toxin and *Escherichia coli* heat-labile enterotoxin hybrids. *Infection and Immunity*, *69* (3), 1528–1535. Available from https://doi.org/10.1128/IAI.69.3.1528-1535.2001.

Bozkir, A., & Hayta, G. (2004). Preparation and evaluation of multiple emulsions water-in-oil-in-water (w/o/w) as delivery system for influenza virus antigens. *Journal of Drug Targeting*, *12*(3), 157–164. Available from https://doi.org/10.1080/10611860410001688018.

Brereton, C. F., Sutton, C. E., Ross, P. J., Iwakura, Y., Pizza, M., Rappuoli, R., Lavelle, E. C., & Mills, K. H. G. (2011). *Escherichia coli* heat-labile enterotoxin promotes protective Th17 responses against infection by driving innate IL-1 and IL-23 production. *Journal of Immunology*, *186*(10), 5896–5906. Available from https://doi.org/10.4049/jimmunol.1003789.

Burny, W., Callegaro, A., Bechtold, V., Clement, F., Delhaye, S., Fissette, L., Janssens, M., Leroux-Roels, G., Marchant, A., Van Den Berg, R. A., Garçon, N., van der Most, R., Didierlaurent, A. M., Carletti, I., Esen, M., Gabor, J., Haelterman, E., Hervé, C., ... Horsmans, Y. (2017). Different adjuvants induce common innate pathways that are associated with enhanced adaptive responses against a model antigen in humans. *Frontiers in Immunology, 8*. Available from https://doi.org/10.3389/fimmu.2017.00943.

Caillet, C., Piras, F., Bernard, M. C., de Montfort, A., Boudet, F., Vogel, F. R., Hoffenbach, A., Moste, C., & Kusters, I. (2010). AF03-adjuvanted and non-adjuvanted pandemic influenza A (H1N1) 2009 vaccines induce strong antibody responses in seasonal influenza vaccine-primed and unprimed mice. *Vaccine, 28*(18), 3076–3079. Available from https://doi.org/10.1016/j.vaccine.2010.02.050.

Cantisani, R., Pezzicoli, A., Cioncada, R., Malzone, C., De Gregorio, E., D'Oro, U., & Piccioli, D. (2015). Vaccine adjuvant MF59 promotes retention of unprocessed antigen in lymph node macrophage compartments and follicular dendritic cells. *Journal of Immunology, 194* (4), 1717–1725. Available from https://doi.org/10.4049/jimmunol.1400623.

Carrasco, F. R., Schmidt, G., Romero, A. L., Sartoretto, J. L., Caparroz-Assef, S. M., Bersani-Amado, C. A., & Cuman, R. K. N. (2009). Immunomodulatory activity of *Zingiber officinale* Roscoe, *Salvia officinalis* L. and *Syzygium aromaticum* L. essential oils: Evidence for humor- and cell-mediated responses. *Journal of Pharmacy and Pharmacology, 61*(7), 961–967. Available from https://doi.org/10.1211/jpp/61.07.0017.

Catanzaro, M., Corsini, E., Rosini, M., Racchi, M., & Lanni, C. (2018). Immunomodulators inspired by nature: A review on curcumin and Echinacea. *Molecules (Basel, Switzerland), 23*(11). Available from https://doi.org/10.3390/molecules23112778.

Cerundolo, V., Barral, P., & Batista, F. D. (2010). Synthetic iNKT cell-agonists as vaccine adjuvants-finding the balance. *Current Opinion in Immunology, 22*(3), 417–424. Available from https://doi.org/10.1016/j.coi.2010.04.003.

Cervantes, J. L., Weinerman, B., Basole, C., & Salazar, J. C. (2012). TLR8: The forgotten relative revindicated. *Cellular and Molecular Immunology, 9*(6), 434–438. Available from https://doi.org/10.1038/cmi.2012.38.

Chahal, J. S., Khan, O. F., Cooper, C. L., McPartlan, J. S., Tsosie, J. K., Tilley, L. D., Sidik, S. M., Lourido, S., Langer, R., Bavari, S., Ploegh, H. L., & Anderson, D. G. (2016). Dendrimer-RNA nanoparticles generate protective immunity against lethal Ebola, H1N1 influenza, and *Toxoplasma gondii* challenges with a single dose. *Proceedings of the National Academy of Sciences of the United States of America, 113*(29), E4133–E4142. Available from https://doi.org/10.1073/pnas.1600299113.

Chang, J. C. C., Diveley, J. P., Savary, J. R., & Jensen, F. C. (1998). Adjuvant activity of incomplete Freund's adjuvant. *Advanced Drug Delivery Reviews, 32*(3), 173–186. Available from https://doi.org/10.1016/S0169-409X(98)00009-X.

Chu, D. W. S., Hwang, S. J., Lim, F. S., Oh, H. M. L., Thongcharoen, P., Yang, P. C., Bock, H. L., Dramé, M., Gillard, P., Hutagalung, Y., Tang, H., Teoh, Y. L., & Ballou, R. W. (2009). Immunogenicity and tolerability of an AS03A-adjuvanted prepandemic influenza vaccine: A phase III study in a large population of Asian adults. *Vaccine, 27*(52), 7428–7435. Available from https://doi.org/10.1016/j.vaccine.2009.07.102.

Ciabattini, A., Olivieri, R., Lazzeri, E., & Medaglini, D. (2019). Role of the microbiota in the modulation of vaccine immune responses. *Frontiers in Microbiology, 10*. Available from https://doi.org/10.3389/fmicb.2019.01305.

Cianci, R., Franza, L., Massaro, M. G., Borriello, R., De Vito, F., & Gambassi, G. (2020). The interplay between immunosenescence and microbiota in the efficacy of vaccines. *Vaccines, 8*(4), 1–16. Available from https://doi.org/10.3390/vaccines8040636.

Cioncada, R., Maddaluno, M., Vo, H. T. M., Woodruff, M., Tavarini, S., Sammicheli, C., Tortoli, M., Pezzicoli, A., De Gregorio, E., Carroll, M. C., D'Oro, U., Piccioli, D., & Bayry, J. (2017). Vaccine adjuvant MF59 promotes the intranodal differentiation of antigen-loaded and activated monocyte-derived dendritic cells. *PLoS One*, *12*(10), e0185843. Available from https://doi.org/10.1371/journal.pone.0185843.

Cleland, J. L., Kensil, C. R., Lim, A., Jacobsen, N. E., Basa, L., Spellman, M., Wheeler, D. A., Wu, J. Y., & Powell, M. F. (1996). Isomerization and formulation stability of the vaccine adjuvant QS-21. *Journal of Pharmaceutical Sciences*, *85*(1), 22–28. Available from https://doi.org/10.1021/js9503136.

Clements, J. D., & Norton, E. B. (2018). The mucosal vaccine adjuvant LT(R192G/L211A) or dmLT. *MSphere*, *3*(4). Available from https://doi.org/10.1128/MSPHERE.00215-18.

Coccia, M., Collignon, C., Hervé, C., Chalon, A., Welsby, I., Detienne, S., Van Helden, M. J., Dutta, S., Genito, C. J., Waters, N. C., Van Deun, K., Smilde, A. K., Van Den Berg, R. A., Franco, D., Bourguignon, P., Morel, S., Garçon, N., Lambrecht, B. N., Goriely, S., ... Didierlaurent, A. M. (2017). Cellular and molecular synergy in AS01-adjuvanted vaccines results in an early IFNγ response promoting vaccine immunogenicity. *npj Vaccines*, *2*(1). Available from https://doi.org/10.1038/s41541-017-0027-3.

Cohet, C., van der Most, R., Bauchau, V., Bekkat-Berkani, R., Doherty, T. M., Schuind, A., Tavares Da Silva, F., Rappuoli, R., Garçon, N., & Innis, B. L. (2019). Safety of AS03-adjuvanted influenza vaccines: A review of the evidence. *Vaccine*, *37*(23), 3006–3021. Available from https://doi.org/10.1016/j.vaccine.2019.04.048.

Coler, R. N., Day, T. A., Ellis, R., Piazza, F. M., Beckmann, A. M., Vergara, J., Rolf, T., Lu, L., Alter, G., Hokey, D., Jayashankar, L., Walker, R., Snowden, M. A., Evans, T., Ginsberg, A., Reed, S. G., Ashman, J., Sagawa, Z. K., ... Tait, D. (2018). The TLR-4 agonist adjuvant, GLA-SE, improves magnitude and quality of immune responses elicited by the ID93 tuberculosis vaccine: First-in-human trial. *npj Vaccines*, *3*(1). Available from https://doi.org/10.1038/s41541-018-0057-5.

Coughlin, R. T., Fattom, A., Chu, C., White, A. C., & Winston, S. (1995). Adjuvant activity of QS-21 for experimental *E. coli* 018 polysaccharide vaccines. *Vaccine*, *13*(1), 17–21. Available from https://doi.org/10.1016/0264-410X(95)80005-X.

Couzin, J. (2005). CELL BIOLOGY: The ins and outs of exosomes. *Science (New York, N.Y.)*, *308*(5730), 1862–1863. Available from https://doi.org/10.1126/science.308.5730.1862.

Cunningham, A. L., Lal, H., Kovac, M., Chlibek, R., Hwang, S. J., Díez-Domingo, J., Godeaux, O., Levin, M. J., McElhaney, J. E., Puig-Barberà, J., Vanden Abeele, C., Vesikari, T., Watanabe, D., Zahaf, T., Ahonen, A., Athan, E., Barba-Gomez, J. F., Campora, L., De Looze, F., ... Heineman, T. C. (2016). Efficacy of the herpes zoster subunit vaccine in adults 70 years of age or older. *New England Journal of Medicine*, *375*(11), 1019–1032. Available from https://doi.org/10.1056/NEJMoa1603800.

Dalsgaard, K. (1974). Saponin adjuvants - III. Isolation of a substance from *Quillaja saponaria* molina with adjuvant activity in foot-and-mouth disease vaccines. *Archiv Für Die Gesamte Virusforschung*, *44*(3), 243–254. Available from https://doi.org/10.1007/BF01240612.

Dalsgaard, K. (1978). A study of the isolation and characterization of the saponin Quil A. Evaluation of its adjuvant activity, with a special reference to the application in the vaccination of cattle against foot-and-mouth disease. *Acta Veterinaria Scandinavica. Supplementum*, *69*, 7–40.

Datoo, M. S., Natama, M. H., Somé, A., Traoré, O., Rouamba, T., Bellamy, D., Yameogo, P., Valia, D., Tegneri, M., Ouedraogo, F., Soma, R., Sawadogo, S., Sorgho, F., Derra, K., Rouamba, E., Orindi, B., Ramos Lopez, F., Flaxman, A., Cappuccini, F., ... Tinto, H. (2021). Efficacy of a low-dose candidate malaria vaccine, R21 in adjuvant Matrix-M, with seasonal administration to children in Burkina Faso: A randomized controlled trial. *The Lancet*, *397*(10287), 1809–1818. Available from https://doi.org/10.1016/S0140-6736(21)00943-0.

Datta, S. K., Sabet, M., Nguyen, K. P. L., Valdez, P. A., Gonzalez-Navajas, J. M., Islam, S., Mihajlov, I., Fierer, J., Insel, P. A., Webster, N. J., Guiney, D. G., & Raz, E. (2010). Mucosal adjuvant activity of cholera toxin requires Th17 cells and protects against inhalation anthrax. *Proceedings of the National Academy of Sciences of the United States of America, 107*(23), 10638–10643. Available from https://doi.org/10.1073/pnas.1002348107.

Davenport, F. M., & Hennessy, A. V. (1956). A serologic recapitulation of past experiences with influenza A; antibody response to monovalent vaccine. *The Journal of Experimental Medicine, 104*(1), 85–97. Available from https://doi.org/10.1084/jem.104.1.85.

Davitt, C. J. H., Longet, S., Albutti, A., Aversa, V., Nordqvist, S., Hackett, B., McEntee, C. P., Rosa, M., Coulter, I. S., Lebens, M., Tobias, J., Holmgren, J., & Lavelle, E. C. (2019). Alpha-galactosylceramide enhances mucosal immunity to oral whole-cell cholera vaccines. *Mucosal Immunology, 12*(4), 1055–1064. Available from https://doi.org/10.1038/s41385-019-0159-z.

Davitt, C. J. H., McNeela, E. A., Longet, S., Tobias, J., Aversa, V., McEntee, C. P., Rosa, M., Coulter, I. S., Holmgren, J., & Lavelle, E. C. (2016). A novel adjuvanted capsule based strategy for oral vaccination against infectious diarrhoeal pathogens. *Journal of Controlled Release, 233*, 162–173. Available from https://doi.org/10.1016/j.jconrel.2016.05.001.

De, S. N. (1959). Enterotoxicity of bacteria-free culture-filtrate of *Vibrio cholerae*. *Nature, 183*(4674), 1533–1534. Available from https://doi.org/10.1038/1831533a0.

De Jong, S. E., Olin, A., & Pulendran, B. (2020). The impact of the microbiome on immunity to vaccination in humans. *Cell Host and Microbe, 28*(2), 169–179. Available from https://doi.org/10.1016/j.chom.2020.06.014.

De Serrano, L. O., & Burkhart, D. J. (2017). Liposomal vaccine formulations as prophylactic agents: Design considerations for modern vaccines. *Journal of Nanobiotechnology, 15*(1). Available from https://doi.org/10.1186/s12951-017-0319-9.

Del Giudice, G., Rappuoli, R., & Didierlaurent, A. M. (2018). Correlates of adjuvanticity: A review on adjuvants in licensed vaccines. *Seminars in Immunology, 39*, 14–21. Available from https://doi.org/10.1016/j.smim.2018.05.001.

Den Brok, M. H., Büll, C., Wassink, M., De Graaf, A. M., Wagenaars, J. A., Minderman, M., Thakur, M., Amigorena, S., Rijke, E. O., Schrier, C. C., & Adema, G. J. (2016). Saponin-based adjuvants induce cross-presentation in dendritic cells by intracellular lipid body formation. *Nature Communications, 7*. Available from https://doi.org/10.1038/ncomms13324.

Desselberger, U. (2018). The Mammalian intestinal microbiome: Composition, interaction with the immune system, significance for vaccine efficacy, and potential for disease therapy. *Pathogens, 7*(3). Available from https://doi.org/10.3390/pathogens7030057.

Detienne, S., Welsby, I., Collignon, C., Wouters, S., Coccia, M., Delhaye, S., Van Maele, L., Thomas, S., Swertvaegher, M., Detavernier, A., Elouahabi, A., Goriely, S., & Didierlaurent, A. M. (2016). Central role of CD169 + lymph node resident macrophages in the adjuvanticity of the QS-21 component of AS01. *Scientific Reports, 6*. Available from https://doi.org/10.1038/srep39475.

Di Bonito, P., Chiozzini, C., Arenaccio, C., Anticoli, S., Manfredi, F., Olivetta, E., Ferrantelli, F., Falcone, E., Ruggieri, A., & Federico, M. (2017). Antitumor HPV E7-specific CTL activity elicited by in vivo engineered exosomes produced through DNA inoculation. *International Journal of Nanomedicine, 12*, 4579–4591. Available from https://doi.org/10.2147/IJN.S131309.

Didierlaurent, A. M., Laupèze, B., Di Pasquale, A., Hergli, N., Collignon, C., & Garçon, N. (2017). Adjuvant system AS01: Helping to overcome the challenges of modern vaccines. *Expert Review of Vaccines, 16*(1), 55–63. Available from https://doi.org/10.1080/14760584.2016.1213632.

Didierlaurent, A. M., Morel, S., Lockman, L., Giannini, S. L., Bisteau, M., Carlsen, H., Kielland, A., Vosters, O., Vanderheyde, N., Schiavetti, F., Larocque, D., Van Mechelen, M., & Garçon, N. (2009). AS04, an aluminum salt- and TLR4 agonist-based adjuvant system, induces a transient localized innate immune response leading to enhanced adaptive immunity. *Journal of Immunology*, *183*(10), 6186–6197. Available from https://doi.org/10.4049/jimmunol.0901474.

Di Sotto, A., Vitalone, A., & Di Giacomo, S. (2020). Plant-derived nutraceuticals and immune system modulation: An evidence-based overview. *Vaccines*, *8*(3), 1–34. Available from https://doi.org/10.3390/vaccines8030468.

Domnich, A., Arata, L., Amicizia, D., Puig-Barberà, J., Gasparini, R., & Panatto, D. (2017). Effectiveness of MF59-adjuvanted seasonal influenza vaccine in the elderly: A systematic review and *meta*-analysis. *Vaccine*, *35*(4), 513–520. Available from https://doi.org/10.1016/j.vaccine.2016.12.011.

Du, X., Zhao, B., Li, J., Cao, X., Diao, M., Feng, H., Chen, X., Chen, Z., & Zeng, X. (2012). Astragalus polysaccharides enhance immune responses of HBV DNA vaccination via promoting the dendritic cell maturation and suppressing Treg frequency in mice. *International Immunopharmacology*, *14*(4), 463–470. Available from https://doi.org/10.1016/j.intimp.2012.09.006.

Duan, Q., Xia, P., Nandre, R., Zhang, W., & Zhu, G. (2019). Review of newly identified functions associated with the heat-labile toxin of enterotoxigenic *Escherichia coli*. *Frontiers in Cellular and Infection Microbiology*, *9*. Available from https://doi.org/10.3389/fcimb.2019.00292.

Eggermont, A. M., Suciu, S., Rutkowski, P., Marsden, J., Testori, A., Corrie, P., Aamdal, S., Ascierto, P. A., Patel, P., & Spatz, A. (2010). Randomized phase III trial comparing postoperative adjuvant ganglioside GM2-KLH/QS-21 vaccination vs observation in stage II (T3-T4N0M0) melanoma: Final results of study EORTC 18961. *Journal of Clinical Oncology*, *28*, 8505. Available from https://doi.org/10.1200/jco.2010.28.15_suppl.8505.

Ellebedy, A. H., Nachbagauer, R., Jackson, K. J. L., Dai, Y.-N., Han, J., Alsoussi, W. B., Davis, C. W., Stadlbauer, D., Rouphael, N., Chromikova, V., McCausland, M., Chang, C. Y., Cortese, M., Bower, M., Chennareddy, C., Schmitz, A. J., Zarnitsyna, V. I., Lai, L., Rajabhathor, A., … Ahmed, R. (2020). Adjuvanted H5N1 influenza vaccine enhances both cross-reactive memory B cell and strain-specific naive B cell responses in humans. *Proceedings of the National Academy of Sciences*, *117*(30), 17957–17964. Available from https://doi.org/10.1073/pnas.1906613117.

Elmowalid, G. A., Abd El-Hamid, M. I., Abd El-Wahab, A. M., Atta, M., Abd El-Naser, G., & Attia, A. M. (2019). Garlic and ginger extracts modulated broiler chicks innate immune responses and enhanced multidrug resistant *Escherichia coli* O78 clearance. *Comparative Immunology, Microbiology and Infectious Diseases*, *66*101334. Available from https://doi.org/10.1016/j.cimid.2019.101334.

Eremina, N. V., Kazey, V. I., Mishugin, S. V., Leonenkov, R. V., Pushkar, D. Y., Mett, V. L., & Gudkov, A. V. (2020). First-in-human study of anticancer immunotherapy drug candidate mobilan: Safety, pharmacokinetics and pharmacodynamics in prostate cancer patients. *Oncotarget*, *11*(14), 1273–1288. Available from https://doi.org/10.18632/oncotarget.27549.

Eriksson, A. M., Schön, K. M., & Lycke, N. Y. (2004). The cholera toxin-derived CTA1-DD vaccine adjuvant administered intranasally does not cause inflammation or accumulate in the nervous tissues. *Journal of Immunology*, *173*(5), 3310–3319. Available from https://doi.org/10.4049/jimmunol.173.5.3310.

Evans, T. G., McElrath, M. J., Matthews, T., Montefiori, D., Weinhold, K., Wolff, M., Keefer, M. C., Kallas, E. G., Corey, L., Gorse, G. J., Belshe, R., Graham, B. S., Spearman, P. W.,

Schwartz, D., Mulligan, M. J., Goepfert, P., Fast, P., Berman, P., Powell, M., ... Francis, D. (2001). QS-21 promotes an adjuvant effect allowing for reduced antigen dose during HIV-1 envelope subunit immunization in humans. *Vaccine, 19*(15–16), 2080–2091. Available from https://doi.org/10.1016/S0264-410X(00)00415-1.

Fernández-Tejada, A., Chea, E. K., George, C., Pillarsetty, N., Gardner, J. R., Livingston, P. O., Ragupathi, G., Lewis, J. S., Tan, D. S., & Gin, D. Y. (2014). Development of a minimal saponin vaccine adjuvant based on QS-21. *Nature Chemistry, 6*(7), 635–643. Available from https://doi.org/10.1038/nchem.1963.

Fernández-Tejada, A., Walkowicz, W. E., Tan, D. S., & Gin, D. Y. (2017). *Semisynthesis of analogues of the saponin immunoadjuvant QS-21. Methods in molecular biology* (1494, pp. 45–71). Humana Press, Inc. Available from https://doi.org/10.1007/978-1-4939-6445-1_4.

Foged, C., Arigita, C., Sundblad, A., Jiskoot, W., Storm, G., & Frokjaer, S. (2004). Interaction of dendritic cells with antigen-containing liposomes: Effect of bilayer composition. *Vaccine, 22*(15–16), 1903–1913. Available from https://doi.org/10.1016/j.vaccine.2003.11.008.

Fox, C. B. (2017). *Vaccine Adjuvants Methods and Protocols - Springer Protocols Christopher B. Fox (editor)* (1st ed.). Springer Science + Business Media (copyright holder).

Francica, J. R., Flynn, B. J., Foulds, K. E., Noe, A. T., Werner, A. P., Moore, I. N., Gagne, M., Johnston, T. S., Tucker, C., Davis, R. L., Flach, B., O'Connell, S., Andrew, S. F., Lamb, E., Flebbe, D. R., Nurmukhambetova, S. T., Donaldson, M. M., Todd, J. P. M., Zhu, A. L., ... Seder, R. A. (2021). Vaccination with SARS-CoV-2 spike protein and AS03 adjuvant induces rapid anamnestic antibodies in the lung and protects against virus challenge in non-human primates. *BioRxiv*. Available from https://doi.org/10.1101/2021.03.02.433390.

Freund, J. (1951). The effect of paraffin oil and mycobacteria on antibody formation and sensitization; a review. *American Journal of Clinical Pathology, 21*(7), 645–656. Available from https://doi.org/10.1093/ajcp/21.7.645.

Freund, J., Casals, J., & Hosmer, E. P. (1937). Sensitization and antibody formation after injection of tubercle bacilli and paraffin oil. *Proceedings of the Society for Experimental Biology and Medicine, 37*(3), 509–513. Available from https://doi.org/10.3181/00379727-37-9625.

Freytag, L. C., & Clements, J. D. (2005). Mucosal adjuvants. *Vaccine, 23*(15), 1804–1813. Available from https://doi.org/10.1016/j.vaccine.2004.11.010.

Friedman-Klabanoff, D. A. J., Berry, A. A., Travassos, M. A., Cox, C., Zhou, Y., Mo, A. X., Nomicos, E. Y. H., Deye, G. A., Pasetti, M. F., & Laurens, M. B. (2021). Low dose recombinant full-length circumsporozoite protein-based *Plasmodium falciparum* vaccine is well-tolerated and highly immunogenic in phase 1 first-in-human clinical testing. *Vaccine, 39*(8), 1195–1200. Available from https://doi.org/10.1016/j.vaccine.2020.12.023.

Fujii, S. I., Shimizu, K., Smith, C., Bonifaz, L., & Steinman, R. M. (2003). Activation of natural killer T cells by α-galactosylceramide rapidly induces the full maturation of dendritic cells in vivo and thereby acts as an adjuvant for combined CD4 and CD8 T cell immunity to a coadministered protein. *Journal of Experimental Medicine, 198*(2), 267–279. Available from https://doi.org/10.1084/jem.20030324.

Fukuyama, Y., Okada, K., Yamaguchi, M., Kiyono, H., Mori, K., Yuki, Y., & Reisert, J. (2015). Nasal administration of cholera toxin as a mucosal adjuvant damages the olfactory system in mice. *PLoS One, 10*(9), e0139368. Available from https://doi.org/10.1371/journal.pone.0139368.

Galson, J. D., Trück, J., Kelly, D. F., & Van Der Most, R. (2016). Investigating the effect of AS03 adjuvant on the plasma cell repertoire following pH1N1 influenza vaccination. *Scientific Reports, 6*. Available from https://doi.org/10.1038/srep37229.

Garçon, N., & Di Pasquale, A. (2017). From discovery to licensure, the Adjuvant System story. *Human Vaccines and Immunotherapeutics*, *13*(1), 19–33. Available from https://doi.org/10.1080/21645515.2016.1225635.

Garçon, N., Vaughn, D. W., & Didierlaurent, A. M. (2012). Development and evaluation of AS03, an Adjuvant System containing α-tocopherol and squalene in an oil-in-water emulsion. *Expert Review of Vaccines*, *11*(3), 349–366. Available from https://doi.org/10.1586/erv.11.192.

Gautam, M., Diwanay, S., Gairola, S., Shinde, Y., Patki, P., & Patwardhan, B. (2004). Immunoadjuvant potential of *Asparagus racemosus* aqueous extract in experimental system. *Journal of Ethnopharmacology*, *91*(2–3), 251–255. Available from https://doi.org/10.1016/j.jep.2003.12.023.

Giannini, S. L., Hanon, E., Moris, P., Van Mechelen, M., Morel, S., Dessy, F., Fourneau, M. A., Colau, B., Suzich, J., Losonksy, G., Martin, M. T., Dubin, G., & Wettendorff, M. A. (2006). Enhanced humoral and memory B cellular immunity using HPV16/18 L1 VLP vaccine formulated with the MPL/aluminum salt combination (AS04) compared to aluminum salt only. *Vaccine*, *24*(33–34), 5937–5949. Available from https://doi.org/10.1016/j.vaccine./2006.06.005.

Gilewski, T., Adluri, S., Ragupathi, G., Zhang, S., Yao, T. J., Panageas, K., Moynahan, M., Houghton, A., Norton, L., & Livingston, P. O. (2000). Vaccination of high-risk breast cancer patients with mucin-1 (MUC1) keyhole limpet hemocyanin conjugate plus QS-21. *Clinical Cancer Research*, *6*(5), 1693–1701.

Gilewski, T. A., Ragupathi, G., Dickler, M., Powell, S., Bhuta, S., Panageas, K., Koganty, R. R., Chin-Eng, J., Hudis, C., Norton, L., Houghton, A. N., & Livingston, P. O. (2007). Immunization of high-risk breast cancer patients with clustered sTn-KLH conjugate plus the immunologic adjuvant QS-21. *Clinical Cancer Research*, *13*(10), 2977–2985. Available from https://doi.org/10.1158/1078-0432.CCR-06-2189.

Gin, D. Y., & Slovin, S. F. (2011). Enhancing immunogenicity of cancer vaccines: QS-21 as an immune adjuvant. *Current Drug Therapy*, *6*(3), 207–212. Available from https://doi.org/10.2174/157488511796391988.

Givord, C., Welsby, I., Detienne, S., Thomas, S., Assabban, A., Lima Silva, V., Molle, C., Gineste, R., Vermeersch, M., Perez-Morga, D., Leo, O., Collignon, C., Didierlaurent, A. M., & Goriely, S. (2018). Activation of the endoplasmic reticulum stress sensor IRE1α by the vaccine adjuvant AS03 contributes to its immunostimulatory properties. *npj Vaccines*, *3*(1). Available from https://doi.org/10.1038/s41541-018-0058-4.

Glenny, A. T., Pope, C. G., Waddington, H., & Wallace, U. (1926). Immunological notes. XVII-XXIV. *The Journal of Pathology and Bacteriology*, *29*(1), 31–40. Available from https://doi.org/10.1002/path.1700290106.

Greenwood, B. (2014). The contribution of vaccination to global health: Past, present and future. *Philosophical Transactions of the Royal Society B: Biological Sciences*, *369*(1645), 20130433. Available from https://doi.org/10.1098/rstb.2013.0433.

Gregoriadis, G., & Ryman, B. E. (1972). Fate of protein-containing liposomes injected into rats: An approach to the treatment of storage diseases. *European Journal of Biochemistry*, *24*(3), 485–491. Available from https://doi.org/10.1111/j.1432-1033.1972.tb19710.x.

Gregoriadis, G., Leathwood, P. D., & Ryman, B. E. (1971). Enzyme entrapment in liposomes. *FEBS Letters*, *14*(2), 95–99. Available from https://doi.org/10.1016/0014-5793(71)80109-6.

Grun, J. L., & Maurer, P. H. (1989). Different T helper cell subsets elicited in mice utilizing two different adjuvant vehicles: The role of endogenous interleukin 1 in proliferative responses.

Cellular Immunology, *121*(1), 134–145. Available from https://doi.org/10.1016/0008-8749(89)90011-7.

Guan, S., He, J., Guo, W., Wei, J., Lu, J., & Deng, X. (2011). Adjuvant effects of salidroside from *Rhodiola rosea* L. on the immune responses to ovalbumin in mice. *Immunopharmacology and Immunotoxicology*, *33*(4), 738–743. Available from https://doi.org/10.3109/08923973.2011.567988.

Guerra Mendoza, Y., Garric, E., Leach, A., Lievens, M., Ofori-Anyinam, O., Pirçon, J. Y., Stegmann, J. U., Vandoolaeghe, P., Otieno, L., Otieno, W., Owusu-Agyei, S., Sacarlal, J., Masoud, N. S., Sorgho, H., Tanner, M., Tinto, H., Valea, I., Mtoro, A. T., Njuguna, P., ... Schuerman, L. (2019). Safety profile of the RTS,S/AS01 malaria vaccine in infants and children: Additional data from a phase III randomized controlled trial in sub-Saharan Africa. *Human Vaccines and Immunotherapeutics*, *15*(10), 2386–2398. Available from https://doi.org/10.1080/21645515.2019.1586040.

Guiducci, C., Gong, M., Cepika, A. M., Xu, Z., Tripodo, C., Bennett, L., Crain, C., Quartier, P., Cush, J. J., Pascual, V., Coffman, R. L., & Barrat, F. J. (2013). RNA recognition by human TLR8 can lead to autoimmune inflammation. *Journal of Experimental Medicine*, *210*(13), 2903–2919. Available from https://doi.org/10.1084/jem.20131044.

Hagan, T., Cortese, M., Rouphael, N., Boudreau, C., Linde, C., Maddur, M. S., Das, J., Wang, H., Guthmiller, J., Zheng, N. Y., Huang, M., Uphadhyay, A. A., Gardinassi, L., Petitdemange, C., McCullough, M. P., Johnson, S. J., Gill, K., Cervasi, B., Zou, J., ... Pulendran, B. (2019). Antibiotics-driven gut microbiome perturbation alters immunity to vaccines in humans. *Cell*, *178*(6), 1313–1328. Available from https://doi.org/10.1016/j.cell.2019.08.010, e13.

Hall, J. A., Bouladoux, N., Sun, C. M., Wohlfert, E. A., Blank, R. B., Zhu, Q., Grigg, M. E., Berzofsky, J. A., & Belkaid, Y. (2008). Commensal DNA limits regulatory T cell conversion and is a natural adjuvant of intestinal immune responses. *Immunity*, *29*(4), 637–649. Available from https://doi.org/10.1016/j.immuni.2008.08.009.

Hamborg, M., Rose, F., Jorgensen, L., Bjorklund, K., Pedersen, H. B., Christensen, D., & Foged, C. (2014). Elucidating the mechanisms of protein antigen adsorption to the CAF/NAF liposomal vaccine adjuvant systems: Effect of charge, fluidity and antigen-to-lipid ratio. *Biochimica et Biophysica Acta - Biomembranes*, *1838*(8), 2001–2010. Available from https://doi.org/10.1016/j.bbamem.2014.04.013.

Han, J., Zhao, D., Li, D., Wang, X., Jin, Z., & Zhao, K. (2018). Polymer-based nanomaterials and applications for vaccines and drugs. *Polymers*, *10*(1). Available from https://doi.org/10.3390/polym10010031.

Harrer, T., Plettenberg, A., Arastéh, K., Van Lunzen, J., Fätkenheuer, G., Jaeger, H., Janssens, M., Burny, W., Collard, A., Roman, F., Loeliger, A., Koutsoukos, M., Bourguignon, P., Lavreys, L., & Voss, G. (2014). Safety and immunogenicity of an adjuvanted protein therapeutic HIV-1 vaccine in subjects with HIV-1 infection: A randomized placebo-controlled study. *Vaccine*, *32*(22), 2657–2665. Available from https://doi.org/10.1016/j.vaccine.2013.10.030.

Harris, V. C., Haak, B. W., Handley, S. A., Jiang, B., Velasquez, D. E., Hykes, B. L., Droit, L., Berbers, G. A. M., Kemper, E. M., Van Leeuwen, E. M. M., Boele Van Hensbroek, M., & Wiersinga, W. J. (2018). Effect of antibiotic-mediated microbiome modulation on rotavirus vaccine immunogenicity: A human, randomized-control proof-of-concept trial. *Cell Host and Microbe*, *24*(2), 197–207. Available from https://doi.org/10.1016/j.chom.2018.07.005, e4.

Häse, C. C., Thai, L. S., Boesman-Finkelstein, M., Mar, V. L., Burnette, W. N., Kaslow, H. R., Stevens, L. A., Moss, J., & Finkelstein, R. A. (1994). Construction and characterization of recombinant *Vibrio cholerae* strains producing inactive cholera toxin analogs. *Infection and*

Immunity, *62*(8), 3051–3057. Available from https://doi.org/10.1128/iai.62.8.3051-3057.1994.

Heineman, T. C., Cunningham, A., & Levin, M. (2019). Understanding the immunology of Shingrix, a recombinant glycoprotein E adjuvanted herpes zoster vaccine. *Current Opinion in Immunology*, *59*, 42–48. Available from https://doi.org/10.1016/j.coi.2019.02.009.

Hem, S. L., & White, J. L. (1984). Characterization of aluminum hydroxide for use as an adjuvant in parenteral vaccines. *Journal of Parenteral Science and Technology*, *38*(1), 2–10.

Hemmi, H., Kaisho, T., Takeuchi, O., Sato, S., Sanjo, H., Hoshino, K., Horiuchi, T., Tomizawa, H., Takeda, K., & Akira, S. (2002). Small anti-viral compounds activate immune cells via the TLR7 MyD88–dependent signaling pathway. *Nature Immunology*, 196–200. Available from https://doi.org/10.1038/ni758.

Henriksen-Lacey, M., Christensen, D., Bramwell, V. W., Lindenstrøm, T., Agger, E. M., Andersen, P., & Perrie, Y. (2010). Liposomal cationic charge and antigen adsorption are important properties for the efficient deposition of antigen at the injection site and ability of the vaccine to induce a CMI response. *Journal of Controlled Release*, *145*(2), 102–108. Available from https://doi.org/10.1016/j.jconrel.2010.03.027.

Henriksen-Lacey, M., Christensen, D., Bramwell, V. W., Lindenstrøm, T., Agger, E. M., Andersen, P., & Perrie, Y. (2011). Comparison of the depot effect and immunogenicity of liposomes based on dimethyldioctadecylammonium (DDA), 3β-[N-(N′,N′- dimethylaminoethane)carbomyl] cholesterol (DC-Chol), and 1,2-dioleoyl-3- trimethylammonium propane (DOTAP): Prolonged liposome retention mediates stronger Th1 responses. *Molecular Pharmaceutics*, *8*(1), 153–161. Available from https://doi.org/10.1021/mp100208f.

Henry, C., Palm, A. K. E., Krammer, F., & Wilson, P. C. (2018). From original antigenic sin to the universal influenza virus vaccine. *Trends in Immunology*, *39*(1), 70–79. Available from https://doi.org/10.1016/j.it.2017.08.003.

Herbert, W. J. (1968). The mode of action of mineral-oil emulsion adjuvants on antibody production in mice. *Immunology*, *14*(3), 301–318.

Hermans, I. F., Silk, J. D., Gileadi, U., Salio, M., Mathew, B., Ritter, G., Schmidt, R., Harris, A. L., Old, L., & Cerundolo, V. (2003). NKT cells enhance CD4 + and CD8 + T cell responses to soluble antigen in vivo through direct interaction with dendritic cells. *Journal of Immunology*, *171*(10), 5140–5147. Available from https://doi.org/10.4049/jimmunol.171.10.5140.

Hirota, K., Turner, J. E., Villa, M., Duarte, J. H., Demengeot, J., Steinmetz, O. M., & Stockinger, B. (2013). Plasticity of TH 17 cells in Peyer's patches is responsible for the induction of T cell-dependent IgA responses. *Nature Immunology*, *14*(4), 372–379. Available from https://doi.org/10.1038/ni.2552.

Holmgren, J., Lycke, N., & Czerkinsky, C. (1993). Cholera toxin and cholera B subunit as oral-mucosal adjuvant and antigen vector systems. *Vaccine*, *11*(12), 1179–1184. Available from https://doi.org/10.1016/0264-410X(93)90039-Z.

Honda-Okubo, Y., Saade, F., & Petrovsky, N. (2012). Advax^{TM}, a polysaccharide adjuvant derived from delta inulin, provides improved influenza vaccine protection through broad-based enhancement of adaptive immune responses. *Vaccine*, *30*(36), 5373–5381. Available from https://doi.org/10.1016/j.vaccine.2012.06.021.

Hosomi, K., & Kunisawa, J. (2020). Impact of the intestinal environment on the immune responses to vaccination. *Vaccine*, *38*(44), 6959–6965. Available from https://doi.org/10.1016/j.vaccine.2020.08.079.

Hu, K. F., Elvander, M., Merza, M., Åkerblom, L., Brandenburg, A., & Morein, B. (1998). The immunostimulating complex (ISCOM) is an efficient mucosal delivery system for respiratory syncytial virus (RSV) envelope antigens inducing high local and systemic antibody responses. *Clinical and Experimental Immunology, 113*(2), 235–243. Available from https://doi.org/10.1046/j.1365-2249.1998.00650.x.

Hussain, M. J., Wilkinson, A., Bramwell, V. W., Christensen, D., & Perrie, Y. (2014). Th1 immune responses can be modulated by varying dimethyldioctadecylammonium and distearoyl-sn-glycero-3-phosphocholine content in liposomal adjuvants. *Journal of Pharmacy and Pharmacology, 66*(3), 358–366. Available from https://doi.org/10.1111/jphp.12173.

Ibrahim, E. E. S., Gamal, W. M., Hassan, A. I., Mahdy, S. E. D., Hegazy, A. Z., & Abdel-Atty, M. M. (2015). Comparative study on the immunopotentiator effect of ISA 201, ISA 61, ISA 50, ISA 206 used in trivalent foot and mouth disease vaccine. *Veterinary World, 8*(10), 1189–1198. Available from https://doi.org/10.14202/vetworld.2015.1189-1198.

Jafarzadeh, A., & Nemati, M. (2018). Therapeutic potentials of ginger for treatment of multiple sclerosis: A review with emphasis on its immunomodulatory, anti-inflammatory and anti-oxidative properties. *Journal of Neuroimmunology, 324*, 54–75. Available from https://doi.org/10.1016/j.jneuroim.2018.09.003.

Janeway, C. A., Jr. (1989). Pillars article: Approaching the asymptote? Evolution and revolution in immunology. Cold spring harb symp quant biol. *Journal of Immunology, 54*, 4475–4487.

Jertborn, M., Nordstrom, I., Kilander, A., Czerkinsky, C., & Holmgren, J. (2001). Local and systemic immune responses to rectal administration of recombinant cholera toxin B subunit in humans. *Infection and Immunity, 69*(6), 4125–4128. Available from https://doi.org/10.1128/IAI.69.6.4125-4128.2001.

Kaczmarek, J. C., Patel, A. K., Kauffman, K. J., Fenton, O. S., Webber, M. J., Heartlein, M. W., DeRosa, F., & Anderson, D. G. (2016). Polymer–lipid nanoparticles for systemic delivery of mRNA to the lungs. *Angewandte Chemie - International Edition, 55*(44), 13808–13812. Available from https://doi.org/10.1002/anie.201608450.

Kashala, O., Amador, R., Valero, M. V., Moreno, A., Barbosa, A., Nickel, B., Daubenberger, C. A., Guzman, F., Pluschke, G., & Patarroyo, M. E. (2002). Safety, tolerability and immunogenicity of new formulations of the *Plasmodium falciparum* malaria peptide vaccine SPf66 combined with the immunological adjuvant QS-21. *Vaccine, 20*(17–18), 2263–2277. Available from https://doi.org/10.1016/S0264-410X(02)00115-9.

Kauffman, K. J., Webber, M. J., & Anderson, D. G. (2016). Materials for non-viral intracellular delivery of messenger RNA therapeutics. *Journal of Controlled Release, 240*, 227–234. Available from https://doi.org/10.1016/j.jconrel.2015.12.032.

Kazmin, D., Nakaya, H. I., Lee, E. K., Johnson, M. J., Van Der Most, R., Van Den Berg, R. A., Ballou, W. R., Jongert, E., Wille-Reece, U., Ockenhouse, C., Aderem, A., Zak, D. E., Sadoff, J., Hendriks, J., Wrammert, J., Ahmed, R., & Pulendran, B. (2017). Systems analysis of protective immune responses to RTS,S malaria vaccination in humans. *Proceedings of the National Academy of Sciences of the United States of America, 114*(9), 2425–2430. Available from https://doi.org/10.1073/pnas.1621489114.

Keller, S., Ridinger, J., Rupp, A. K., Janssen, J. W. G., & Altevogt, P. (2011). Body fluid derived exosomes as a novel template for clinical diagnostics. *Journal of Translational Medicine, 9*. Available from https://doi.org/10.1186/1479-5876-9-86.

Kensil, C. R., Patel, U., Lennick, M., & Marciani, D. (1991). Separation and characterization of saponins with adjuvant activity from *Quillaja saponaria* Molina cortex. *Journal of Immunology, 146*(2), 431–437.

Kester, K. E., Cummings, J. F., Ofori-Anyinam, O., Ockenhouse, C. F., Krzych, U., Moris, P., Schwenk, R., Nielsen, R. A., Debebe, Z., Pinelis, E., Juompan, L., Williams, J., Dowler, M., Stewart, V. A., Wirtz, R. A., Dubois, M. C., Lievens, M., Cohen, J., Ballou, W. R., ...

Heppner, D. G. (2009). Randomized, double-blind, phase 2a trial of falciparum malaria vaccines RTS,S/AS01B and RTS,S/AS02A in malaria-naive adults: Safety, efficacy, and immunologic associates of protection. *Journal of Infectious Diseases*, *200*(3), 337–346. Available from https://doi.org/10.1086/600120.

Kim, S. K., Ragupathi, G., Musselli, C., Choi, S. J., Park, Y. S., & Livingston, P. O. (1999). Comparison of the effect of different immunological adjuvants on the antibody and T-cell response to immunization with MUC1-KLH and GD3-KLH conjugate cancer vaccines. *Vaccine*, *18*(7–8), 597–603. Available from https://doi.org/10.1016/S0264-410X(99)00316-3.

Kong, N. C. T., Beran, J., Kee, S. A., Miguel, J. L., Sánchez, C., Bayas, J. M., Vilella, A., Calbo-Torrecillas, F., López De Novales, E., Srinivasa, K., Stoffel, M., & Hoet, B. (2008). A new adjuvant improves the immune response to hepatitis B vaccine in hemodialysis patients. *Kidney International*, *73*(7), 856–862. Available from https://doi.org/10.1038/sj.ki.5002725.

Korsholm, K. S., Hansen, J., Karlsen, K., Filskov, J., Mikkelsen, M., Lindenstrøm, T., Schmidt, S. T., Andersen, P., & Christensen, D. (2014). Induction of CD8 + T-cell responses against subunit antigens by the novel cationic liposomal CAF09 adjuvant. *Vaccine*, *32*(31), 3927–3935. Available from https://doi.org/10.1016/j.vaccine.2014.05.050.

Kotta, S., Khan, A. W., Ansari, S. H., Sharma, R. K., & Ali, J. (2015). Formulation of nanoemulsion: A comparison between phase inversion composition method and high-pressure homogenization method. *Drug Delivery*, *22*(4), 455–466. Available from https://doi.org/10.3109/10717544.2013.866992.

Kozlowski, P. A., Cu-Uvin, S., Neutra, M. R., & Flanigan, T. P. (1997). Comparison of the oral, rectal, and vaginal immunization routes for induction of antibodies in rectal and genital tract secretions of women. *Infection and Immunity*, *65*(4), 1387–1394. Available from https://doi.org/10.1128/iai.65.4.1387-1394.1997.

Kuate, S., Cinatl, J., Doerr, H. W., & Überla, K. (2007). Exosomal vaccines containing the S protein of the SARS coronavirus induce high levels of neutralizing antibodies. *Virology*, *362*(1), 26–37. Available from https://doi.org/10.1016/j.virol.2006.12.011.

Kurzatkowski, W., Kartoĝlu, u, Staniszewska, M., Górska, P., Krause, A., & Wysocki, M. J. (2013). Structural damages in adsorbed vaccines affected by freezing. *Biologicals: Journal of the International Association of Biological Standardization*, *41*(2), 71–76. Available from https://doi.org/10.1016/j.biologicals.2011.10.011.

Lal, H., Cunningham, A. L., Godeaux, O., Chlibek, R., Diez-Domingo, J., Hwang, S. J., Levin, M. J., McElhaney, J. E., Poder, A., Puig-Barberà, J., Vesikari, T., Watanabe, D., Weckx, L., Zahaf, T., & Heineman, T. C. (2015). Efficacy of an adjuvanted herpes zoster subunit vaccine in older adults. *New England Journal of Medicine*, *372*(22), 2087–2096. Available from https://doi.org/10.1056/NEJMoa1501184.

Lasic, D. D. (1998). Novel applications of liposomes. *Trends in Biotechnology*, *16*(7), 307–321. Available from https://doi.org/10.1016/S0167-7799(98)01220-7.

Laupèze, B., Hervé, C., Di Pasquale, A., & Tavares Da Silva, F. (2019). Adjuvant Systems for vaccines: 13 years of post-licensure experience in diverse populations have progressed the way adjuvanted vaccine safety is investigated and understood. *Vaccine*, *37*(38), 5670–5680. Available from https://doi.org/10.1016/j.vaccine.2019.07.098.

Lemoine, C., Thakur, A., Krajišnik, D., Guyon, R., Longet, S., Razim, A., Górska, S., Pantelić, I., Ilić, T., Nikolić, I., Lavelle, E. C., Gamian, A., Savić, S., & Milicic, A. (2020). Technological approaches for improving vaccination compliance and coverage. *Vaccines*, *8*(2), 1–29. Available from https://doi.org/10.3390/vaccines8020304.

Leroux-Roels, G., Marchant, A., Levy, J., Van Damme, P., Schwarz, T. F., Horsmans, Y., Jilg, W., Kremsner, P. G., Haelterman, E., Clément, F., Gabor, J. J., Esen, M., Hens, A., Carletti, I., Fissette, L., Tavares Da Silva, F., Burny, W., Janssens, M., Moris, P., ... Van Mechelen, M. (2016). Impact of adjuvants on CD4 + T cell and B cell responses to a protein antigen vaccine: Results from a phase II, randomized, multicenter trial. *Clinical Immunology*, *169*, 16–27. Available from https://doi.org/10.1016/j.clim.2016.05.007.

Li, H., Nookala, S., & Re, F. (2007). Aluminum hydroxide adjuvants activate caspase-1 and induce IL-1β and IL-18 release. *Journal of Immunology*, *178*(8), 5271–5276. Available from https://doi.org/10.4049/jimmunol.178.8.5271.

Li, H., Willingham, S. B., Ting, J. P. Y., & Re, F. (2008). Cutting edge: Inflammasome activation by alum and alum's adjuvant effect are mediated by NLRP3. *Journal of Immunology*, *181*(1), 17–21. Available from https://doi.org/10.4049/jimmunol.181.1.17.

Li, J.-K., Balic, J. J., Yu, L., & Jenkins, B. (2017). *TLR agonists as adjuvants for cancer vaccines. Regulation of inflammatory signaling in health and disease* (pp. 192–212). Springer.

Li, W., Deng, Y., Chu, Q., & Zhang, P. (2019). Gut microbiome and cancer immunotherapy. *Cancer Letters*, *447*, 41–47. Available from https://doi.org/10.1016/j.canlet.2019.01.015.

Liang, J. G., Su, D., Song, T. Z., Zeng, Y., Huang, W., Wu, J., Xu, R., Luo, P., Yang, X., Zhang, X., Luo, S., Liang, Y., Li, X., Huang, J., Wang, Q., Huang, X., Xu, Q., Luo, M., Huang, A., ... Liang, P. (2021). S-Trimer, a COVID-19 subunit vaccine candidate, induces protective immunity in nonhuman primates. *Nature Communications*, *12*(1). Available from https://doi.org/10.1038/s41467-021-21634-1.

Liao, L., He, Y., Li, L., Meng, H., Dong, Y., Yi, F., & Xiao, P. (2018). A preliminary review of studies on adaptogens: Comparison of their bioactivity in TCM with that of ginseng-like herbs used worldwide. *Chinese Medicine*, *13*(1). Available from https://doi.org/10.1186/s13020-018-0214-9.

Lind, P. E. (1968). Cellular localization Of 125 I-labelled *Salmonella* adelaide flagellin following injection into the rat in the presence of Freund's complete adjuvant. *Australian Journal of Experimental Biology and Medical Science*, *46*(2), 189–208. Available from https://doi.org/10.1038/icb.1968.16.

Lindblad, E. B. (2004). Aluminum adjuvants - In retrospect and prospect. *Vaccine*, *22*(27–28), 3658–3668. Available from https://doi.org/10.1016/j.vaccine.2004.03.032.

Lindblad, E. B., & Duroux, L. (2016). *Mineral adjuvants. Immunopotentiators in modern vaccines* (2nd ed., pp. 347–375). Elsevier. Available from https://doi.org/10.1016/B978-0-12-804019-5.00018-9.

Lindqvist, M., Persson, J., Thörn, K., & Harandi, A. M. (2009). The mucosal adjuvant effect of α-galactosylceramide for induction of protective immunity to sexually transmitted viral infection. *Journal of Immunology*, *182*(10), 6435–6443. Available from https://doi.org/10.4049/jimmunol.0900136.

Lofano, G., Mancini, F., Salvatore, G., Cantisani, R., Monaci, E., Carrisi, C., Tavarini, S., Sammicheli, C., Paccani, S. R., Soldaini, E., Laera, D., Finco, O., Nuti, S., Rappuoli, R., De Gregorio, E., Bagnoli, F., & Bertholet, S. (2015). Oil-in-water emulsion MF59 increases germinal center B cell differentiation and persistence in response to vaccination. *Journal of Immunology*, *195*(4), 1617–1627. Available from https://doi.org/10.4049/jimmunol.1402604.

Longet, S., Abautret-Daly, A., Davitt, C. J. H., McEntee, C. P., Aversa, V., Rosa, M., Coulter, I. S., Holmgren, J., Raghavan, S., & Lavelle, E. C. (2019). An oral alpha-galactosylceramide adjuvanted *Helicobacter pylori* vaccine induces protective IL-1R- and IL-17R-dependent Th1 responses. *npj Vaccines*, *4*(1). Available from https://doi.org/10.1038/s41541-019-0139-z.

Luabeya, A. K. K., Kagina, B. M. N., Tameris, M. D., Geldenhuys, H., Hoff, S. T., Shi, Z., Kromann, I., Hatherill, M., Mahomed, H., Hanekom, W. A., Andersen, P., Scriba, T. J., Schoeman, E., Krohn, C., Day, C. L., Africa, H., Makhethe, L., Smit, E., ... Brown, Y. (2015). First-in-human trial of the post-exposure tuberculosis vaccine H56: IC31 in *Mycobacterium tuberculosis* infected and non-infected healthy adults. *Vaccine*, *33*(33), 4130–4140. Available from https://doi.org/10.1016/j.vaccine.2015.06.051.

Luchner, M., Reinke, S., & Milicic, A. (2021). Tlr agonists as vaccine adjuvants targeting cancer and infectious diseases. *Pharmaceutics*, *13*(2), 1–16. Available from https://doi.org/10.3390/pharmaceutics13020142.

Lundgren, A., Bourgeois, L., Carlin, N., Clements, J., Gustafsson, B., Hartford, M., Holmgren, J., Petzold, M., Walker, R., & Svennerholm, A. M. (2014). Safety and immunogenicity of an improved oral inactivated multivalent enterotoxigenic *Escherichia coli* (ETEC) vaccine administered alone and together with dmLT adjuvant in a double-blind, randomized, placebo-controlled Phase I study. *Vaccine*, *32*(52), 7077–7084. Available from https://doi.org/10.1016/j.vaccine.2014.10.069.

Lynn, D. J., & Pulendran, B. (2018). The potential of the microbiota to influence vaccine responses. *Journal of Leukocyte Biology*, *103*(2), 225–231. Available from https://doi.org/10.1189/jlb.5MR0617-216R.

Ma, J., Bulger, P. A., Davis, D. V. R., Perilli-Palmer, B., Bedore, D. A., Kensil, C. R., Young, E. M., Hung, C. H., Seals, J. R., Pavia, C. S., & Coughlin, R. T. (1994). Impact of the saponin adjuvant QS-21 and aluminum hydroxide on the immunogenicity of recombinant OspA and OspB of *Borrelia burgdorferi*. *Vaccine*, *12*(10), 925–932. Available from https://doi.org/10.1016/0264-410X(94)90036-1.

MacLennan, R., Schofield, F. D., Pittman, M., Hardegree, M. C., & Barile, M. F. (1965). Immunization against neonatal tetanus in New Guinea. Antitoxin response of pregnant women to adjuvant and plain toxoids. *Bulletin of the World Health Organization*, *32*(5), 683–697.

Macy, D. W. (1997). Vaccine adjuvants. *Seminars in Veterinary Medicine and Surgery-Small Animal*, *12*(3), 206–211. Available from https://doi.org/10.1016/S1096-2867(97)80034-5.

Magwira, C. A., & Taylor, M. B. (2018). Composition of gut microbiota and its influence on the immunogenicity of oral rotavirus vaccines. *Vaccine*, *36*(24), 3427–3433. Available from https://doi.org/10.1016/j.vaccine.2018.04.091.

Mahase, E. (2021). Covid-19: Novavax vaccine efficacy is 86% against UK variant and 60% against South African variant. *BMJ (Clinical Research Edition)*, *372*, n296. Available from https://doi.org/10.1136/bmj.n296.

Mannhalter, J. W., Neychev, H. O., Zlabinger, G. J., Ahmad, R., & Eibl, M. M. (1985). Modulation of the human immune response by the non-toxic and non-pyrogenic adjuvant aluminum hydroxide: Effect on antigen uptake and antigen presentation. *Clinical and Experimental Immunology*, *61*(1), 143–151.

Marciani, D. J. (2003). Vaccine adjuvants: Role and mechanisms of action in vaccine immunogenicity. *Drug Discovery Today*, *8*(20), 934–943. Available from https://doi.org/10.1016/S1359-6446(03)02864-2.

Marciani, D. J. (2014). New Th2 adjuvants for preventive and active immunotherapy of neurodegenerative proteinopathies. *Drug Discovery Today*, *19*(7), 912–920. Available from https://doi.org/10.1016/j.drudis.2014.02.015.

Marciani, D. J. (2018). Elucidating the mechanisms of action of saponin-derived adjuvants. *Trends in Pharmacological Sciences*, *39*(6), 573–585. Available from https://doi.org/10.1016/j.tips.2018.03.005.

Marciani, D. J., Pathak, A. K., Reynolds, R. C., Seitz, L., & May, R. D. (2001). Altered immunomodulating and toxicological properties of degraded *Quillaja saponaria* Molina saponins. *International Immunopharmacology*, *1*(4), 813–818. Available from https://doi.org/10.1016/S1567-5769(01)00016-9.

Martins, K. A. O., Bavari, S., & Salazar, A. M. (2015). Vaccine adjuvant uses of poly-IC and derivatives. *Expert Review of Vaccines*, *14*(3), 447–459. Available from https://doi.org/10.1586/14760584.2015.966085.

Marty-Roix, R., Vladimer, G. I., Pouliot, K., Weng, D., Buglione-Corbett, R., West, K., MacMicking, J. D., Chee, J. D., Wang, S., Lu, S., & Lien, E. (2016). Identification of QS-21 as an inflammasome-activating molecular component of saponin adjuvants. *Journal of Biological Chemistry*, *291*(3), 1123–1136. Available from https://doi.org/10.1074/jbc.M115.683011.

McDevitt, H. O., Askonas, B. A., Humphrey, J. H., Schechter, I., & Sela, M. (1966). The localization of antigen in relation to specific antibody-producing cells. I. Use of a synthetic polypeptide [(T, G)-A–L] labelled with iodine-125. *Immunology*, *11*(4), 337–351.

McGuire, E. P., Fong, Y., Toote, C., Cunningham, C. K., McFarland, E. J., Borkowsky, W., Barnett, S., Itell, H. L., Kumar, A., Gray, G., Julianna McElrath, M., Tomaras, G. D., Permar, S. R., & Fouda, G. G. (2018). HIV-exposed infants vaccinated with an MF59/recombinant gp120 vaccine have higher magnitude anti-V1V2 IgG responses than adults immunized with the same vaccine. *Journal of Virology*, *92*(1). Available from https://doi.org/10.1128/JVI.01070-17.

Morein, B., Sundquist, B., Höglund, S., Dalsgaard, K., & Osterhaus, A. (1984). Iscom, a novel structure for antigenic presentation of membrane proteins from enveloped viruses. *Nature*, *308*(5958), 457–460. Available from https://doi.org/10.1038/308457a0.

Morel, S., Didierlaurent, A., Bourguignon, P., Delhaye, S., Baras, B., Jacob, V., Planty, C., Elouahabi, A., Harvengt, P., Carlsen, H., Kielland, A., Chomez, P., Garçon, N., & Van Mechelen, M. (2011). Adjuvant System AS03 containing α-tocopherol modulates innate immune response and leads to improved adaptive immunity. *Vaccine*, *29*(13), 2461–2473. Available from https://doi.org/10.1016/j.vaccine.2011.01.011.

Morishita, M., Takahashi, Y., Matsumoto, A., Nishikawa, M., & Takakura, Y. (2016). Exosome-based tumor antigens–adjuvant co-delivery utilizing genetically engineered tumor cell-derived exosomes with immunostimulatory CpG DNA. *Biomaterials*, *111*, 55–65. Available from https://doi.org/10.1016/j.biomaterials.2016.09.031.

Moutia, M., Habti, N., & Badou, A. (2018). In vitro and in vivo immunomodulator activities of *Allium sativum* L. *Evidence-Based Complementary and Alternative Medicine*, *2018*, 1–10. Available from https://doi.org/10.1155/2018/4984659.

Mutsch, M., Zhou, W., Rhodes, P., Bopp, M., Chen, R. T., Linder, T., Spyr, C., & Steffen, R. (2004). Use of the inactivated intranasal influenza vaccine and the risk of Bell's Palsy in Switzerland. *New England Journal of Medicine*, *350*(9), 896–903. Available from https://doi.org/10.1056/NEJMoa030595.

Naud, P. S., Roteli-Martins, C. M., De Carvalho, N. S., Teixeira, J. C., De Borba, P. C., Sanchez, N., Zahaf, T., Catteau, G., Geeraerts, B., & Descamps, D. (2014). Sustained efficacy, immunogenicity, and safety of the HPV-16/18 AS04-adjuvanted vaccine: Final analysis of a long-term follow-up study up to 9.4 years post-vaccination. *Human Vaccines and Immunotherapeutics*, *10*(8), 2147–2162. Available from https://doi.org/10.4161/hv.29532.

Newman, M. J., Wu, J. Y., Coughlin, R. T., Murphy, C. I., Seals, J. R., Wyand, M. S., & Kensil, C. R. (1992). Immunogenicity and toxicity testing of an experimental HIV-1 vaccine in nonhuman primates. *AIDS Research and Human Retroviruses*, *8*(8), 1413–1418. Available from https://doi.org/10.1089/aid.1992.8.1413.

Nikfarjam, S., Rezaie, J., Kashanchi, F., & Jafari, R. (2020). Dexosomes as a cell-free vaccine for cancer immunotherapy. *Journal of Experimental and Clinical Cancer Research*, *39*(1). Available from https://doi.org/10.1186/s13046-020-01781-x.

Oda, K., Matsuda, H., Murakami, T., Katayama, S., Ohgitani, T., & Yoshikawa, M. (2000). Adjuvant and hemolytic activities of 47 saponins derived from medicinal and food plants. *Biological Chemistry*, *381*(1). Available from https://doi.org/10.1515/bc.2000.009.

Oda, K., Matsuda, H., Murakami, T., Katayama, S., Ohgitani, T., & Yoshikawa, M. (2003). Relationship between adjuvant activity and amphipathic structure of soyasaponins. *Vaccine*, *21*(17–18), 2145–2151. Available from https://doi.org/10.1016/S0264-410X(02)00739-9.

Ogonuki, H., Hashizume, S., & Takashashi, B. (1967). Preparation and laboratory tests of oil-adjuvant cholera vaccine. *Bulletin of the World Health Organization*, *37*(5), 729–736.

Oh, J. Z., Ravindran, R., Chassaing, B., Carvalho, F. A., Maddur, M. S., Bower, M., Hakimpour, P., Gill, K. P., Nakaya, H. I., Yarovinsky, F., Sartor, R. B., Gewirtz, A. T., & Pulendran, B. (2014). TLR5-mediated sensing of gut microbiota is necessary for antibody responses to seasonal influenza vaccination. *Immunity*, *41*(3), 478–492. Available from https://doi.org/10.1016/j.immuni.2014.08.009.

O'Hagan, D. T., Ott, G. S., De Gregorio, E., & Seubert, A. (2012). The mechanism of action of MF59 - An innately attractive adjuvant formulation. *Vaccine*, *30*(29), 4341–4348. Available from https://doi.org/10.1016/j.vaccine.2011.09.061.

O'Hagan, D. T., Ott, G. S., Van Nest, G., Rappuoli, R., & Del Giudice, G. (2013). The history of MF59® adjuvant: A phoenix that arose from the ashes. *Expert Review of Vaccines*, *12* (1), 13–30. Available from https://doi.org/10.1586/erv.12.140.

Orr, M. T., Khandhar, A. P., Seydoux, E., Liang, H., Gage, E., Mikasa, T., Beebe, E. L., Rintala, N. D., Persson, K. H., Ahniyaz, A., Carter, D., Reed, S. G., & Fox, C. B. (2019). Reprogramming the adjuvant properties of aluminum oxyhydroxide with nanoparticle technology. *npj Vaccines*, *4*(1). Available from https://doi.org/10.1038/s41541-018-0094-0.

Osterhaus, A., de Vries, P., Morein, B., Akerblom, L., & Heeney, J. (1992). Comparison of protection afforded by whole virus ISCOM vs MDP adjuvanted formalin-inactivated SIV vaccines from IV cell-free or cell-associated homologous challenge. *AIDS Research and Human Retroviruses*, *8*(8), 1507–1510. Available from https://doi.org/10.1089/aid.1992.8.1507.

Osterhaus, A., Weijer, K., Uytdehaag, F., Jarrett, O., Sundquist, B., & Morein, B. (1985). Induction of protective immune response in cats by vaccination with feline leukemia virus iscom. *Journal of Immunology*, *135*(1), 591–596.

Osterhaus, A., Weijer, K., UytdeHaag, F., Knell, P., Jarrett, O., Åkerblom, L., & Morein, B. (1989). Serological responses in cats vaccinated with FeLV ISCOM and an inactivated FeLV vaccine. *Vaccine*, *7*(2), 137–141. Available from https://doi.org/10.1016/0264-410X(89)90053-4.

Paavonen, J., Naud, P., Salmerón, J., Wheeler, C. M., Chow, S. N., Apter, D., Kitchener, H., Castellsague, X., Teixeira, J. C., Skinner, S. R., Hedrick, J., Jaisamrarn, U., Limson, G., Garland, S., Szarewski, A., Romanowski, B., Aoki, F. Y., Schwarz, T. F., Poppe, W., … Dubin, G. (2009). Efficacy of human papillomavirus (HPV)-16/18 AS04-adjuvanted vaccine against cervical infection and precancer caused by oncogenic HPV types (PATRICIA): Final analysis of a double-blind, randomized study in young women. *The Lancet*, *374*(9686), 301–314. Available from https://doi.org/10.1016/S0140-6736(09)61248-4.

Paget, C., Mallevaey, T., Speak, A. O., Torres, D., Fontaine, J., Sheehan, K. C. F., Capron, M., Ryffel, B., Faveeuw, C., Leite de Moraes, M., Platt, F., & Trottein, F. (2007). Activation of invariant NKT cells by Toll-like receptor 9-stimulated dendritic cells requires type I interferon and charged glycosphingolipids. *Immunity*, *27*(4), 597–609. Available from https://doi.org/10.1016/j.immuni.2007.08.017.

Panossian, A. (2017). Understanding adaptogenic activity: Specificity of the pharmacological action of adaptogens and other phytochemicals. *Annals of the New York Academy of Sciences*, *1401*(1), 49–64. Available from https://doi.org/10.1111/nyas.13399.

Pearse, M. J., & Drane, D. (2005). ISCOMATRIX® adjuvant for antigen delivery. *Advanced Drug Delivery Reviews*, *57*(3), 465–474. Available from https://doi.org/10.1016/j.addr.2004.09.006.

Pedersen, G. K., Andersen, P., & Christensen, D. (2018). Immunocorrelates of CAF family adjuvants. *Seminars in Immunology*, *39*, 4–13. Available from https://doi.org/10.1016/j.smim.2018.10.003.

Penn-Nicholson, A., Geldenhuys, H., Burny, W., Van Der Most, R., Day, C. L., Jongert, E., Moris, P., Hatherill, M., Ofori-Anyinam, O., Hanekom, W., Bollaerts, A., Demoitie, M. A., Kany Luabeya, A. K., De Ruymaeker, E., Tameris, M., Lapierre, D., & Scriba, T. J. (2015). Safety and immunogenicity of candidate vaccine M72/AS01E in adolescents in a TB endemic setting. *Vaccine*, *33*(32), 4025–4034. Available from https://doi.org/10.1016/j.vaccine.2015.05.088.

Peres, C., Matos, A. I., Conniot, J., Sainz, V., Zupančič, E., Silva, J. M., Graça, L., Sá Gaspar, R., Préat, V., & Florindo, H. F. (2017). Poly(lactic acid)-based particulate systems are promising tools for immune modulation. *Acta Biomaterialia*, *48*, 41–57. Available from https://doi.org/10.1016/j.actbio.2016.11.012.

Perrie, Y., Kastner, E., Kaur, R., Wilkinson, A., & Ingham, A. J. (2013). A case-study investigating the physicochemical characteristics that dictate the function of a liposomal adjuvant. *Human Vaccines and Immunotherapeutics*, *9*(6), 1374–1381. Available from https://doi.org/10.4161/hv.24694.

Petrovska, L., Lopes, L., Simmons, C. P., Pizza, M., Dougan, G., & Chain, B. M. (2003). Modulation of dendritic cell endocytosis and antigen processing pathways by *Escherichia coli* heat-labile enterotoxin and mutant derivatives. *Vaccine*, *21*(13–14), 1445–1454. Available from https://doi.org/10.1016/S0264-410X(02)00696-5.

Petrovsky, N. (2015). Comparative safety of vaccine adjuvants: A summary of current evidence and future needs. *Drug Safety*, *38*(11), 1059–1074. Available from https://doi.org/10.1007/s40264-015-0350-4.

Pillet, S., Aubin, É., Trépanier, S., Poulin, J. F., Yassine-Diab, B., Ter Meulen, J., Ward, B. J., & Landry, N. (2018). Humoral and cell-mediated immune responses to H5N1 plant-made virus-like particle vaccine are differentially impacted by alum and GLA-SE adjuvants in a Phase 2 clinical trial. *npj Vaccines*, *3*(1). Available from https://doi.org/10.1038/s41541-017-0043-3.

Pitcovski, J., Bazak, Z., Wasserman, E., Elias, O., Levy, A., Peretz, T., Fingerut, E., & Frankenburg, S. (2006). Heat labile enterotoxin of *E. coli*: A potential adjuvant for transcutaneous cancer immunotherapy. *Vaccine*, *24*(5), 636–643. Available from https://doi.org/10.1016/j.vaccine.2005.08.052.

Pizza, M., Giuliani, M. M., Fontana, M. R., Monaci, E., Douce, G., Dougan, G., Mills, K. H. G., Rappuoli, R., & Del Giudice, G. (2001). Mucosal vaccines: Nontoxic derivatives of LT and CT as mucosal adjuvants. *Vaccine*, *19*(17–19), 2534–2541. Available from https://doi.org/10.1016/S0264-410X(00)00553-3.

Pollack, K. E., Meneveau, M. O., Melssen, M. M., Lynch, K. T., Koeppel, A. F., Young, S. J., Turner, S., Kumar, P., Sol-Church, K., Mauldin, I. S., & Slingluff, C. L. (2020). Incomplete

Freund's adjuvant reduces arginase and enhances th1 dominance, tlr signaling and cd40 ligand expression in the vaccine site microenvironment. *Journal for ImmunoTherapy of Cancer*, *8*(1). Available from https://doi.org/10.1136/jitc-2020-000544.

Poltorak, A., He, X., Smirnova, I., Liu, M. Y., Van Huffel, C., Du, X., Birdwell, D., Alejos, E., Silva, M., Galanos, C., Freudenberg, M., Ricciardi-Castagnoli, P., Layton, B., & Beutler, B. (1998). Defective LPS signaling in C3H/HeJ and C57BL/10ScCr mice: Mutations in Tlr4 gene. *Science (New York, N.Y.)*, *282*(5396), 2085–2088. Available from https://doi.org/10.1126/science.282.5396.2085.

Powell, M. F., Cleland, J. L., Eastman, D. J., Lim, A., Newman, M. J., Nunberg, J. H., Weissburg, R. P., Vennari, J. C., Wrin, T., & Berman, P. W. (1994). Immunogenicity and HIV-1 virus neutralization of MN recombinant glycoprotein 120/HIV-1 QS21 vaccine in baboons. *AIDS Research and Human Retroviruses*, *10*(2), S105–S108.

Press, J. B., Reynolds, R. C., May, R. D., & Marciani, D. J. (2000). *Structure/function relationships of immunostimulating saponins. Studies in natural products chemistry* (24, pp. 131–174). Elsevier. Available from https://doi.org/10.1016/S1572-5995(00)80045-9.

Pulendran, B., S. Arunachalam, P., & O'Hagan, D. T. (2021). Emerging concepts in the science of vaccine adjuvants. *Nature Reviews. Drug Discovery*, *20*(6), 454–475. Available from https://doi.org/10.1038/s41573-021-00163-y.

Putkonen, P., Björling, E., Åkerblom, L., Thorstensson, R., Lovgren, K., Benthin, L., Chiodi, F., Morein, B., Biberfeld, G., Norrby, E., & Wigzell, H. (1994). Long-standing protection of macaques against cell-free HIV-2 with a HIV-2 iscom vaccine. *Journal of Acquired Immune Deficiency Syndromes*, *7*(6), 551–559.

Qadri, F., Akhtar, M., Bhuiyan, T. R., Chowdhury, M. I., Ahmed, T., Rafique, T. A., Khan, A., Rahman, S. I. A., Khanam, F., Lundgren, A., Wiklund, G., Kaim, J., Löfstrand, M., Carlin, N., Bourgeois, A. L., Maier, N., Fix, A., Wierzba, T., Walker, R. I., ... Svennerholm, A. M. (2020). Safety and immunogenicity of the oral, inactivated, enterotoxigenic *Escherichia coli* vaccine ETVAX in Bangladeshi children and infants: A double-blind, randomized, placebo-controlled phase 1/2 trial. *The Lancet Infectious Diseases*, *20*(2), 208–219. Available from https://doi.org/10.1016/S1473-3099(19)30571-7.

Raghu, R., Sharma, D., Ramakrishnan, R., Khanam, S., Chintalwar, G. J., & Sainis, K. B. (2009). Molecular events in the activation of B cells and macrophages by a non-microbial TLR4 agonist, G1-4A from *Tinospora cordifolia*. *Immunology Letters*, *123*(1), 60–71. Available from https://doi.org/10.1016/j.imlet.2009.02.005.

Ramon, G. (1924). Sur la toxine et surranatoxine diphtheriques. *Annales de l'Institut Pasteur*, *38*, 1–10.

Ramon, G. (1925). Sur l'augmentation anormale de l'antitoxine chez les chevaux producteurs de serum antidiphterique. *Bulletin de la Société centrale de médecine vétérinaire*, *101*, 227–234.

Ratan, Z. A., Youn, S. H., Kwak, Y. S., Han, C. K., Haidere, M. F., Kim, J. K., Min, H., Jung, Y. J., Hosseinzadeh, H., Hyun, S. H., & Cho, J. Y. (2021). Adaptogenic effects of *Panax ginseng* on modulation of immune functions. *Journal of Ginseng Research*, *45*(1), 32–40. Available from https://doi.org/10.1016/j.jgr.2020.09.004.

Razim, A., Pyclik, M., Pacyga, K., Górska, S., Xu, J., Olszewski, M. A., Gamian, A., & Mic, A. (2021). Silicone oil-based nanoadjuvants as candidates for a new formulation of intranasal vaccines. *Vaccines*, *9*(3), 1–20. Available from https://doi.org/10.3390/vaccines9030234.

Reinke, S., Thakur, A., Gartlan, C., Bezbradica, J. S., & Milicic, A. (2020). Inflammasome-mediated immunogenicity of clinical and experimental vaccine adjuvants. *Vaccines*, *8*(3), 1–25. Available from https://doi.org/10.3390/vaccines8030554.

Reyna-Margarita, H. R., Irais, C. M., Mario-Alberto, R. G., Agustina, R. M., Luis-Benjamín, S. G., & David, P. E. (2019). Plant phenolics and lectins as vaccine adjuvants. *Current Pharmaceutical Biotechnology*, *20*(15), 1236–1243. Available from https://doi.org/10.2174/1389201020666190716110705.

Richmond, P., Hatchuel, L., Dong, M., Ma, B., Hu, B., Smolenov, I., Li, P., Liang, P., Han, H. H., Liang, J., & Clemens, R. (2021). Safety and immunogenicity of S-Trimer (SCB-2019), a protein subunit vaccine candidate for COVID-19 in healthy adults: A phase 1, randomized, double-blind, placebo-controlled trial. *The Lancet*, *397*(10275), 682–694. Available from https://doi.org/10.1016/S0140-6736(21)00241-5.

Rimmelzwaan, G. F., Claas, E. C. J., Van Amerongen, G., De Jong, J. C., & Osterhaus, A. D. M. E. (1999). ISCOM vaccine induced protection against a lethal challenge with a human H5N1 influenza virus. *Vaccine*, *17*(11–12), 1355–1358. Available from https://doi.org/10.1016/S0264-410X(98)00390-9.

Rivera, E., Daggfeldt, A., & Hu, S. (2003). Ginseng extract in aluminum hydroxide adjuvanted vaccines improves the antibody response of pigs to porcine parvovirus and *Erysipelothrix rhusiopathiae*. *Veterinary Immunology and Immunopathology*, *91*(1), 19–27. Available from https://doi.org/10.1016/S0165-2427(02)00269-6.

Rivera, E., Hu, S., & Concha, C. (2003). Ginseng and aluminum hydroxide act synergistically as vaccine adjuvants. *Vaccine*, *21*(11–12), 1149–1157. Available from https://doi.org/10.1016/S0264-410X(02)00518-2.

Rose, F., Wern, J. E., Gavins, F., Andersen, P., Follmann, F., & Foged, C. (2018). A strong adjuvant based on glycol-chitosan-coated lipid-polymer hybrid nanoparticles potentiates mucosal immune responses against the recombinant *Chlamydia trachomatis* fusion antigen CTH522. *Journal of Controlled Release*, *271*, 88–97. Available from https://doi.org/10.1016/j.jconrel.2017.12.003.

Ryan, E. J., McNeela, E., Pizza, M., Rappuoli, R., O'Neill, L., & Mills, K. H. G. (2000). Modulation of innate and acquired immune responses by *Escherichia coli* heat-labile toxin: Distinct pro- and anti-inflammatory effects of the nontoxic AB complex and the enzyme activity. *Journal of Immunology*, *165*(10), 5750–5759. Available from https://doi.org/10.4049/jimmunol.165.10.5750.

Saade, F., Honda-Okubo, Y., Trec, S., & Petrovsky, N. (2013). A novel hepatitis B vaccine containing Advax^{TM}, a polysaccharide adjuvant derived from delta inulin, induces robust humoral and cellular immunity with minimal reactogenicity in preclinical testing. *Vaccine*, *31*(15), 1999–2007. Available from https://doi.org/10.1016/j.vaccine.2012.12.077.

Sabbatini, P., Tsuji, T., Ferran, L., Ritter, E., Sedrak, C., Tuballes, K., Jungbluth, A. A., Ritter, G., Aghajanian, C., Bell-McGuinn, K., Hensley, M. L., Konner, J., Tew, W., Spriggs, D. R., Hoffman, E. W., Venhaus, R., Pan, L., Salazar, A. M., Diefenbach, C. M., ... Gnjatic, S. (2012). Phase I trial of overlapping long peptides from a tumor self-antigen and poly-ICLC shows rapid induction of integrated immune response in ovarian cancer patients. *Clinical Cancer Research*, *18*(23), 6497–6508. Available from https://doi.org/10.1158/1078-0432.CCR-12-2189.

Salio, M., Speak, A. O., Shepherd, D., Polzella, P., Illarionov, P. A., Veerapen, N., Besra, G. S., Platt, F. M., & Cerundolo, V. (2007). Modulation of human natural killer T cell ligands on TLR-mediated antigen-presenting cell activation. *Proceedings of the National Academy of Sciences of the United States of America*, *104*(51), 20490–20495. Available from https://doi.org/10.1073/pnas.0710145104.

Salk, J. E., Contakos, M., Laurent, A. M., Sorensen, M., Rapalski, A. J., Simmons, I. H., & Sandberg, H. (1953). Use of adjuvants in studies on influenza immunization: 3. Degree of

TABLE 10.1 (Continued)

	• Yeast, most commonly *Pichia pastoris,* and *Saccharomyces cerevisiae* • Plants (a new approach), typically *Nicotiana benthamiana* using a technique known as agroinfiltration, using recombinant *Agrobacterium tumefaciens* • Insect cells, for example *D melanogaster* Schneider-2 cells, or using baculovirus infection of lines including Sf9 and Sf21 • Mammalian cells, most commonly HEK293 (human in origin) and CHO (Chinese hamster ovary) For ease of purification, a subunit vaccine is usually secreted into the medium. This is not always possible, for example in the manufacture of Hepatitis B Surface Antigen VLPs, which assemble inside cells.Expression can be constitutive if the vaccine subunit is not toxic to the expression system. If it is toxic the subunit can be expressed following induction for a short period during the USP, for example using a methanol-inducible promoter in *P. Pastoris* It should be noted that methanol induction may not be suitable for large-scale manufacture, due to the heat produced during production. ***Which expression system to choose?*** • Research work should investigate several systems to establish which can produce a correctly folded, stable, immunogenic vaccine. Relative yields should be compared as these have a high impact on cost of manufacture. • If glycosylation is important for potency/ stability (as it is for monoclonal antibodies) then mammalian cells are the logical choice. • Microbial/fungal systems tend to be cheaper, partly because QC costs are much lower – for example, they cannot support the replication of viruses which infect humans.
Conjugate vaccines	Conjugate vaccines consist of strain-specific bacterial polysaccharide chemically conjugated to a carrier protein (expressed and purified separately prior to conjugation), for example tetanus toxoid.Polysaccharides have previously been isolated from bacteria grown in fermenters, leading to issues with bacterial cell wall contaminants and product heterogeneity. More

(*Continued*)

TABLE 10.1 (Continued)

	recently, oligosaccharides have been chemically synthesised, and linked to synthetically made carrier peptides.
RNA	There are 2 types of RNA vaccine: mRNA and saRNA (self-amplifying). Production may involve the following steps: • Produce DNA plasmid template (typically in *E. coli,* requiring fermentation to amplify the DNA); • Produce RNA copies in cell-free system using *in vitro* transcription e.g. with T7 purified RNA polymerase; • Remove DNA and purify RNA; • Formulate e.g. in liposomes for effective delivery *in vivo*
DNA	The manufacturing process mimics the first part of the RNA vaccine production process (see above)
Peptides (typically for cancer vaccines, in combination with T-cell inducing adjuvant)	There are 4 possible production approaches listed below, although most peptide vaccines are made by chemical (usually solid-phase) synthesis as this has become a rapid, cost effective approach. • Solid-phase synthesis • Liquid-phase synthesis • Enzymatic synthesis • Recombinant production (expression from engineered cell line)

Particle Vaccines (VLP manufacture) and Chapter 7: Approaches to Cancer Vaccination (manufacture of RNA vaccines).

Some examples of approaches to vaccine manufacture are given in the following publications:

- Vaccine production in yeast (Bill, 2015)
- Synthetic conjugate vaccines (bacterial polysaccharides) (Mettu et al., 2020)
- RNA vaccines (Kis et al., 2019; Schlake et al., 2012)
- A soluble protein malaria vaccine candidate manufactured in insect cells (Jing et al., 2018)
- Adenovirus vector vaccine manufacture (Fedosyuk et al., 2019)
- Attenuated whole virus vaccine manufacture, with Rabivax as an example (Kulkarni et al., 2017)

- GMP manufacture in plants (Kopertekh & Schiemann, 2019; Lamprecht et al., 2016)

Vaccine purification

Vaccine purification or the downstream process (DSP) makes use of the vaccine's size, charge, and sometimes binding affinity, compared to the characteristics of contaminants present at the end of the USP. Examples of purification strategies include the following:

- Separation based on size: centrifugation (including density gradient), size exclusion chromatography, and filtration;
- Separation based on charge or hydrophobicity: anion/cation exchange chromatography and hydrophobic interaction chromatography;
- Separation based on affinity: Monoclonal antibodies bind to Protein A and Protein G via the Fc component, and affinity tags can be added to a subunit vaccine which bind to affinity reagents, for example, an antibody/nanobody;
- Specific steps can be used to remove contaminants if required, for example, benzonase can be used to digest DNA arising from lysed cells, and for some product types (e.g., recombinant proteins manufactured using a mammalian cell bank) specific steps need to be included in the manufacturing process to inactivate/remove potential viral contaminants, such as the use of a low pH hold step or nano-filtration.

Following completion of purification and typically after a single sterile filtration step, the vaccine is known as the "Drug Substance" (DS or active pharmaceutical ingredient). After final formulation, final sterile filtration, and filling into vials/syringes, the vaccine becomes a "Drug Product" (DP). It is common to carry out formulation studies during process development to optimize vaccine stability, and to decide on the intended long-term storage temperature of the DP. All buffer components and excipients should be GMP grade where available, and permitted for administration into humans—there is a searchable FDA list of acceptable inactive ingredients in drug formulations (FDA approved ingredients, 2021) Fig. 10.2 shows the GMP process from starting materials, through the USP and DSP to produce the DS, ending will fill and finish, to produce a DP that is ready for deployment in the clinic. Fig. 10.3 outlines post-GMP activities, including QC, stability, and in vivo toxicity testing, which must be completed before regulatory permission is sought for a clinical trial.

In process controls and final release testing

An integral part of GMP manufacture is QC testing during the process, and of the final vaccine product, to ensure that it is safe to administer to humans

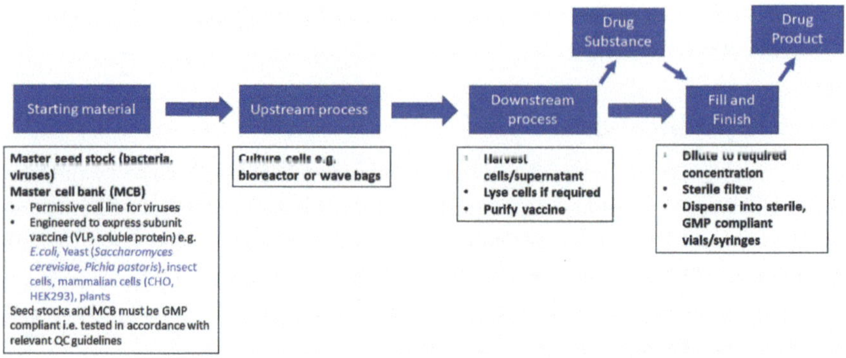

FIGURE 10.2 The GMP process. [This diagram indicates that Master stocks (of virus, bacteria, or cell banks) are used directly in GMP manufacture: this may be the case, but it is also possible that working stocks/cell banks are prepared from the Master stocks as an intermediate].

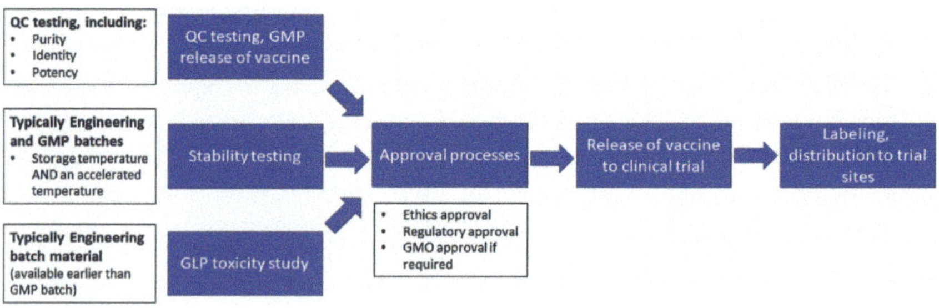

FIGURE 10.3 Postmanufacturing activities.

and has the anticipated biological activity. "Safe" means below defined levels of impurities that could cause toxicity, for contaminants including pathogens, endotoxin, DNA, and chemicals. Each GMP manufacturing process will require in process controls and product specifications for the MCB, DS, and DP. The DP is also tested periodically to ensure that it is stable, and assigned an expiry date based on ongoing stability data from representative engineering batch(es). QC tests are carried out by the CDMO or outsourced to specialist testing organizations. CDMOs have quality assurance departments to ensure adherence to GMP regulations, including thorough documentation of testing and any deviations from expected results including out of specification results.

Product specifications are set in advance of manufacture, and consist of a set of QC tests to establish product identity, purity, and potency, together with acceptance criteria. Required tests and acceptable limits for contaminants including endotoxin are defined in regulatory guidelines, and acceptance criteria are established during process development and testing of the engineering batch. Each vaccine will require a unique specification, which depends on the type of vaccine and the manufacturing process used, in particular the cells used to produce/express the vaccine. It is essential for vaccine developers to seek the advice of an expert in finalizing product specifications, for example a QP (in Europe). If there is any doubt as to whether a particular test or acceptance criterion will be permissible, regulatory authorities can often be approached ahead of GMP manufacture for scientific advice.

Any vaccine manufacturing process which utilizes a cell line capable of supporting replication of viruses pathogenic to humans (including mammalian, avian, and insect cell lines) must be tested in viral clearance studies. These studies test the ability of isolated purification steps from the DSP to remove contaminating viruses. Known amounts of model viruses are spiked into purification intermediates, and virus recovery measured following each purification step. The total log reduction in virus for the purification steps tested is compared to the theoretical value estimated to be present in the MCB.

Tables 10.2 and 10.3 are illustrative examples of QC tests required at different stages of GMP manufacture. Product-specific specifications must in all cases be developed in collaboration with qualified individuals, and in accordance with the relevant regulatory guidelines. The QC tests described are suitable for early-phase vaccine products in clinical development: later phase/licensed vaccines are likely to require additional characterization. QC tests need to be qualified for early stage trials, or validated, for later stages in development. During GMP manufacture of viruses (killed/attenuated/viral vectors), Master and Working Viral Seed Stocks (if used) would undergo QC testing similar to that of the DS. RNA is becoming a more commonly used vaccination strategy in the wake of Covid-19, and regulatory guidance for this type of vaccine is being developed as a consequence (WHO guidelines on QC testing of RNA vaccines, 2020)

TABLE 10.2 Examples of Master Cell Bank (MCB) specifications.

Type of test	Examples	Possible specification
Identity/ Characterisation (product)	• Detection of product (for example Western blot, ELISA) • Detection of product DNA, by PCR and/or sequencing	Positive for product
Identity (host cell line)	• Random Amplified Polymorphic DNA (RAPD) Assay • Species-specific PCR (for species matching cell bank and others)	Specific for relevant species
Purity and Safety: Sterility	Growth in broths supporting bacterial/fungal growth, typically for 14 days	Sterile

(Continued)

TABLE 10.2 (Continued)

Purity and Safety: Adventitious agents	Mycoplasma: by PCR, *in vitro* indicator cells or Agar and Broth	Negative
	Spiroplasma: typically by PCR	Negative
	For vaccines made in mammalian/avian/ insect cells: • Toxicity *in vitro* (co-cultivation with relevant cell lines) • Toxicity *in vivo* e.g. embryonated eggs, suckling mice, guinea pigs • PCR tests for specific viruses e.g. porcine/ bovine viruses if the starting material was exposed to trypsin and/or bovine serum, and viruses capable of infecting host cell line • F-PERT or Q-PERT (**F**luorescent or **Q**uantitative product-enhanced reverse transcriptase) assay for reverse transcriptase activity, either in MCB or in secondary detector cell line following incubation with culture supernatant • Transmission electron microscopy to look for viral particles, bacteria, fungi – usually of 200 cells	Negative

TABLE 10.3 Examples of drug substance and drug product specifications.

Type of test	Examples	Possible specification
Identity (product)	Western blot (subunit vaccine) DNA sequence (live vaccine) Sequence of RT-PCR product (RNA vaccine)	Positive for vaccine; no mutations in DNA/RNA sequence
Purity	SDS-PAGE HPLC-SEC Electron microscopy (VLP)	Depends on vaccine: >90% for soluble protein
Extraneous agents:Sterility	Growth in broths supporting bacterial/fungal growth, typically for 14 days	Sterile (no growth)
Extraneous agents:Endotoxin (LPS, or pyrogen)	Usually detected with Limulus amebocyte lysate (LAL) reagent, an extract of blood cells (amoebocytes) from the Atlantic horseshoe crab Limulus polyphemus	Depends on vaccine, and on dose. Typically <10EU/ml

(Continued)

TABLE 10.3 (Continued)

Type of test	Examples	Possible specification
	which binds to endotoxin. There are 3 possible readouts, a) clotting (non-quantitative), b) turbidity, and c) colorimetric, in which the enzymatic reaction cause a yellow colour to develop upon cleavage of the chromophore p-nitroaniline. The endotoxin level is determined by comparison with a standard curve.	
Extraneous (adventitious) agents	Mycoplasma: by PCR, *in vitro* indicator cells or Agar and Broth	Negative
	Spiroplasma: typically by PCR	Negative
	For vaccines made in mammalian/avian/insect cells, to detect contaminating viruses: • Toxicity *in vitro* (co-cultivation with relevant cell lines) • Toxicity *in vivo* e.g. embryonated eggs, suckling mice, guinea pigs • PCR tests for specific viruses • Q-PERT assay to detect retroviruses (by measuring reverse transcriptase activity) • Replication Competent Assay (RCA) for viral vector vaccines (which are replication incompetent) For vaccines that are themselves viruses, the vaccine product can be removed using, for example, a specific antibody to permit testing for contaminating pathogens	Negative
Quantification: Concentration	• Protein concentration for subunit vaccine, measured by UV light absorbance	Within predetermined limits, vaccine dependent

(*Continued*)

TABLE 10.3 (Continued)

Type of test	Examples	Possible specification
	(OD280) or BSA (colorimetric assay using Bradford reagent, compared to standard curve); • Viral particles/ml for whole virus vaccine e.g. measured by DNA/protein content • *In vitro* infectivity titre (live virus)	
Quantification: Potency	Product specific. Examples: • ELISA assay with antibody recognising a neutralising epitope; • Immunogenicity *in vivo;* • *In vitro* infectivity titre (live virus) • CFU (colony forming units) – live bacteria	Above predetermined limit, vaccine dependent
Process residuals (contaminants)	Host cell DNA by quantitative PCR	≤10ng per dose
	Host cell protein, usually by ELISA assay	Typically <10ng/ml
	Specific cell line contaminant: depends on contaminant e.g. specific proteins detected by W blot/ELISA	Vaccine dependent
	Specific process contaminant e.g. detergent, resin used in chromatography or DNA template/enzymes for RNA vaccine. Assay is contaminant-specific.	Vaccine dependent
Characterisation: Appearance	Physical inspection: Colour, opacity	Product dependent
Extraneous agents:Visible particles	Physical inspection	Free of visible partilces
Extraneous agents:Sub-visible particles	A variety of techniques exist including DLS (dynamic light scattering), and microscopy	< 6000 / 10μm < 600 / 25μm

(*Continued*)

TABLE 10.3 (Continued)

Type of test	Examples	Possible specification
Characterisation: pH	pH meter	Product dependent, within specified range. Ideally close to pH7.4 (physiological pH)
Characterisation: Osmolality	Measured using an osmometer - either a freezing point depression or a vapour pressure depression osmometer	Product dependent, within specified range

References

Annex 2: WHO good manufacturing practices for biological products. (2016). <https://www.who.int/biologicals/areas/vaccines/Annex_2_WHO_Good_manufacturing_practices_for_biological_products.pdf?ua = 1>.

Bill, R. M. (2015). Recombinant protein subunit vaccine synthesis in microbes: A role for yeast? *Journal of Pharmacy and Pharmacology*, *67*(3), 319–328. Available from https://doi.org/10.1111/jphp.12353.

EudraLex – Volume 4 – GMP guidelines. (2017). <https://ec.europa.eu/health/documents/eudralex/vol-4_en>.

European pharmacopoeia, 10th edition. (2019). <https://www.edqm.eu/en/european-pharmacopoeia-ph-eur-10th-edition>.

FDA approved ingredients. (2021). <https://www.accessdata.fda.gov/scripts/cder/iig/index.cfm>.

Fedosyuk, S., Merritt, T., Peralta-Alvarez, M. P., Morris, S. J., Lam, A., Laroudie, N., Kangokar, A., Wright, D., Warimwe, G. M., Angell-Manning, P., Ritchie, A. J., Gilbert, S. C., Xenopoulos, A., Boumlic, A., & Douglas, A. D. (2019). Simian adenovirus vector production for early-phase clinical trials: A simple method applicable to multiple serotypes and using entirely disposable product-contact components. *Vaccine*, *37*(47), 6951–6961. Available from https://doi.org/10.1016/j.vaccine.2019.04.056.

ICH guidelines. (2021). <https://www.ich.org/page/ich-guidelines>.

International pharmacopoeia. (2020). <https://apps.who.int/phint/en/p/docf/>.

Jing, J., Richard, D. T., Emma, J. B., Philip, A.-M., Max, S., David, J. P., Pawan, D., Silk, E. S., Jennifer, M. M., Rebecca, A. D., Fay, L. N., Jordan, R. B., Kathryn, A. H., Lars, P., Thomas, J., Tanja, B., Ioana, N. B., Harracho, M. P., Abdessamad, T.-A., … Simon, J. , D. (2018). Production, quality control, stability, and potency of cGMP-produced *Plasmodium falciparum* RH50.1 protein vaccine expressed in *Drosophila* S2 cells. *NPJ Vaccines*. Available from https://doi.org/10.1038/s41541-018-0071-7.

Kis, Z., Shattock, R., Shah, N., & Kontoravdi, C. (2019). Emerging technologies for low-cost, rapid vaccine manufacture. *Biotechnology Journal*, *14*(1), e1800376.

Kopertekh, L., & Schiemann, J. (2019). Transient production of recombinant pharmaceutical proteins in plants: Evolution and perspectives. *Current Medicinal Chemistry*, *26*(3), 365–380. Available from https://doi.org/10.2174/0929867324666170718114724.

Kulkarni, P. S., Sahai, A., Gunale, B., & Dhere, R. M. (2017). Development of a new purified vero cell rabies vaccine (Rabivax-S) at the Serum Institute of India Pvt Ltd. *Expert Review of Vaccines*, *16*(4), 303–311. Available from https://doi.org/10.1080/14760584.2017.1294068.

Lamprecht, R. L., Kennedy, P., Huddy, S. M., Bethke, S., Hendrikse, M., Hitzeroth, I. I., & Rybicki, E. P. (2016). Production of human papillomavirus pseudovirions in plants and their use in pseudovirion-based neutralisation assays in mammalian cells. *Scientific Reports*, *6*. Available from https://doi.org/10.1038/srep20431.

Mettu, R., Chen, C.-Y., & Wu, C.-Y. (2020). Synthetic carbohydrate-based vaccines: Challenges and opportunities. *Journal of Biomedical Science*, *27*(1), 9.

Schlake, T., Thess, A., Fotin-Mleczek, M., & Kallen, K.-J. (2012). Developing mRNA-vaccine technologies. *RNA Biology*, *9*(11), 1319–1330.

WHO guidelines on QC testing of RNA vaccines. (2020). <https://www.who.int/docs/default-source/biologicals/ecbs/reg-considerations-on-rna-vaccines_1st-draft_pc_tz_22122020.pdf?sfvrsn = c13e1e20_3>.

Chapter 11

Testing vaccines: essential steps in clinical trial design

Julia L. Marshall
The University of Oxford, Oxford, United Kingdom

Introduction

A vaccine clinical trial is a research study in humans which aims to investigate or confirm clinical, pharmacological, or immunological effects of a vaccine to establish the vaccine's safety and/or efficacy. This chapter outlines the types of vaccine trials and the processes for delivering a trial including the principles and application of good clinical practice (GCP), trial design, documentation, regulatory requirements, and safety considerations. Assessing vaccine efficacy is covered in Chapter 4, Vaccine development: from the laboratory to the field.

Vaccine development pathway

Once a promising vaccine candidate has been chosen from preclinical studies, a vaccine goes through several phases of clinical trials in humans before and after licensure (see Fig. 11.1). While preclinical testing is a crucial step in vaccine development, human clinical trials are crucial. When tested in humans, promising preclinical vaccine candidates may not be protective or may cause unacceptable adverse event (AE) profiles.

Fig. 11.1 outlines the phases of vaccine development. The primary objective of Phase I studies are safety and dose finding. These studies are usually the first time the vaccine has been given to humans and careful monitoring starting with low doses, is required. Phase II trials aim to establish the effect of the vaccine on the body by assessing protective immune markers. They are also designed to detect any responses against the vaccine product itself (e.g., vaccine vector antibodies) (Buttner & Muller, 2011; Medicines and Healthcare Products Regulatory Agency [MHRA], 2012). A "first in human" trial might be designed to establish initial safety and dose while also detecting immune responses, and hence might be classified as a combined Phase I/II

Vaccinology and Methods in Vaccine Research. DOI: https://doi.org/10.1016/B978-0-323-91146-7.00002-0

Licensing ↓ (between Phase III and Phase IV)

Preclinical	Phase I	Phase II	Phase III	Phase IV
Animal models provide information on dose, safety and efficacy	First in human studies. Safety and dose finding. Careful dose escalation.	Dose and pilot efficacy. End point immune markers or clinical disease (e.g., challenge studies). Continue adverse event (AE) monitoring.	Extensive efficacy clinical trial. Naturally acquired infection or possibly CHIM trial. Confirm dosing, ongoing AE assessment. Noninferiority trials to current vaccine (if applicable)	Post licensure. To detect rare or long-term adverse events. Compare side effects of different licensed vaccines. Effectiveness in different population groups.
	3–80 healthy volunteers	20–200 healthy volunteers or volunteers exposed to pathogen	100 to several thousand volunteers exposed to pathogen	Thousands
	Case series. Usually no randomization or blinding	Case control or blinded randomized control trials.	Large-scale blinded randomized control trials, often multicenter.	Observational (post marketing surveillance) or randomized control trials

FIGURE 11.1 Vaccine development pathway. Phases of vaccine development.

trial. Trial phases can be further subdivided into (a) and (b). For instance, a Phase IIa trial is an earlier phase proof of concept trial perhaps investigating several doses, whereas Phase IIb is a later stage trial which might focus more on efficacy of an established dose (Buttner & Muller, 2011; MHRA, 2012).

Phase III trials are large trials which aim to confirm vaccine efficacy by determining if the vaccine prevents natural infection. Depending on the incidence of the disease in the community, Phase III vaccine trials can be costly and time consuming. Challenge trials (controlled human infection models, CHIM) can be a useful method to detect an efficacy signal at an early stage of vaccine development (Gordon et al., 2017). CHIM trials, if carefully designed to protect the safety of the volunteer, are a powerful tool to fast track vaccine development, bypassing the need for costly time consuming Phase II and Phase III field studies, especially where the incidence of naturally acquired infection is low (WHO, 2016).

Ensuring the trial is designed correctly to answer the specific question or objective is essential. Examples of appropriate trials designs for each phase are shown in Fig. 11.1 (Buttner & Muller, 2011). Except for some Phase IV studies, all clinical vaccine studies are interventional experimental trials, using an Investigational Medical Product (IMP) and recording outcome. Randomized double-blinded control trials are the gold standard for Phase III efficacy trials. Variations on this design can include cross-over trials, ring vaccination trials during epidemics such as Ebola (Henao-Restrepo, 2015), or other cluster randomized control designs, to assess herd immunity in communities (Loeb et al., 2010).

Introduction to good clinical practice

GCP is a set of internationally agreed scientific and ethical principles that determine the framework for conducting trials in humans. It is designed to ensure the safety and wellbeing of trial participants and to protect their rights. A human trial by its very nature is conducted to answer a question, the outcome of which is unknown. Due to these uncertainties, the health of participants may be put at risk. Clinical trials are an accepted necessary part of medication and vaccine development, but the overall potential benefit for society needs to be carefully balanced against the risk to the individual. In the past, research has been conducted without consent and sometimes under duress (e.g., Nuremburg trials; Tuskegee syphilis study). As a direct result of the unethical practices carried out by medical professionals under the Nazi regime during World War II, the Nuremberg Code was established in 1947. This was the first international code of practice, highlighting the requirement for informed consent. Following this, in 1964, the World Medical Association developed a set of ethical principles around medical research, "The Declaration of Helsinki" (World Medical Association, 2013). A revised version of this declaration makes up the underlying principles of international guidance around correct Clinical Trial conduct and most countries legally require adherence to these principles.

The primary tenants of the declaration are minimizing risk to the participant; safeguarding participants; informed consent; adhering to an approved research protocol; and ensuring data integrity (Council for International Organizations of Medical Sciences [CIOMS], 2016).

The WHO provides international guidelines on the ethical running of human trials, based on internationally accepted ethical principles of beneficence, justice, and autonomy. At time of publication, the current guidelines are the 2016 International ethical guidelines for health-related research involving humans, published by The Council for International Organizations of Medical Sciences (CIOMS), in association with the WHO, which is available for free download (CIOMS, 2016). The CIOMS also provides GCP training, which is available at the same website. For most countries, it is a legal requirement to seek approval from a local ethics committee before trial commencement and proof of this is usually a requirement for publication. The ethics committee's role is to ensure that the trial protocol adheres to these internationally accepted ethical principles. In addition, if the trial is testing a new medication or vaccine planned for human use, most countries require prior approval from their national medicines regulatory authority. There may also be other approvals required before starting the trial, such as approval for collection and storage of human tissue or ensuring compliance with data protection law. It is important to check all local legal requirements during trial planning stages.

Consent

Enrolling volunteers who have freely and willingly provided fully informed consent to participate in the trial is a cornerstone of GCP. Several criteria need to be met to ensure integrity of this process:

- Written information (participant information sheet, PIS) that is both comprehensible to the volunteer and comprehensive, that is, covers all trial procedures and risks.
    - Further explanation or translation of documents should be provided as required.
- Sufficient time to read the information and consider any implications of participation.
    - Often a minimum period of at least 48 hours is stipulated between providing the volunteer with trial information and asking them to sign the consent.
- The PIS and all other participant facing documents must have been reviewed and approved by an ethics committee to ensure the information is accurate, understandable, and not coercive.
- The trial has been verbally explained to the participant and an opportunity has been given to address any questions raised.
- The study staff undertaking the consent process must be appropriately qualified.
    - Staff must be trained in GCP, receive specific trial training, and have sufficient clinical knowledge to address questions which may arise.
- Study staff must check volunteer understanding and assess volunteer capacity to consent.
    - Consider a written questionnaire to check comprehension of the trial.
    - If the volunteer does not have capacity (i.e., if they are a minor or do not have mental capacity), a guardian may consent on their behalf but this procedure must be compliant with national law.
- Written consent must be obtained.
    - Consider the need for a witness to co-sign consent if literacy may be an issue.
    - A copy is usually kept and filed as part of study documentation and a second copy given to the volunteer.
- Ensure the volunteer is aware that they are free to withdraw their consent at any point in the trial.
    - If early withdrawal may cause safety issues for the participant, this must be explained and an appropriate contingency plan included in the protocol design to protect participant safety without the need for further research procedures.
- Continued consent should with confirmed with the volunteer throughout the trial process.

Good clinical practice compliant documentation

As a requirement of GCP, the trial must have a GCP-compliant protocol. This document outlines in detail all aspects of how the trial is to be run. See Table 11.1 for common sections contained in this document. An important

TABLE 11.1 Example sections of a trial protocol.

Common protocol sections	Description
General information	Includes study title, authorization numbers from regulatory bodies and ethics committees, sponsor, study staff and roles, protocol number, and date.
Background information	Including scientific justification for the trial.
Trial objective(s)	Often divided into primary, secondary, and other/ exploratory. Each objective should be accompanied by an endpoint including how and when this endpoint will be measured. The primary objective should reflect the research question. Its endpoint with outcome measures is required for regulatory oversight and determines overall trial conclusions.
Trial design	Overview of planned trial visits (with diagram), procedures that will occur at each visit, volunteer and staff blinding, criteria for dose escalation (if applicable).
Selection of participants	Inclusion and exclusion criteria of participants, withdrawal criteria.
Trial treatments (e.g., vaccine, placebo)	Needs to include information about dosing, preparation, GMP manufacture, receipting, storage, and destruction.
Assessment of safety	Solicited adverse events, safety holding rules, data safety monitoring committee oversight.
Assessment of efficacy	Often defined in primary objectives and outcome measures. Might include clinical (i.e., developing target disease or infection) and/or laboratory assays.
Statistics	Justification of sample size and power. Explanation of any planned interim analyses. Planned inclusion/ exclusion of data from withdrawn participants. Randomization plan.
Data management	Data handling, record keeping, volunteer confidentiality, e.g., outline what will be "source data," i.e., primary data source, length of time data is kept and how it is kept secure.
Financing and insurance	Proof that the trial is adequately financed to carry out all planned procedures outlined in the protocol. Indemnity insurance and insurance for participants in case of medical need.
Publication policy	Outline plans for publication including that consent will be sought from participants for publication of deidentified ("pseudonymized") data
Appendices	

aspect of running a trial is ensuring that all logistical aspects are well thought through as this helps to protect participant safety and enables collection of high-quality data, the two tenants of GCP. A detailed protocol and related documents such as standard operating procedures (SOPs) are therefore key parts of trial planning.

Participant facing documents

Every written piece of information about the trial that participants will receive needs to be approved by the ethics committee before use. These documents may include the PIS, consent form, and any advertising materials required to aid trial recruitment (including posters or radio text).

Participant information sheet

PIS provides all information about the planned trial needed for the volunteer to make an informed decision about their participation in the trial. The language used must be at a level that could be understood by the target audience. For instance, in the United Kingdom, this might be at a level understandable by an average adult who has obtained formal education to age 15. If necessary, the PIS should be provided in a range of languages.

Consent form

This must reference the PIS as this outlines what the volunteer is consenting to. In addition, it is good practice to reference all pertinent research procedures that the volunteer is consenting to as part of the trial. The consent form might also include other aspects of the trial such as consent to store data and research samples.

Other documents

These might include a letter to the participant's family doctor to verify medical information (to confirm volunteer eligibility and safety) and to advise the doctor of medical procedures that the volunteer has undergone. Other required documents include case report forms, contracts, and SOPs.

Case report forms

These are the forms used to collect trial data and are either paper or electronic. If information is collected directly onto these forms, then the case report form is the "source data." Other source data might also be used to input into the case report forms, for example, blood test results. In that case, the source data are the laboratory report. The origin of all source data needs to be outlined in the protocol (see Table 11.1). These case report forms need to be carefully designed to maximize the quality of data collected, i.e., it is good practice to ensure clear precise questions and minimize the variability of possible answers while still ensuring accurate data collection.

Contracts

These are situation specific but usually include agreements for site use; financial arrangements; contracts with local laboratories or other service providers; legal compliance; material transfer agreements (if research samples are being processed at another institution); confidentiality agreements; and delegation of responsibility (see "Clinical trial roles and responsibilities" section).

Standard operating procedures

These are often site-specific documents that outline in detail how particular procedures described in the protocol are to be carried out. This is to ensure research procedures are consistent between members of staff and between participants.

Clinical trial roles and responsibilities

Every trial has a sponsor who has overall responsibility for the trial. For vaccine trials, this could be the pharmaceutical company or a research institute such as a University. Noncommercial institutes often delegate responsibility for running the trial to a Chief Investigator, though they usually retain responsibility for financing, safety reporting, and legal aspects of the trial. For academic institutes, the Chief Investigator is usually the person who has conceived the research question, designed the trial, sourced the grant funding, and then sought Sponsor approval to run their trial. Depending on the size of the trial, the Chief Investigator can then delegate responsibility to specific research staff. If the trial is running across multiple sites, the Chief Investigator may delegate responsibility to a Principal Investigator who will then lead the trial at their site. Assigned responsibilities are outlined in the delegation log. This is filed with training records, including GCP training, and evidence of the relevant qualifications, to prove each delegated staff member is qualified for the role they have been assigned. Staff who have not had formal trial-specific training nor been formally delegated to a role within the trial are not able to perform any trial procedures on the volunteer.

Ensuring quality

Collecting quality data that accurately reflect the effect of the vaccine on the trial population is an important requirement of GCP. To ensure research quality, there are several integral requirements. These include a well-designed protocol, appropriately trained staff, using SOPs, and accurate data collection and recording; as described earlier in this chapter. To ensure compliance, trials are subject to regular monitoring and may be audited by the regulatory authorities. Any breach of protocol must be recorded and reported, and processes put in places to avoid a repeat incident. If the breach is assessed as being a serious breach of protocol or GCP, the regulatory authority will mandate urgent reporting and may advise pausing the trial. Depending on the regulatory authority, a serious breach may be defined as something which puts the safety of participants or the scientific integrity of the trial at significant risk (MHRA, 2004).

Data safety monitoring committee

Vaccine trials often have a data safety monitoring committee (DSMB) that is made up of a panel of specialists who are independent to the trial. The DSMB may also be called an independent data monitoring committee. A DSMB committee may be a compulsory regulatory requirement. The committee is tasked with ensuring real-time quality, compliance, and safety of the trial.

DSMB committees usually meet at regular preagreed intervals to review safety data, and in addition may be called upon for advice between meetings for any clinical concerns. Depending on their preagreed role, they may give advice on pausing the trial, advising the regulatory authorities, adding in additional safety assessments, or confirming escalation to the next planned vaccine dose in a trial.

Trial steering committee

Some trials also have a trial steering committee (TSC) that is made up of members of the research team and external advisors. This committee is tasked with overseeing the general progress of the trial, thus helping to ensure data integrity. Creation of a TSC may be particularly helpful if the vaccine trial is of a complex design requiring input from a variety of specialists, or if it is carried out over multiple sites, especially if international.

Safety

Screening

Following consent, all potential participants must undergo screening to ensure that volunteers recruited into the trial meet the protocol's inclusion and exclusion criteria. The inclusion and exclusion criteria should be chosen on the basis of the following factors:

- Maximizing safety of participants
- Reflect target population

- Known contraindications to vaccine based on published literature, the Investigator's Brochure and/or Summary of Product Characteristics
- Safe and effective running of the trial including logistics and volunteer ability to participate fully in the trial.

Screening participants for a clinical trial, therefore, usually involves a medical history, physical examination, and investigations such as blood or urine tests or radiological investigations. With volunteer consent, the screening process may also include seeking information from other sources such as medical history from the volunteer's doctor and previous participation in other trials. Initial assessment of volunteer capacity to participate in the trial should have also been completed as part of the consent process.

Medical cover

Appropriate medical cover to minimize risk to participants throughout the trial must be carefully planned. If there is extremely limited safety data on the trial vaccine or if there is a higher risk for serious systemic side effects or anaphylaxis (e.g., due to intravenous instillation), continuous volunteer monitoring in an inpatient unit might be appropriate. At a minimum, administration of the vaccine should take place in a clinical facility sufficiently equipped with emergency resuscitation equipment, with an inpatient facility nearby in case of a medical emergency. Volunteers should not be discharged home until the clinician is happy that there is no immediate evolving AE. Safety ring fencing needs to be in place. This could include ensuring the volunteer has access to a nearby 24-hour research clinic (if close to the volunteer's homes, e.g., in the same village) or a dedicated 24-hour contact phone number to an appropriately qualified trial clinician. It is important to ensure volunteers are aware of the importance of accessing appropriate medical care to reduce the risk of causing harm. Following vaccination, volunteers should also be formally followed up. The frequency and nature of this follow-up depends on the risk assessment and the safety data collection requirements (see "Adverse events" section). Follow-up visits could be phone calls, home visits, or clinic appointments.

Adverse events

AEs must be collected in any trial that assesses vaccine safety. AEs are defined as any untoward sign, symptom, or disease occurring or worsening after delivery of the agent under investigation.

Solicited and unsolicited adverse events

It is important to outline in the protocol the AEs that are going to be specifically collected (or "solicited"). The choice of solicited AEs depends on the clinical situation, route of vaccination, and the known ("expected") AEs for that vaccine (e.g., as outlined in the Investigational Brochure). If this is the first time, the vaccine is being tested in humans and hence there is a paucity of "expected" AE data, solicited AEs could be chosen on the basis of known AEs to similar vaccines, and from AEs seen in preclinical studies of that vaccine.

If this is a later phase trial where the AEs of the vaccine are well described, the Chief Investigator might decide not to collect these expected AEs, but only record AEs that are unexpected or serious (see "Adverse event assessment" section).

Table 11.2 shows a typical list of solicited AEs collected for an intradermal injected vaccine or aerosol vaccination.

Subjective AEs can be collected through volunteer self-reporting on diary cards, or at clinic visits. Objective AEs that are commonly recorded include vital signs, local reactions, or specific investigations such as blood tests or lung function tests. For an example of blood tests collected during a trial, see Table 11.3 (Folegatti et al., 2020). To facilitate reporting, volunteers can be given thermometers, tape measures, or other equipment to record AEs at home, depending on volunteer literacy levels and infrastructure of the trial setting.

TABLE 11.2 Example list of solicited adverse events following intradermal vaccination over the deltoid (ID) or aerosol vaccination. (A) local injection site, (B) local respiratory, or (C) systemic AEs after either ID or aerosol vaccination.

a. Local AEs after ID vaccination	b. Respiratory AEs after aerosol vaccination	c. Systemic AEs
Warmth	Chest pain	Fever
Itchiness	Coughing phlegm	Feverishness
Pain	Coughing blood	Malaise
Swelling	Shortness of breath	Arthralgia
Scaling	Wheeze	Nausea
Redness	Sore throat	Fatigue
	Cough	Headache
		Myalgia

Source: Adapted from Manjaly Thomas, Z. R., Satti, I., Marshall, J. L., Harris, S. A., Ramon, R. L., Hamidi, A., Minhinnick, A., Riste, M., Stockdale, L., Lawrie, A. M., Vermaak, S., Wilkie, M., Bettinson, H., & McShane, H. (2019). Alternate aerosol and systemic immunization with a recombinant viral vector for tuberculosis, MVA85A: A phase I randomized controlled trial. *PLoS Medicine, 16*(4). https://doi.org/10.1371/journal.pmed.1002790.

TABLE 11.3 Laboratory parameters measured for Group 1 during the Phase I/II Oxford AstraZeneca ChAdOx1 nCoV-19 SARs-CoV-2 trial.

Test	Timepoint
Full blood count, sodium, potassium, urea, creatinine, albumin, liver function tests	Day 0 (vaccination), 3, 7, 28

Apart from solicited AEs, it is common to also collect "unsolicited" AEs. This is especially true in Phase I trials, where AEs related to the vaccine are largely unknown and hence the Chief Investigator might plan to collect every AE. In addition, Phase IV trials might only collect any unsolicited AE that meets a certain severity threshold (see "Severity" section).

Recording adverse events

AEs can be recorded (on case report forms) by study staff as part of inpatient assessment or at outpatient visits, phone interviews, or on take-home diaries. There are advantages and disadvantages of each method and the decision about frequency of AE recording is dependent on the kinetics of expected AEs and the likelihood of volunteers retrospectively remembering these AEs. For example, immediate vaccine reactions expected to occur in the first hour or days following vaccinations might be recorded differently to longer term AEs.

Adverse event assessment

Three main assessments need to be made for each AE: relatedness, severity, and seriousness.

Causality (relatedness)

The Chief Investigator, or an appropriately qualified delegated candidate, is tasked with determining if AEs are related to the IMP, in this case the vaccine. This should be done both in real time for AEs of concern (see below) and at the end of the trial once all data are collected and quality controlled. This second review is important as it is at this stage that subtle patterns might be picked up, especially if the vaccine causes an AE that was not previously expected. Ideally a second independent clinician should also make this assessment.

Criteria for determining relatedness need to be established in the protocol and should be based on the following criteria:

- known or expected AE;
- proximal time course to vaccination; and
- AEs that are not clearly due to another cause.

On the basis of these criteria, the likelihood of relatedness is assigned (no relationship, unlikely, possibly, probably, or definitely). Any AEs considered at least possibly related to the vaccine are classified as "adverse reactions" and should be reported. Adverse reactions should be followed up until resolved or stabilized.

Severity

Each AE is also graded on severity. Grading criteria can vary depending on the nature of the AE and the norms of clinical practice in the local area. However, to ensure consistency between trials, it is good practice to use internationally accepted graded criteria, using a marker of severity that is as objective as possible (see Table 11.4) (Sibille et al., 2010). Grading criteria for laboratory AEs can be adapted from the US Food and Drug Administration toxicity grading scale adjusted for local laboratory ranges (US Food & Drug Administration, 2007). Grading criteria for observations and other planned AE measurements should also be written into the protocol.

Seriousness (serious adverse events)

Of primary importance is detecting any AEs caused by the vaccine that could do serious harm to participants. Therefore any serious adverse event (SAE), defined by the criteria in Fig. 11.2, must be reported to the relevant regulatory bodies overseeing the trial within the timeframe as per their guidelines and as defined in the protocol. As part of this process, the SAE is assessed for relatedness to the vaccine (see above). The SAE is of particular concern if it was also "unexpected," meaning it is not consistent with the reference safety information for that vaccine, or as predefined in the protocol. In this case, that AE is classified as a suspected unexpected serious adverse reaction (SUSAR) and the timeframe for reporting is urgent. The trial may be paused to allow for an appropriate review. Such a review may involve unblinding,

TABLE 11.4 Severity grading criteria for local and systemic adverse events (AEs).

Grade 0	None
Grade 1	Mild: Transient or mild discomfort; no medical treatment required, does not interfere with daily activity
Grade 2	Moderate: Moderate discomfort; interferes with daily activity; no treatment required except simple analgesia
Grade 3	Severe: Significant discomfort; prevents daily activity or requires treatment, hospitalization possible
Grade 4	Life threatening

Adjusted from the criteria used in clinical trial "Aerosol BCG challenge trial in healthy UK adults." Satti, I., Marshall, J. et al. Clinicaltrials.gov NCT02709278.

Serious adverse event (SAE) is defined as any AE that:

1. Results in death,
2. Is life-threatening at the time of the event
3. Requires inpatient hospitalization or prolongation of existing hospitalization,
4. Results in persistent or significant disability/incapacity,
5. Is a congenital anomaly/birth defect.
6. Other important medical event, for example,requires medical intervention to prevent one of the above criteria

FIGURE 11.2 Definition of a serious adverse event (SAE) as per the UK regulatory authority. Note: The definition of an SAE may vary subtly between regulatory authorities so it is important to check the criteria locally.

seeking further medical information about the case in question and reviewing safety data from other participants. A decision to continue is then taken, and if continued, the trial may be recommenced with alterations, for example with a lower dose or with more intense safety monitoring requirements. Specific "holding rules" are often built into the trial protocol, especially in Phase I trials, to ensure any non-SAE patterns are picked up early, ideally preventing the occurrence of a SUSAR.

Not all SAEs need to be reported immediately to the regulatory authorities if they are expected and this is made explicit in the protocol. For instance, death from Ebola might be expected during an Ebola vaccine trial that is being conducted during an epidemic. While this event would be recorded as part of the collected efficacy outcome measures for the vaccine, as these events would be expected during the trial, the regulatory authority *may not* require these SAEs to be reported urgently as safety events, if determined to be unrelated to the vaccine.

Reporting requirements

The protocol needs to outline the plan for AE reporting. Apart from the urgent reporting discussed above, local regulatory authorities and ethics committees usually require periodic reporting during the trial, for example, submission of a yearly drug safety update report and annual progress reports. The vaccine manufacturer will also require reports of any AEs of interest, with a requirement for a regular update of the investigator's brochure. In addition, the sponsor, grant funders, and other interested stake holders will stipulate timeframes for reports, which may also include interim efficacy data and recruitment updates.

Other trial considerations

Reimbursement

Reimbursement for trial participation is an important ethical consideration. It is unethical to require significant amounts of time from a participant without reimbursement to cover potential loss of wages. Any costs incurred by the volunteer for participating in the study, for example, bus fares to attend clinic visits, should also be covered. However, if reimbursement levels are too high, this could unduly influence a volunteer into participating in a trial. Reimbursement must, therefore, be at a level suitable for that community and reflective of trial commitments. For some trials, it may be sufficient to provide a free meal on clinic attendance or a gift voucher. However, for trials requiring inpatient admission or extensive time, monetary reimbursement is appropriate. Ethical dilemmas may arise where the same trial is run over multiple sites where the income levels of volunteers vary greatly. For instance, reimbursement of £10 per hour at a UK trial site would be considered undue inducement at a trial site in Kenya. Yet volunteers across sites must be seen to be treated the same. These issues need to be carefully considered during trial design. Input from the local community can be helpful, and all reimbursements need to be approved by the local ethics committee.

Publication considerations

For publication, most journals require proof that a favorable opinion from the relevant ethics committee and medicines regulatory authority was received before starting the trial. The International Committee of Medical Journal Editors (ICMJE) also stipulate that the trial must have been registered on a publicly accessible database before enrollment to avoid duplication of trial research and selective publication of results (The International Committee of Medical Journal Editors, 2021). Examples of free databases recognized by the ICMJE include the WHO International Clinical Trials Registry Platform (ICTRP) and Clinicaltrials.gov (US National Library of Medicine, 2020; WHO, 2021). Prominent medical journals also require submission of a CONSORT (CONsolidated Standards Of Reporting Trials) checklist for randomized trials to ensure the trial has been correctly designed and reported (CONSORT Transparent Reporting of Trials, 2021). Other publication stipulations could include confirmation that participants gave consent for publication of data, and author compliance with the journal's open access data sharing policy.

Confidentiality and data control

It is essential that the personal information of participants is protected. Confidentiality is an essential part of this process and trial information is usually deidentified by assigning a volunteer number to a participant in lieu of identifiable information such name or date of birth (officially this is "pseudonymization" as the volunteer number is unique to that individual and can be traced back to the participant). Identifiable information does usually need to be collected, for example, for contact purposes, to pay the volunteer or to contact their family doctor. However, this information needs to be securely stored and accessible to a limited number of staff. Trials must also comply with local data laws which may dictate what data may be collected, how it is stored and the length of time data can be stored for.

Final word

The COVID-19 pandemic which started in 2019 has once again highlighted the risk to humanity of emerging pathogens. The rapid speed with which the scientific community developed effective vaccines against the SARS-CoV-2 virus could not have occurred without clinical trials and the willingness of participants around the world to volunteer for them. A well-designed GCP-compliant clinical trial, as described in this chapter, is critical to the continued development of vaccines against these emerging and endemic pathogens.

References

Buttner, P., & Muller, R. (2011). *Experimental designs. Epidemiology* (pp. 215–257). Oxford University Press.

CONSORT Transparent Reporting of Trials. (2021). http://www.consort-statement.org/.

Council for International Organizations of Medical Sciences. (2016). *International ethical guidelines for health-related research involving humans*. https://cioms.ch/wp-content/uploads/2017/01/WEB-CIOMS-EthicalGuidelines.pdf.

Folegatti, P. M., Bittaye, M., Flaxman, A., Lopez, F. R., Bellamy, D., Kupke, A., Mair, C., Makinson, R., Sheridan, J., Rohde, C., Halwe, S., Jeong, Y., Park, Y. S., Kim, J. O., Song, M., Boyd, A., Tran, N., Silman, D., Poulton, I., ... Gilbert, S. (2020). Safety and immunogenicity of a candidate Middle East respiratory syndrome coronavirus viral-vectored vaccine: a dose-escalation, open-label, non-randomized, uncontrolled, phase 1 trial. *The Lancet Infectious Diseases*, *20*(7), 816–826. Available from https://doi.org/10.1016/S1473-3099(20)30160-2.

Gordon, S. B., Rylance, J., Luck, A., Jambo, K. C., Ferreira, D. M., Manda-Taylor, L., Bejon, P., Ngwira, B., Littler, K., Seager, Z., Gibani, M., Gmeiner, M., Roestenberg, M., & Mlombe, Y. (2017). A framework for Controlled Human Infection Model (CHIM) studies in Malawi: Report of a wellcome trust workshop on CHIM in low income countries held in Blantyre, Malawi. *Wellcome Open Research*, *2*, 70. Available from https://doi.org/10.12688/wellcomeopenres.12256.1.

Henao-Restrepo, A. M. (2015). The ring vaccination trial: a novel cluster randomized controlled trial design to evaluate vaccine efficacy and effectiveness during outbreaks, with special reference to Ebola. Ebola ça suffit ring vaccination trial consortium. *BMJ: British Medical Journal*, 351.

Loeb, M., Russell, M. L., Moss, L., Fonseca, K., Fox, J., Earn, D. J. D., Aoki, F., Horsman, G., Van Caeseele, P., Chokani, K., Vooght, M., Babiuk, L., Webby, R., & Walter, S. D. (2010). Effect of influenza vaccination of children on infection rates in Hutterite communities: A randomized trial. *JAMA - Journal of the American Medical Association*, *303*(10), 943–950. Available from https://doi.org/10.1001/jama.2010.250.

Medicines and Healthcare Products Regulatory Agency. (2004). *The medicines for human use*. https://www.legislation.gov.uk/uksi/2004/1031/contents/made.

Medicines and Healthcare Products Regulatory Agency. (2012). *Good clinical practice for clinical trials*. Available from https://www.gov.uk/guidance/good-clinical-practice-for-clinical-trials.

Sibille, M., Patat, A., Caplain, H., & Donazzolo, Y. (2010). A safety grading scale to support dose escalation and define stopping rules for healthy subject first-entry-into-man studies: Some points to consider from the French Club Phase I* working group. *British Journal of Clinical Pharmacology*, *70*(5), 736–748. Available from https://doi.org/10.1111/j.1365-2125.2010.03741.x.

The International Committee of Medical Journal Editors. (2021). *Clinical trials*. Available from http://www.icmje.org/recommendations/browse/publishing-and-editorial-issues/clinical-trial-registration.html.

US Food & Drug Administration. (2007). *Guidance for industry: Toxicity grading scale for healthy adult and adolescent volunteers enrolled in preventive vaccine clinical trials*. Available from https://www.fda.gov/media/73679/download.

US National Library of Medicine. (2020). *Clinicaltrials.gov*. Available from https://clinicaltrials.gov/.

World Health Organization. (2016). *Human challenge trials for vaccine development: Regulatory considerations*. Available from https://www.who.int/biologicals/expert_committee/Human_challenge_Trials_IK_final.pdf.

World Health Organization. (2021). *International clinical trials registry platform*. Available from https://www.who.int/clinical-trials-registry-platform.

World Medical Association. (2013). World Medical Association declaration of Helsinki: Ethical principles for medical research involving human subjects. *JAMA - Journal of the American Medical Association*, *310*(20), 2191–2194. Available from https://doi.org/10.1001/jama.2013.281053.

Index

Note: Page numbers followed by "*f*" and "*t*" refer to figures and tables, respectively.

A

E

F

G

H

L

M

N

W

Z

FSC
www.fsc.org
MIX
Papier | Fördert
gute Waldnutzung
FSC® C083411

Kolibri 360 GmbH
Ferdinand-Jühlke-Straße 7
99095 Erfurt, Deutschland
produktsicherheit@kolibri360.de